Advances in Aquatic Ecology

— Volume 9 —

The Editors

Dr.V.B.Sakhare is Head, Post Graduate Department of Zoology, Yogeshwari Mahavidyalaya, Ambajogai. He has 15 years' experience as an outstanding teacher and researcher. He is recipient of fellowship of Indian Association of Aquatic Biologists, Hyderabad. He has done pioneering work in the field of Reservoir Fisheries and Limnology.Dr.Sakhare has successfully organized *National Conference on Emerging Trends in Fisheries and Aquaculture* (*ETFA-2012*), *National Workshop on Techniques of Scientific Writing (TSW-2014)*, *National Conference on Current Perspectives in Limnology* (*NCCPL-2009*) and *Regional Workshop* on *Water Quality Assessment* (*Implications in Potability, Productivity and Pollution Control*).

Dr.Sakhare has been editing an international journal '*Ecology and Fisheries'* (ISSN 0974-6323).Dr.Sakhare has authored/edited few books such as '*Fish and Fisheries of Indian Reservoirs*','*Applied Fisheries*', '*Reservoir Fisheries and Limnology*','*Reservoir Fisheries and Ecology: A Literary Survey*', '*Methodology for Water Analysis*', '*Aquatic Ecology*', '*Aquatic Biology and Aquaculture*', '*Inland Fisheries*', '*Applied Ecology*', '*Perspectives in Ecology*', and '*Advances in Aquatic Ecology* (*Vol.* I, II, III, IV, V, VI, VII,VIII).

Dr.Sakhare has supervised a research project funded by University Grants Commission, New Delhi and he is a recognized post graduate teacher and research guide of Dr.Babasaheb Ambedkar Marathwada University, Aurangabad, Solapur University, Solapur and J.J.T.University, Rajasthan. Under his guidance three students have completed Ph.D. He has published more than 45 research articles and reviews in peer reviewed journals and about 70 marathi articles in newspapers and magazines.

Dr.Sakhare has chaired a number of sessions of different seminars/symposia. He has been invited to different colleges/institutes to deliver lectures on different topics in aquatic ecology and reservoir fisheries.

Dr.B.Vasanthkumar is presently working as Associate Professor and Head of the Department of Zoology at Government Degree Arts and Science College,Karwar, Karnataka. He is recipient of summer research fellowship of Indian Academy of Science for year 2014.He has published more than 50 research papers in the national and international journals and more than 50 popular science articles in different aspects of ecology and environment. He has also published 12 text books for under graduate students of Karnataka University, Dharwad. He is co-author of reference books like 'Aquatic Ecosystem and its management' '*Applied Ecology*' '*Advances in Aquatic Ecology (Volume 7 and 8)*' and *'Emerging Trends in Fisheries and Aquaculture'*. Presently Dr.Vasanthkunar is working as Principal Investigator with Major Research Project funded by University Grants Commission, New Delhi.

Dr.J.S.Mohite is Principal of Yeshwantrao Chavan Mahavidyalaya, Tuljapur (Maharashtra).He received his Ph.D. from Swami Ramanand Teerth Marathwada University, Nanded.His research work is related to limnology and fisheries and has published several research papers in various national and international journals of repute. Dr.Mohite has worked as University Senate Member and Joint secretary of Marathwada Principals' Association.

Advances in Aquatic Ecology

— Volume 9 —

Editors
Dr. Vishwas B. Sakhare
Head,
Post Graduate Department of Zoology
Yogeshwari Mahavidyalaya,
Ambajogai – 431 517, Maharashtra
INDIA

Dr. B. Vasanthkumar
Head
Department of Zoology
Government Degree Arts and Science College,
Karwar – 581 301
Karnataka
INDIA

Dr. J.S. Mohite
Principal
Yashwantrao Chavan Mahavidyalaya,
Tuljapur – 413 601, Maharashtra
INDIA

2015
Daya Publishing House®
A Division of
Astral International Pvt. Ltd.
New Delhi – 110 002

Cataloging in Publication Data–DK
Courtesy: D.K. Agencies (P) Ltd. <docinfo@dkagencies.com>

Advances in aquatic ecology / editors, Dr. Vishwas B. Sakhare, Dr. B. Vasanthkumar, Dr. J.S. Mohite.
     volume 9   cm
     Includes bibliographical references and index.
     ISBN 978-93-5130-691-7 (International Edition)

     1. Fisheries–Environmental aspects–India. 2. Aquatic ecology–India. I. Sakhare, V. B. (Vishwas Balasaheb), 1974–, editor. II. Vasanthkumar, B., 1968–, editor. III. Mohite, J. S., editor.

     DDC 577.60954   23

Published by   : **Daya Publishing House®**
                     A Division of
                     Astral International Pvt. Ltd.
                     – ISO 9001:2008 Certified Company –
                     4760-61/23, Ansari Road, Darya Ganj
                     New Delhi-110 002
                     Ph. 011-43549197, 23278134
                     E-mail: info@astralint.com
                     Website: www.astralint.com

Laser Typesetting   : **Classic Computer Services, Delhi - 110 035**

Printed at   : **Thomson Press India Limited**

PRINTED IN INDIA

Preface

The present book entitled '*Advances in Aquatic Ecology (Volume 9)*' comprises chapters by well known experts and research workers in their respective fields. We are thankful to all the contributors who responded promptly by making available their articles and it is the overwhelming response which has catalytic impact upon s to complete the task of publishing this book. This volume is an assemblage of up to date information of rapid advances and developments taking place in the field of aquatic ecology. With its application oriented and interdisciplinary approach, we hope that the students, teachers, researchers, scientists, policy makers and environmental lawyers in India and abroad will find this volume much more useful. The articles in the book have been contributed by eminent scientists/academicians active in the areas of aquatic ecology.

Editors are grateful to the following under mentioned distinguished scientists and other fellow colleagues for their constant encouragements, suggestions, valuable guidelines and necessary help. These respected, distinguished, beloved scientists and well wishers are Dr.B.R.Chavan, Principal of Yogeshwari Mahavidyalaya, Ambajogai, Dr.Kalpana Kerwadikar, Principal, Government Degree College, Karwar, Dr.P.K.Joshi of Dnyanopasak College, Parbhani, Dr.K.Vijaykumar, Registrar of Srikrishna Devararaya University, Bellary, Dr.Ansuman Das of Fishery Survey of India, Mumbai, Dr.V.Ravi of Centre of Advanced Studies in Marine Biology, Annamalai University, Parangipettai, Dr.Chhaya Panse, Vice Principal of Maharshi Dayanand College, Mumbai, V.Rajani, S. Aswathy Krishnan S.Navami and Ayona Jayadev of All Saint's College, Thiruvananthapuram, Dr.J.B.Solanki of Junagadh Agricultural University, Veraval, Dr.S.P.Chavan of Swami Ramanand Teerth Marathwada University, Nanded, Dr.P. Nimisha and S. Sheeba of Sree Narayana College, Kollam, H. Krishna Ram of University of Mysore, Mysore, Dr.S.Ramakrishna, Bangalore

University, Bangalore, Dr.H.A. Sayeswara, Sahyadri College, Shivamogga, Dr.Chandrakant Bharambe of Vidnyan Mahavidyalaya, Malkapur, Prof.V.S.Hamde of Yogeshwari Mahavidyalaya, Ambajogai, Dr.B.R.Shinde and Dr.V.R. Borane of Jigamata Arts, Science and Commerce College, Nandurbar, Dr.S.Jeyakumar of Aditanar College of Arts and Science, Virapandianpatnam, Dr.S.Mala of Govindammal Aditan College for Women, Tiruchendur, Dr.R.P.Mali of Yeshwant Mahavidyalaya, Nanded, Dr.Basha Mohidden of S.K.University, Anantapur, Dr.Sandhya Rani Gaur, Dr.Dushyant Kumar Damle, Dr.Neha Chandrawanshi and Dr.T.K.Thakur of Indira Gandhi Agricultural University, Raipur, Dr.G. Sattanathan of Government College for Women, Kumbakonam, Dr.R.Rajesh of Centre of Advanced Study in Marine Biology, Annamalai University, Parangipettai, Dr.A.C.Kumbhar of Shankarrao Mohie Patil Mahavidyalaya, Akluj, Dr.Lazarus Lanka of Devchand College, Arjunanagar and many others.

The editors wish to express their deep sense of gratitude to shri Anil Mittal of Astral International Private Limited, New Delhi for publishing this volume.

We are grateful to our family members for helping us by several ways and encouraging us for publication of this volume.

Dr. V.B. Sakhare
Dr. B. Vasanthkumar
Dr.J.S. Mohite

Contents

List of Contributors

Ahamed, M. Rafi
S.K.P. Government Degree College (U.G and P.G), Guntakal – 515 801

Anusikha, A.
Fishery Survey of India, Botawala Chambers, Sir P.M. Road, Fort, Mumbai – 400 001

Ashashree, H.M.
Department of Zoology, Sahyadri Science College (Autonomous), Kuvempu University, Shivamogga – 577 203

Aswathy Krishnan G.
P.G. Department of Environmental Sciences, All Saints' College, Thiruvananthapuram – 695 007

Bhadane, Rekha S.
Department of Zoology, L.V.H. Arts Science and Commerce College, Panchavati, Nashik – 422 005

Borane, V. R.
Department of Zoology, Jijamata Arts, Science and Commerce College, Nandurbar – 425 412

Chandrawanshi, Neha
Department of Fisheries, Indira Gandhi Krishi Vishwavidyalaya, Raipur – 492 006

Chavan, S.P.
Aquatic Parasitology and Inland Fisheries Research Laboratory, Swami Ramanand Teerth Marathwada University, Nanded – 431 006

Chougule, S.H.
Department of Zoology, Shankarrao Mohite Mahavidyalaya, Akluj – 413 101

Damle, Dushyant Kumar
Department of Fisheries, Indira Gandhi Agricultural University, Raipur – 492 006

Darekar, P.V.
Department of Zoology, Shankarrao Mohite Mahavidyalaya, Akluj – 413 101

Das, Ansuman
Fishery Survey of India, Botawala Chambers, Sir P.M. Road, Fort, Mumbai – 400 001

Deshmukh, A.L.
Department of Zoology and Research Centre, Shankarrao Mohite Mahavidyalaya, Akluj – 413 101

Dodia, A.R.
College of Fisheries, Junagadh Agricultural University, Veraval – 362 265

Gangadhar, B.K.
Department of Zoology, Government Arts and Science College, Karwar – 581 301

Gaur, Sandhya R.
Department of Fisheries, Indira Gandhi Agricultural University, Raipur – 492 006

Ghorpade, B.N.
Department of Zoology and Research Centre, Shankarrao Mohite Mahavidyalaya, Akluj – 413 101

Hamde, V.S.
Department of Microbiology, Yogeshwari Mahavidyalaya, Ambajogai – 431 517

Hashmi, Seema
Department of Zoology and Microbiology, Milliya College, Beed – 431 122

Ilyas, Mohd
Post Graduate Department of Zoology, Milliya Arts, Science and Management Science College, Beed – 431 122

Jagtap, A.R.
Post Graduate and Research Department of Zoology, Yeshwant Mahavidyalaya, Nanded – 431 601

Jayadev, Ayona
P.G. Department of Environmental Sciences, All Saints' College, Thiruvananthapuram – 695 007

Jayaprakash
Department of Zoology, Bangalore University, Bengaluru – 560 056

Jeyakumar, S.
Department of Advanced Zoology and Biotechnology, Aditanar College of Arts and Science, Virapandianpatnam

Kadam, Mangal Sitaram
Post Graduate and Research Department of Zoology, Yeshwant College, Nanded – 431 602

Kannewad, Pandurang
Aquatic Parasitology and Inland Fisheries Research Laboratory, Swami Ramanand Teerth Marathwada University, Nanded – 431 606

Kumbhar, A.C.
Department of Zoology, Shankarrao Mohite Mahavidyalaya, Akluj – 413 101

Lanka, Lazarus
Department of Zoology, Devchand College, Arjunnagar

Mala, S.
Department of Advanced Zoology and Biotechnology, Govindammal Aditan College for Women, Tiruchendur – 628 215

Mali, R.P.
Post Graduate and Research Department of Zoology, Yeshwant Mahavidyalaya, Nanded – 431 601

Malthane, G.A.
Department of Zoology, Vidnyan Mahavidyalaya, Malkapur – 443 101

Mohapatra, Jayalaxmi
Fishery Survey of India, Botawala Chambers, Sir P.M. Road, Fort, Mumbai – 400 001

Mohidden, Basha
S.K. University, Anantapur – 515 003

Navami, S.S.
Department of Environmental Sciences, All Saints' College, Thiruvananthapuram – 695 007

Nimisha, P.
Department of Zoology, Sree Narayana College, Kollam – 691 001

Niture, S.D.
Department of Zoology, Shivaji Mahavidyalaya, Udgir – 413 517

Panse, Chhaya
Vice Principal, Maharshi Dayanand College, Parel, Mumbai – 400 012

Parmar, H.L.
College of Fisheries, Junagadh Agricultural University, Veraval – 362 265

Parmar, P.V.
College of Fisheries, Junagadh Agricultural University, Veraval – 362 265

Patil, B.V.
Department of Zoology, Vidnyan Mahavidyalaya, Malkapur – 443 101

Poul, Shivaji
Department of Zoology, Madhavrao Patil Mahavidhyalaya, Palam – 431 720

Purushothama, R.
Department of Environmental Science, Sahyadri Science College (Autonomous), Kuvempu University, Shivamogga – 577 203

Rajani, V.
P.G. Department of Environmental Sciences, All Saints' College, Thiruvananthapuram – 695 007

Rajesh, R.
Centre of Advanced Study in Marine Biology, Annamalai University, Parangipettai – 608 502

Ram, H. Krishna
Department Of Studies in Zoology, University of Mysore, Mysore – 570 006

Ramakrishna, S.
Department of Zoology, Bangalore University, Bengaluru – 560 056

Rathod, J.L.
Department of Studies and Research in Marine Biology, Karnataka University P.G. Centre, Kodibag, Karwar – 581 303

Ravi, V.
Centre of Advanced Study in Marine Biology, Faculty of Marine Sciences, Annamalai University, Parangipettai – 608 502

Roopa, S.V.
Department of Studies and Research in Marine Biology, Karnataka University P.G Centre, Kodibag, Karwar – 581 303

Sakhare, Vishwas B.
Post Graduate Department of Zoology, Yogeshwari Mahavidyalaya, Ambajogai – 431 517

Saleem, Quazi
Department of Zoology and Microbiology, Milliya College, Beed – 431 122

Sattanathan, G.
PG and Research Department of Zoology, Government College for Women (Autonomous), Kumbakonam – 612 001

Sayeswara, H.A.
Department of Zoology, Sahyadri Science College (Autonomous), Kuvempu University, Shivamogga – 577 203

Shaikh, F.I.
Post Graduate Department of Zoology, Milliya Arts, Science and Management Science College, Beed – 431 122

Shaikh, Imran
Post Graduate Department of Zoology, Milliya Arts, Science and Management Science College, Beed – 431 122

Shaikh, Iqbal
Department of Zoology, Vidnyan Mahavidyalaya, Malkapur – 443 101

Sheeba, S.
Department of Zoology, Sree Narayana College, Kollam – 691 001

Shinde, B.R.
Department of Zoology, Jijamata Arts, Science and Commerce College, Nandurbar – 425 412

Solanki, J.B.
College of Fisheries, Junagadh Agricultural University, Veraval – 362 265

Sreedhara Nayaka, B.M.
Karnataka State Pollution Control Board, Hassan – 573 201

Thakur, T.K.
Department of Fisheries, Indira Gandhi Agricultural University, Raipur – 492 006

Vasanthkumar, B.
Department of Zoology, Government Arts and Science College, Karwar – 581 301

Chapter 1

Review of Food and Feeding Habits of Freshwater Fishes of India

☆ V.B. Sakhare

Food is the basic prerequisite for growth, development, survival and existence of all organisms. Ross (1986) identified that in aquatic environments food is the main factor and that its partition defines fundamental groups within the community, which get together in guilds according to the trophic similarity. It plays an important role in the growth, migration and spawning behaviour of the fish. As the nature of food depends upon the nature of several biotic and abiotic factors, the problem is interesting from specific, as well as ecological point of view (Bhuiyan *et al.,* 2006).The study of the food and feeding habits of freshwater fish species is a subject of continuous research because it constitutes the basis for the development of a successful fisheries management. Freshwater fishes consume a wide variety of foods. The identification of stomach contents allows us to know about food consumption, feeding and assimilation rates, cannibalism and even habitat segregation (Gomos *et al.,* 2002).The food and feeding habits of fishes vary with time of day, size of fish, and season of the year. Fishes are also known to change the food habits as they grow, accompanied by correlative changes in the digestive system.

The food of young ones is generally different from that of the adult. Young ones with small and short intestine prefer zooplankton, and are able to digest rotifers, cladocerans another microscopic animals easily. The phytoplankton and algae are not easily digested.

The food of fishes according to Schaperclause (1933) may be of three groups such as main, occasional and emergency food. A classification was also made by Nikolskii (1963) based on the relationship between fishes and their food and categorized them according to the extent of variation in the types of food consumed. Das and Moitra (1963) also classified fishes according to the food consumed.

Catla catla

It is a fast growing species among the Indian major carps. It grows to a length up to 45 cm, weighing more than a kilogram in one year and attains 2.2 kg and 6.5 kg weight at the end of second and third years respectively. It is grown in polyculture and matures in the second year. It breeds naturally in rivers during the rainy season, though artificial propagation by hypophysation is possible. It is distributed throughout India, Pakistan, Nepal, Bangladesh and Thailand.

Sakhare and Chalak (2014) studied food and feeding habits of *Catla catla*.Rotifers formed the main item of gut contents forming 25.7 per cent. The major genera of rotifers in the diet of the species were *Brachionus* spp., *Filinia logiseta, Keratella tropica, Lecane bulla, Trichocera orecelus.* Cladocerans were next in the order of dominance froming 21.2 per cent per cent in the gut contents of *Catla catla*.This group was mainly represented by *Ceriodaphnia cornuta, Moina micrura, Alona rectangular* and *Indialona ganapati.* Bacillarophyceae formed 18.7 per cent of the gut contents. This group was represented by *Cymbella turgid, Fragilaria* sp., *Melosira* sp., *Navicula mutica, Synedra ulna* and *Nitzchia* sp. Myxophyceae formed 14.3 per cent of the gut contents of *Catla catla.* Among the myxophyceae, the abundant genera were *Microcystis areuginosa, Nostoc* spp, *Oscillatoria chlorine, Phormidium* sp. and *Anabaena* spp. Aquatic insects formed 8.4 per cent and were represented by *Gryllus* and mosquito larvae. Miscellaneous items and mud contents formed 4.2 and 4.5 per cent respectively.

Pinnularia, diatoms and detritus formed the major food items of this species (CIFRI, 1997).The work on this species from Aliyar reservoir has recorded blue green algae as the dominant item (43.8 to 48.1 per cent) followed by the detritus (21.3 per cent to 24.7 per cent).The zooplanktons were restricted to 12.17 per cent to 24.5 per cent consisting of copepods and rotifers (Selvaraj *et al.,* 1997).

Hora and Pillay (1962) reported *Catla catla* plankton and detritus feeder. Kumar *et al,.* (2007) categorized the fish as planktivorous.

Labeo rohita

This is the most famous major carp found in freshwaters of India. Body elongated with moderately rounded abodomen. Head prominent with blunt snout. It bears a subterminal fringe-lipped mouth bounded by fleshy upper and lower lips. It also contains paired nostrils and paired eyes. A pair of filamentous barbells arises from upper lip. Small tubercles cover the snout, which is oblong, depressed, swollen and projecting beyond the jaws. Lateral line is distinct. The colour of the fish is bluish – black along the back, reddish black along the sides and silvery in the abdominal area.

Detritus formed the bulk (56.1 to 58.8 per cent) of the gut contents. Blue green algae dominated by *Microcystis* (14.6 per cent) and the decayed organic matter (10.2 per cent) occupied the second and third positions (Selvaraj *et al.*1997).

Labeo fimbriatus

It is distributed in freshwaters of India, Pakistan, Nepal and Burma. It is cultured with Indian major carps. Body elongate, its dorsal profile convex than the ventral. Eyes moderate. Mouth moderate and subinferior, lips thick and fringed. It is predominantly a herbivore, feeding on diatoms, blue green algae, green algae, higher aquatic vegetation, insects and decayed organic matter (Talwar and Jhingran, 1991).Mouth is ventrally placed and fimbriated horny lips is highly adapted to bottom browsing. Its stenophagic feeding on sessile diatoms indicates its selectivity in feeding. Irrespective of lentic and lotic environments the fish feeds on similar food (David and Rajagopal, 1975).Plankton formed the basic item of diet. Food comprises myxophyceae, bacillariophyceae, chlorophyceae, plant tissues, copepods, appendages of insects, decayed organic matter, sand and mud, and other miscellaneous items like lower crustacean eggs, annelid setae and some unidentifiable forms. The fish proved to be an absolute bottom feeder mainly subsisting on detritus (42.9 per cent -59 per cent) followed by the decayed organic matter, sand and silt (Selvaraj *et al.,* 1997).

Labeo calbasu

It is one of the major Indian carps. It is an important food fish and at several places is referred as the 'black rohu'. It is also an important game fish and distributed in freshwaters of India, Nepal, Burma, Pakistan, Thailand and Bangladesh. It is bottom feeder (Talwar and Jhingran, 1991; David and Rajagopal 1975). Bacillariophyceae (18.7 per cent) constituted the main food while chlorophyceae with 4.4 per cent (*Spirogyra, Merismopedia, Cosmarium*) ranked next. Insect appendages and plant tissue were occasionally observed in the guts in small quantities. Miscellaneous items like annelidan setae, fungi, oscillatoria, protozoan ciliates (*Paramecium* sp.) were also recorded along with decayed organic matter and mud (David and Rajagopal, 1975).

Khumar and Siddiqui (1989) reported it as bottom dwelling illophagic fish and selective feeder. The food consisted of decayed organic matter, molluscs, diatoms, plant matter, green algae, blue green algae and zooplankton. Juveniles showed a positive selection for zooplanktonic organisms. The adult showed a negative selection for the zooplankton and a positive selection for decayed organic matter and molluscs. The increased feeding intensity of the fish corresponded to a period of maximum available food in the habitat. Maturation of gonads also influenced the feeding intensity of the fish. Post spawning feeding intensity was found to be maximum.

Selvaraj *et al.* (1997) observed only blue green algae (36.2 -55 per cent),detritus (30-42 per cent) and decayed organic matter (15-15.1 per cent) in the gut of *L.calbasu* and concluded that fish is bottom feeder.

Labeo dyocheilus

It is economically important fish and has been categorized as vulnerable by National Bureau of Fish Genetic Resources (NBFGR), Lucknow. The primary food of *Labeo dynocheilus* is green algae and diatoms. Secondary food is sand and debris. Fish take occasionally zooplanktons along with insect and macrophytes as a food. The fish is herbi-omnivorous and had column feeding habit because of its straight mouth position. Gastro-somatic index was highest in February and lowest in August (Verma, 2013).

Cirrhinus mrigala

Body is bright silvery in colour. The body is narrow and linear. Head is small and snout blunt. The mouth is terminal. Lips are thin and nonfringed. Dorsal fin with 12-13 branched rays.The tip of the head is flattened and upper jaw is fringed. Lateral line scales are 40-45 in number. A pair of small barbells are present.

It is mainly illophagous, feeding on the bottom on decayed vegetation. It can also switch to a filter feeding mode. The thin terminal lips are adapted for picking up food material from the substratum (Jhingran and Khan,1979).Larger solid food items are masticated by the pharyngeal teeth. After absorption of yolk-sac, larvae and fry stage feed on zooplankton *i.e.*, rotifers, nauplii, copepods and cladocerans (Hora and Pillay,1962).Khan (1972) showed that fingerlings (up to 100 mm) feed mainly on zooplankton, while phytoplankton becomes more important diet of fish ranging between 100 and 300 mm length. At around 300 mm length, the fish switch to detritus feeding (Jhingran and Khan,1979) and in fish above 560 mm length, semi-decayed organic matter constitute about 65 to 78 per cent of the gut content, while sand and mud make up the rest (Jhingran and Pullin,1985).Food items of animal origin are of little importance in adult fish. The most important phytoplankton groups consumed by juvenile mrigal are chlorophyceae, myxophyceae, bacillarophyceae and euglenophyceae.

The blue green algae, dominated by *Microcystis* constituted the major food item of the fish, closely followed by the detritus. The detritus along with the decayed organic matter constituted almost half of the gut contents (41.8 to 49.4 per cent) indicating the bottom feeding habit of the species (Selvaraj *et al.*, 1997).

Cyprinus carpio

It is popularly known as 'Common carp'.The three varieties of the Prussian strain of common carp, *viz.*, the scale carp (*Cyprinus carpio var.communis*),the mirror carp (*Cyprinus carpio var.specularis*) and the leather carp (*Cyprinus carpio nudus*) were introduced in India during 1939.They were stocked in several high altitude ponds, lakes and reservoirs during the 1950s.Later,in the 1957,the Chinese (Bangkok) strain of the common carp was brought to into the country, primarily for aquacultural purposes, considering its warm water adaptability, easy breeding, omnivorous feeding habits, good growth and hardy nature. The common carp is an important species in aquaculture and enhanced fisheries in India. It is grown either alone or in polyculture, most commonly with catla and rohu.It is virtually cosmopolitan, with populations on every continent except Antartica. Fish is adapted for benthic feeding with a protrusible mouth, large sensory lips, barbels with chemosensory cells, toothless jaw, toothless palate, specialized pharyngeal teeth and a cornified chewing pad.

Sakhare (2010) revealed that the food of *Cyprinus carpio* constituted blue green algae, diatoms, copepods, rotifers, detritus and decayed organic matter. From the gut content analysis it is evident that *Cyprinus carpio* is an omnivorous feeder preferring blue green algae, diatoms, copepods, rotifers, detritus and decayed organic matter.

There was a slight difference in food of different size groups of the fish. The preferred item for small size group was rotifer (62 per cent), copepods (26 per cent), diatoms (6 per cent) and blue green algae (6 per cent), while in the food of large fishes blue green algae (47 per cent), diatoms, (28 per cent) detritus (14 per cent), rotifers (6 per cent) and decayed organic matter (5 per cent) was common.The identified blue green algae were *Microcystis* sp., *Merismopedia* sp., *Lyngybya* sp., *Anabaena* sp., and *Spirulina* sp.Among these *Lyngbya* sp. and *Mersmopedia* sp. were the most dominant species. Diatoms constituted the second most dominant food item in the stomachs of *Cyprinus carpio*. *Melosira* sp., *Tabellaria* sp., *Cyclotella* sp., and *Fragillaria* sp. were the chief representatives. Rotifers were the third most dominant food item. They were represented by *Brachionus* sp., *Filinia* sp. and *Keratella* sp. Detritus and decayed organic matter was also recorded in the gut of carp.

There is certain variability in the feeding intensity.In the female there is a gradual drop in feeding intensity during the maturation phase (January -February and July-September).In the case of the male carp the feeding intensity does not seem to be affected during maturation phase.

According to Menon and Chacko (1956), fishes feeding on filamentous algae, molluscs and worms and in whose gut content, sand grains are found in fair proportion are to be placed under the group of bottom feeders. Due to the occurrence of detritus (including sand), and filamentous algae in the gut content, *Cyprinus carpio* could be categorized as bottom feeders. Vilizzi and Walker (1999b) observed sand in the gut of common carp. Fish is adapted for benthic feeding with a protrusible mouth, large sensory lips, barbels with chemosensory cells, toothless jaw, toothless palate, specialized pharyngeal teeth and a cornified chewing pad (Sibbing *et al.*1986).Common carp can switch feeding modes according to local, seasonal and diurnal peaks in prey abundance, and the availability of prey is reflected in the food items eaten (Hall 1981;Hume *et al.*, 1983a).The modes include benthic feeding, pump-filter feeding and gulping (Lammens and Hoogenboezem, 1991).

The diet of small carp includes chironomides and benthic insects (Hume *et al.*, 1983a).The adults are omnivorous, and their diet includes molluscs, epibenthic cladocerans, copepods, amphipods, chironomids, aquatic and terrestrial insects, detritus, seeds, fragments of dead aquatic plants and filamentous algae (Hall,1981).Common carp are opportunistic feeders on living and non-living organic material. As larvae they are vision-oriented particulate feeders that utilize zooplankton, mostly rotifers, cyclopoid copepods and cladocerans (Hall 1981, Vilizi and Walker 1999b).

According to Dube (2003) common carp is a polyphagus and omnivorous in habit. Initially it feeds on plankton and crustaceans. The fingerlings gradually shift to benthic organisms such as chironomids, tubificids, insect larvae and decayed vegetative matter. The adults are omnivorous, preferring insects, crustaceans and decayed organic matters. In search of benthic food it burrows bottom and embankments of ponds thus making them weak and water turbid. Its feeding habit (browsing nature) could undermine river/reservoir banks leading to the collapse of banks and uprooting vegetation bringing changes to river flows/courses. The foraging behavior of common carp resulted in vegetation removal both by direct consumption and by uprooting due to its proclivity to dig through substrate in search of food. The latter activity also resulted in increased water turbidity rendering the conditions more conducive for its propagation (Lakra snd Singh, 2007).

Saikia and Das (2008) recorded a total of 60 food items of which 22 belonged to chlorophycea, 12 to the cyanobacteria, 10 to the bacillariophyceae and 16 to several zooplankton taxa. According to Piska (1999) common carp is an omnivorous bottom feeding fish. Adult fish feed on bottom dwelling aquatic animals *viz.* insect larvae, worms, molluscs and decayed vegetable matters.It feed on epiphytic plankton also. Young feed on protozoan and small crustaceans.

Ctenopharyngodon idella

It was introduced in India in 1959 by importing from Hongkong and Japan. The body is cylindrical and elongate. Dorsal surface greenish or dark grey in colour and silvery on the belly. Head is broad and snout is rounded. Upper jaw slightly larger than the lower jaw. Mouth is horny. Barbels are absent. Abdomen rounded. Dorsal fin small with less than 9 rays medially placed. There is a notch on the snout. Scales are cycloid and moderate in size. Normally it feeds on aquatic weeds. It needs a minimum food of 25 per cent of its total body weight daily. Its maximum daily feeding ration has been as high as eight times of its total weight.

It is herbivorous and feed primarily on aquatic plants but also take insects and other invertebrates (Ni and Wang, 1999), and readily accept formulated pellets under culture conditions.It feeds on soft and hard aquatic weeds. It also accepts terrestrial grass growing on the bundhs. Fry feeds on organisms like *Cyclops, Diaptomus, Daphnia* etc. However in the later periods, its food habit changes towards aquatic weeds. Its feeding rate is completely different and extremely higher than other carps. It needs a minimum food of 25 per cent of its total body weight daily.Its maximum daily feeding ration has been as high as eight times of its total weight. Its habit of feeding aquatic weeds is beneficial in biologically controlling the aquatic weeds and it also serves as a 'living green manuring machine' besides its own growth. The feeding intensity during post spawning months is high and the adults consume *Hydrilla, Najas, Vallisneria, Utricularia* and soft leaves of *Eichhornia.*

Hypophthalmichtys molitrix

It is an exotic carp introduced in India during 1959 and suitable for culture in confined freshwaters along with Indian major carps. This has an oblong, slightly, laterally compressed body with a pointed head. Snout is bluntly rounded. Lower jaw slightly protruding, small eyes covered with adipose. Dorsal fin small with less than 9 rays and medially placed. Scales are very small about 110 to 124 in numbers on lateral line.

From early fry to late fry stage it feeds on zooplankton as main food with phytoplankton as occasional food but later on phytoplankton becomes the major food with zooplankton as occasional food. Phytoplanktivorous surface feeder from fingerling to adult stage, its main food is euglenoids, microcystes, nostoc, diatoms, desmids and filamentous green algae. Throughout life it feeds as occasional food on zooplankton.

Tilapia mossambica

Tilapias ingests a wide variety of food organisms, including plankton, some aquatic macrophytes, planktonic and benthic aquatic invertebrates, larval fish, detritus,and decomposing organic matter. Tilapias are often considered filter feeders because they can efficiently harvest plankton from the water. The gills of tilapia secrete a mucous that traps plankton (Popma and Masser, 1999).Two mechanisms help tilapia to digest filamentous and planktoic algae and succulent higher plants: physical grinding of plant tissues between two pharyngeal plates of fine teeth; and a stomach pH below 2, which ruptures the cell walls of algae and bacteria (Popma and Masser, 1999).

It is omnivore. The fry feeds exclusively on diatoms and other unicellular planktonic and epiphytic algae. The adults subsist mainly on vegetable food chiefly of chlorophyceae, myxophyceae and bacillariophyceae.When vegetable food are scare, worms, insects, crustaceans, fish larvae and detritus are eaten. Diatoms and chlorophyceae form the major items of food besides weeds in the case of young tilapia. It is observed that 7-10 mm fry feeds almost entirely on zooplankton. However, some algal items may occasionally be found in the stomach.11-60 mm fish feeds almost equally on zooplankton and phytoplankton. Above 60 mm, their main food is phytoplankton, filamentous algae and rarely leaves of higher aquatic plants (Sundararaj and Srikrishnadhas, 2000).

Tor tor

It is a column and bottom feeder and scraps food by the hard jaws. This species is omnivorous and feeds on chlorophyceae (*Spirogyra, Cladophora, Microspora, Hydrodictyon, Ulothrix, Hormedium,* and *Protococcus*), myxophyceae (*Oscillatoria* sp. and *Rivularia* sp.), bacillariophyceae (*Melosira* sp., *Cymbella* sp., *Diatoma* sp., *Navicula* sp., *Gomphonema* sp., *Amphora* sp.) and insects and insect larvae. In later summer and rainy season when food becomes scare in snow fed rivers, these fishes feed to some extent on higher plants and other small fishes. The alimentary canal in rainy season is almost empty and the rectum shows the presence of parts of insect larvae, fish scales and filaments of higher plants (Badola and Singh, 1980).

Sharma *et al.* (2013) carried out gut content analysis of *T.tor* cultured in cages, ponds and Narmada river.Gut content analysis showed that the main component of the gut in *T.tor* from pond and cage constituted plant and insect matter, whereas that from river Narmada showed highest content of macrophytes constituting about 45.5 per cent sand, molluscs and algae also formed the part of the stomach content. Sharma *et al.* (2013) also concluded that T.tor is omnivorous which accepts wide range of available food. Acceptance of different type of food by *T.tor* shows that it can be cultured in different environmental conditions.

Tor khudree

Commonly known as 'Deccan mahseer'.Biju (2003) identified food items belonging to eight categories *i.e.*, 1) semidigested animal matter: includes the semidigested and mutilated flesh of different animals,2) semidigested plant matter represented by broken roots, leaves and stem 3) diatoms mainly represented by

Merismopedia and *Cosmarium* 4) filamentous algae in small quantity 5) crustacean remains comprising mainly crustacean appendages, broken shells etc 6) fish remains includes scales and fish bones,7) sand indicating bottom feeding habit of the fish, and 8) insect larvae.

Semidigested animal matter contributed to 52.3 per cent of the total gut. Other animal protein substitutes such as crustaceans, insect larvae and fish remains were in the ratio of 0.74:5.2:3.2 respectively. Semidigested plant matter contributed to about 24.9 per cent of the gut content, while other items such as diatoms, filamentous algae, sand detritus and miscellaneous substances contributed to about 7.4, 1.24, 0.75, 2.99 and 1.24 per cent respectively in the gut content. Thus, the fish is omnivorous with more preference to carnivorous food. Presence of sand and detritus in the gut content also supports its bottom feeding habit.

Puntius pulchellus

It is one of the important fish fauna of Deccan region of India, now a days rarely found in river Tungabhadra. The identifying characters of this species includes deep body, 2.5 to 3 times of which is equally to standard length. Mouth is narrow having two pairs of berbels.The maxillary pair is equal to orbit and the rostral pair is slightly shorter. The number of lateral line scales are 26-32 and the predorsal scales are 12.Routray *et al.* (2001) studied some aspects of biology and fishery of *Puntius pulchellus* and the gut content analysis revealed that pulchellus is essentially a herbivorous fish having preference for soft vegetation like *Vallisneria* and *Chara*. In India, pulchellus is the only indigenous carp species, which feeds upon aquatic vegetation as well as submerged grasses. In absence of grass leaves, fish consumes grass roots and if this too becomes scare, it feeds upon decaying vegetable matter and bottom debris. Although adults feed upon small sized gastropods, the fingerlings feed upon filamentous algae along with insects.

Puntius conchonius

It is small, active indigenous fish commonly known as Rosy barb. It is distributed in freshwaters of Pakistan, India, Afghanistan, Nepal and Bangladesh. In India it is very common in beels, ponds, small streams and tanks of west Bengal and nearby states. This is one of the important ornamental beautiful fish. The fish can be kept together with other small fishes. This sizable, hardy and very popular Asian minnow is most impressively coloured during the mating period, when the normally silvery male takes on a rich claret flush and the slightly larger female becomes more luminous (Talwar and Jhingran, 1991).The fish has short, deep and compressed body and short head without barbells. Lateral line system incomplete with 24-28 scales. Fish has greenish orange colour on the dorsal side, silvery sides, and a large black blotch on the middle of caudal peduncle. Fins are orange coloured and dorsal fin tips are fringed and in black colour.

Choudhuri (2010) analyzed and identified the food consumed by the fish under natural and artificial conditions. Under natural conditions *P.conchonius* had a great affinity for phytoplankton and aquatic plants. Under artificial conditions (aquarium conditions) food was supplied from outside and *P.conchonius* preferred artificial feed, and tubifex worm. Plant materials were preferred less. So in an aquarium where foods were supplied from outside, the fish became omnivorous. Availability of certain food items made the fish euryphagous and availability of limited food items made it stenophagous. It may perhaps be said that *P.conchonius* is either a stenophagous omnivore depending on the food availability under natural and artificial conditions (Choudhuri, 2010).

Puntius sophore

Das *et al.* (2013) studied feeding habits of *Puntius sophore* through morphometry and observed that the mouth area of the fish did not show significant correlation to food types. On the basis of gut content corresponding to total length of the fish, the food habit of *P. sophore* can be grouped into three total length categories as category I or herbivorous (TL,5-7 cm),II or omnivorous (TL,5-7 cm) and III or omnivorous with more tendency towards zooplankton (TL,7.6 -8.5).

The total phytoplankton in the gut contents of fishes were mainly represented by filamentous algae, diatoms like *Nitzschia* sp., *Melosira* sp., *Synedra* sp., *Desmidium* sp., *Calothrix* sp., *Oscillatoria* sp., *Cloaterium* sp., *Scenedesmus* sp., *Navicula* sp., *Odeogonium* sp. etc.The zooplankton in the gut contents of fish were represented by crsutacea, cladocera, rhizopoda, actinopoda and copepods. Some nematodes and eggs of trematod were also present.

Puntius kolus

It is euryphagic mainly feeds on molluscs, insects. ostracods, bacillariophyceae and on grass seed and decaying plant tissues. Food comprised of bacillariophyceae (*Navicula, Amphora, Pleurosigma, Diatoma, Synedra, Fragilaria, Cymbella, Gomphonenma, Surirella* and *Cocconema*), chlorophyceae (*Spirogyra, Zygnema, Oedogonium, Cosmarium* and *Mougeotia*), plant matter, copepods (*Cyclops, Diaptomus, Nauplii* stages), ostracods (*Cypris* spp), insects (Chironomids), gastropods and bivalves, decayed organic matter, mud and miscellaneous items like *Oscillatoria, Anabaena, Merismopedia, Microcystis, Pediastrum, Scenedesmus, Closterium, Cosmarium, Staurastrum,* ostracod eggs, cladoceran and annelid setae (David and Rajagopal,1975).

Epithemia, Amphora, Synedera, Tabellaria, Eudorina and *Spirogyra* formed the major food items in the species collected from Bhatghar reservoir of Maharashtra (CIFRI, 1997).According to Anon (1982) diet of the species comprise molluscs, mainly gastropods, bivalves (60 per cent), chlorophyceae, mainly *Spirogyra, Pediastrum* (3 per cent), diatoms mainly *Fragilaria, Melosira, Navicula* (5 per cent) and organic detritus (30 per cent).Zooplankton recorded in the gut were mainly the copepods (2 per cent) (Anon,1982).

Puntius sarana

David and Rajagopal (1975) studied food and feeding relationship of *Puntius sarana* from Tugabhadra reservoir. Fish feeds on a variety of aquatic plants like *Hydrilla, Chara, Vallisneria,* bacillariophyceae and chlorophyceae. *Vallisneria* ranged from 4.8 per cent in October to 16.3 per cent in November. Chara was the predominant plant ingested up to 20.3 per cent. *Hydrilla* was maximum in January (22.5 per cent) and minimum in October. Ostracods were recorded in February, March and October as an incidental food. Dismembered appendages in insects were recorded in February, April, October and November. Decayed organic matter ranged from 10 per cent in May to 38.4 per cent in October. Particles of sand and mud would have consumed along with sessile bottom dwelling molluscs. This ranged from 1 per cent in February to 20 per cent in April.

Puntius melanampyx

It is an endemic ornamental fish of the Western Ghats. The colour of the live fish is dull red with three vertical black bands, the first one below dorsal fin to just below lateral line, second slightly behind base of dorsal fin and third just before base of caudal fin. Fins are pinkish edged with black. Anna Mercy *et al.* (2002) studied food and feeding habits of *Puntius melanampyx* from Bharathapuzha and Pamba rivers of Kerala. Mouth of fish is sub-terminal, protrusible and placed at the antero-ventral end of head. Jaws moderately built and gill rakers short and stumpy. Due to the occurrence of detritus including sand and filamentous algae in the gut content, *P.melanampyx* could be categorized as bottom feeder. In general, *P.melanampyx* could be regarded as an omnivore consuming a wide range of food materials like detritus, filamentous algae, plant matter, crustaceans and insects. The distinct preference to benthic flora and fauna is probably a reflection of the behaviour of fish which spend most of the time in benthic zone, as the material in the digestive tract faithfully reflects relative environmental densities of food items falling within the ingestible size range (Anna Mercy *et al.,* 2002).

Puntius chilinoides

It is omnivorous and feeds on insects and their nymphs. Insects and nymphs form major contribution (82 per cent) of the food, while green algae and diatoms come on the second and third place respectively. Fish is bottom and column feeder. It scraps the food materials from the stones and rocks with its hard upper and lower jaws. The mouth is crescentic subterminal and suctorial (Badola and Singh, 1980).

Puntius vittatus

Geetha *et al.* (1990) studied food and feeding habits of *P.vittatus* and observed that the fish is herbivorous feeder, detritus and algae being the most preferred food items. The structure of mouth shows modifications in relation to its feeding habits. The semi-circular mouth of this species is entirely different from that of carnivorous or omnivorous fishes. This is due to the fact that the food capture mechanism of herbivore is more complex than that of the other groups. They have to nibble or scrap algae from a hard substrate and crush hard plant cells or strain an enormous volume of water to retain a small quantity of plankton. Another feature that is associated with feeding is protrusibility.

Salmo-trutta fario

The feed habits and diet composition of the stream dwelling resident brown trout, *Salmo-trutta fario* were investigated by Nusrat *et al.* (2012).Analysis of monthly variations of stomach fullness indicated that feeding intensity was higher between

March and July than that for the spawning season that covered the period from October to December. A total of 4464 individual preys were counted representing Trichoptera, Diptera, Ephemeroptera, Plecoptera, Odonata, Coleoptera, amphipoda, hirudinea, megaloptera, trout eggs, plant seeds and terrestrial ants were also identified in the diet. The index of relative importance revealed that four food items together constituted more than 90 per cent of the diet,with the most important being brachycentridae (51.55 per cent), blepharocera (14.06 per cent), *Baetis* species (10.48 per cent) and *Ephemerella* species (5 per cent).

Mastacembelus armatus

Commonly known as 'Spiny eel', this fish found in muddy waters, especially in ponds and lakes. Body is eel like and greenish olive along back and yellowish beneath. A distinct streak of elongated spots runs along the lateral line from eye to the base of tail.

Seasonal changes in the feeding habits and food items of spiny eel, *M.armatus* were carried by Serajuddin *et al.* (1998).Well developed dentition, absence of gill rakers, strongly built stomach and short intestine, together with the dominance of animal matter in the gut contents, indicates the carnivorous and active predatory habits of *M. armatus*. Macro-crustaceans (especially shrimps) and forage fish were the main food of adults while annelids and aquatic insects were eaten by young specimens. Feeding intensity was high in early maturity and was relatively lower in fish with ripening gonads. Adults consumed more food in summer than winter and the rainy season. Food intake in younger specimens was greater during the post monsoon period and in autumn. The maximum number of empty guts were found in adults during spawning and in winter.

Das and Moitra (1955) pointed out that fish mainly feeds on crustaceans. Dutta (1989, 1990) carried out stomach content analysis of *M. armatus* from Jammu and reported it as selective insectivorous.

Ompak bimaculatus

Commonly known as 'Indian Butter catfish' or 'Pabda' belongs to the family siluridae. The origin of the name of Ompak is traceable to the Malay name Limpok/vernacular Malay name Ompak. The specific name bimaculatus means 'with two spots'. The fish has a wide spread distribution. These are recorded in India, Pakistan, Nepal, Bangladesh, Sri Lanka, China, Thailand, Cambodia, Laos, Vietnam and Indonesia. This fish has an endangered status in the North-eastern part of Western Ghats of India. In south Indian states it is documented as being found in both freshwater and brackishwaters. This fish occurs naturally in streams and rivers. Its movement are described as sluggish to moderate. From the early seedling stage the fish is highly cannibalistic. Larvae show preference towards live foods from fourth day onwards and they accept zooplankton up to 15 days. Mixed zooplankton serves as a good food for early stages of fish. From fry stage onwards their feeding habit was seen to be relatively relaxed, taking catfish, pellets, prawns, earthworms and mussels, beside consuming live, frozen and dried foods. Adults are cannibalistic and predatory in their feeding habit. Their food includes vegetable food and small fish like minnows, crustaceans and molluscs (Debnath *et al.,* 2011).

Fish is omnivorous, feeding mainly on vegetable matter and fish, which dominated with a percentage of 30.04 per cent.Vegatable matter take the precedence with an average percentage of 28.65 per cent.Crustacean adults, crustacean larvae, insect and molluscs were secondary and supplementary foods. Feeding intensity was more in December (48.3 per cent) with minimum (34.4 per cent) in May. The occurrence of empty stomach was found to be more during the peak spawning season (Arthi *et al.,* 2011).

Ompak malabaricus

Arthi *et al.* (2011) studied food and feeding habits of *Ompak malabaricus* of river Amaravathy (Tamil Nadu) and reported the omnivorous feeding habit of fish. *O.malabaricus* mainly feeds on vegetable matter and fishes. In addition also consume crustacean adults, crustacean larvae, and insects as main food. The zooplankton, nekotons, tadpoles, annelid worms and molluscs also observed in the gut. The poorly fed state in adults of *O.malabaricus* was seen more in June. This corresponds with the spawning period of fish.

Wallago attu

Commonly known as 'Indian freshwater shark' is a catfish found in temperate and tropical freshwaters, inhabiting deep flowing waters of rivers and tanks. It serves as a significantly proteinacious dietary contribution to the consumer. The food analysis of *W. attu* by Sakhare and Rawate (2012) revealed that its juveniles prefer aquatic insects (60 per cent), followed by fish (22 per cent), miscellaneous food items (10 per cent) and decayed matter (8 per cent).Therefore, it may be concluded that *W.attu* is a carnivorous fish feeding mainly on fish and insects. The major stomach contents of adults included fish (75 per cent), insects and their larvae (14 per cent), miscellaneous items (8 per cent) and decayed matter (3 per cent). Sugunan (1995) observed fish (82 per cent), insects (6.5 per cent), miscellaneous items (5.4 per cent) and decayed matter in the stomach of fish.

Bhavania australis

It is a typical hill stream loach occupying only in the small streams of highly elevated areas. It has a peculiar body shape with its dorsal profile rounded and having a flattened ventral surface. The body on the dorsal surface has big rounded black spots. It prefers high velocity and highly oxygenated waters.

Biju (2003) studied qualitative and quantitative aspects of feeding habit of this fish and divided food items into different seven groups *i.e.,* 1) Benthic insects (15.50 per cent of the food item), 2) Benthic microinvertebrates (cladocera and copepods forms 8.50 per cent of the consumed food materials), 3) Algae (constituted 40 per cent of the food materials),4)Plant matter (macrovegetation accounted for 8 per cent of the food consumed and comprised fragmentary/semidigested parts of leaves, stem and roots of aquatic plants and other semi digested vegetable matter), 5) Animal matter (accounted for 2.35 per cent of the gut content and comprised of unidentified and semidigested animal matter), 6) Detritus (5 per cent of the food was detritus), 7) Miscellaneous matter (included items like protozoans, fish eggs, shell matter, crustacean and insect appendages and sand particles). Thus, fish is benthic omnivore and mainly feeds on benthic insects, insect larvae, copepods, cladocerans and unicellular and filamentous algae.

Nemacheilus rupicola

Fish has a large terminal mouth. Because of the carnivorous feeding habit the elaborate feeding apparatus in the form of food scrapers are absent. The maxillary and pharyngeal teeth are present. The food items of *N.rupicola* are identified into seven groups *i.e.*, chlorophyceae, xanthophyceae, bacillariophyceae, cladocerans, copepods, ostracods, insects and miscellaneous items are the seven groups of food items obtained from the gut of *N.rupicola*.

Nemacheilus guntheri

It prefers hilly areas with gravel, cobbles and bedrock with a little amount of sand, as the substrate. In midland areas, gravel with pebble forms the main substrate. Body colour dark brown with three rows of whitish spots of different sizes and form; a deep short vertical bar at the base of caudal fin; and a spot on dorsal fin origin. Fish is benthic omnivore that mainly feeds on benthic microinvertebrates (52 per cent), algae (10.50 per cent), plant matter (5.40 per cent), ephemeropteran larvae (8.6 per cent), chironomid larvae (7.51 per cent), detritus (14 per cent) and miscellaneous items (2.30 per cent).

Neolissochilus hexagonolepis

Commonly known as 'Chocolate mahseer' or 'snubnosed mahseer' it is not only considered as a game and food fish but is also recognized as an icon in the water bodies of high lands of Eastern Himalaya region. It was also at one time a major fishery source of the state of Arunachal Pradesh. However, unfortunately, as of now, it is on the verge of extension. An economically important fish, its specific name Hexagonolepis is because of the hexagonal shape of the exposed portions of the scales.

Neolissochilus hexagonolepis is a voracious omnivorous feeder (Talwar and Jhingran, 1991) indicated by the high values of its gastro somatic indices (Jhingran 1975 and Dasgupta 1988).It mainly feeds on gastropod shells, filamentous and planktonic algae and vegetable debris. Sand and mud are also encountered in the stomach. In the early fingerling stage, it mainly feeds on insect larvae, and on aquatic beetles and flies. Aquatic vegetation and marginal grass constitute the main food, found in the guts of advanced fingerlings and adults.

It also prefers dead fish fingerlings, earthworms and shrimps offered by means of hooks. Observations of the gut content analysis revealed that the bulk food consisted of digested plant parts (51.4 per cent), followed by insects (24.5 per cent), other animal matter (7.5 per cent), sand and mud (7.2 per cent) (Sarma, 2009).

Botia birdi

Kant and Vohra (1993) studied the gut contents to find out the quantitative and qualitative composition of the different algal components. According to Kant and

Vohra (1993) the major dietary of this fish is animal food and the fish is carnivore. Insects and their larvae constituted about 62 per cent gut contents form the major part of the food of this fish, followed by 15 per cent daphnids, 4 per cent crustaceans and 2 per cent rotifers. In all 83 per cent diet of the fish is of animal origin. Of the rest 17 per cent of the digested green matter contributes 4 per cent.Only 13 per cent algae is found in the stomach of the fish, of which cyanophyceae is the major constituent (5 per cent), followed by 4 per cent each of desmids and diatoms and green filamentous algae. The algae in this fish is not digested giving an indication that probably this does not form the basic food of the fish or the fish does not relish it much and perhaps these algae enter inadvertently into the suctorial mouth of the fish. Even though algae do not form a direct food of the fish, but it has been proved that the insects and daphnids along with rotifers and crustaceans do feed on the algae thereby forming an indirect food for the fish.

Eutropiichthys vacha

The freshwater catfish, *E.vacha* is an economically important teleost fish. The smaller size has often been placed in aquaria while large size are used as food. Fish is highly predaceous. Aquatic insects, crustaceans, annelids, small forage fish were the main food of adult, while phytoplanktons, crustaceans and macrophytes constitute basic food of juveniles. Feeding intensity was high in early maturation and post spawning stages and was relatively lower in the specimens with ripe gonads. Adult consumed more food in autumn than in winter and rainy season. Food intake in younger specimens was greater during monsoon and post-monsoon season (Abbas, 2010).

The organs concerned with feeding and digestion are modified according to its feeding behaviour. The mouth is sub terminal and slightly over hang by the snout. The gap of the mouth is 19 per cent of the length of the head. Mouth opening is wide and surrounded by four pairs of barbells. These barbells help in seeking out food according to its selectivity. Teeth are present in jaws, mouth and pharynx. Teeth on the jaws are small numerous and villiform. Teeth are backwardly directed and pointed. A pair of upper and lower pharyngeal teeth occurs as pads near to the gill arches. Gill rackers are few and widely spaced. They are elongated. Small prey prevented from escaping through the opercular opening by these elongated filamentous rakers, which act as traps.

Gudusia chapra

It is a commercially important fish in freshwater resources of India, Pakistan and Bangladesh. Recently population of the fish has alarmingly declined in the rivers and reservoirs and its production has been restricted to the lakes, ponds, ditches, inundated fields and other closed water impoundments.

Feeding habit of *Gudusia chapra* from flood plain lakes of West Bengal were studied by Mondal and Kaviraj (2010).The fish showed a planktophagus and omnivorous feeding habit, algae constituting bulk of the food item. Feeding intensity fluctuated between seasons, pre and post spawning months showing higher intensity than the spawning months. Maximum feeding activity was recorded in April in both the sexes. Feeding intensity reduced thereafter, both sexes showing two more peaks, July and December for male and September and December for female. Feeding intensity in male was significantly higher than female during May to August. The fish shows a preference for herbivorous diet. Algae and other plant matters constituted the bulk of the gut content followed by crustacean, rotifers, plant matter, protozoa, insecta and miscellaneous materials. Among algae, chlorophyceae was the dominant group followed by bacillariophyceae. The other important groups were cyanophyceae, eugelnophyceae and dinophyceae. Average percentage of algae, crustaceans, rotifers, plant matters, protozoa, insecta and miscellaneous materials in the gut content were respectively 49.20, 18.29, 15.70, 5.60,5.20,3.60 and 2.60 per cent.

Mystus vittatus

Locally known as 'Tengra', it is a minor bagrid catfish that occurs in the Eastern India especially in West Bengal, Orissa and Assam having food value and a good market demand. It is due to its taste. Chakrabarty *et al.* (2007) studied food and feeding habits of *Mystus vittatus*. The post larvae (5-25mm in length) feed on zooplankton consisting mainly of cladocerans, copepods and protozoans. A small quantity of *Microcystis, Volvox* and other colonial algae were observed to pass through the gut mostly in undigested condition. Most of the juveniles (26-45 mm in length) consumed zooplankton, mainly copepods, followed by aquatic insects. Phytoplankton, mud, detritus and vegetable matter were also encountered in the diet. The food of adults (46-118 mm in length),constituted mainly of zooplankton and small aquatic insects in bulk, followed by less quantity of worms, mud, detritus and vegetable matter. Most of the adults examined were "full" with feed. Empty stomachs having trace of food were rarely encountered, thereby suggesting that intensity of feeding of the species was fairly high. The study revealed an exclusively phytophagous habit in post larvae stages, *M.vittatus* changes over to an essentially plankton and insect diet in juvenile stages. The adults fed on plankton and aquatic insects mainly, and worm, mud, detritus, vegetable matter in small quantity as supplementary diets. The fish seemed to explore all ecological niches in aquatic biotopes for food.

In highly polluted Hussainsagar lake of Hyderabad (Andhra Pradesh), fish mainly feed on insect larvae and insects. The other food items include both plant and animal matter. Intensive feeding was observed during December to February (Siva Reddy and Babu Rao,1987).Insect larvae and insects formed the major food item. Insect larvae occurred in all the guts. They were mainly chironomid larvae and pupae (diptera), nymphs of dragon flies, larvae of tubifera, lepidoptera and trichiptera. Insects consisted of corixa and notonects.

Mystus seenghala

Gut content analysis in fish indicates that the gut content consists of about 25 to 30 per cent plant food matter, 50-60 per cent of animal matter and detritus. The percentage of animal matter in the diet increased by 5-10 per cent with increase in body length from smaller juvenile 5-10 cm to about 25-35 cm. The quality and quantity of food varies according to habitat (Yeragi and Yeragi, 2014).They also concluded that *Mystus seenghala* is omnivorous.

Gut content analysis of *Mystus seenghala* was carried out by Babare *et al.* (2013).They observed that about 80-90 per cent gut content is animal food matter (weed fishes such as *Puntius ticto, Chela phulo* and *Ambassis nama*).The percentage of animal matter in the diet increased by 5-10 per cent with increase in body size from small (20-25 cm) to large (60-65cm).

Mystus montanus

Food and feeding habits of *M.montanus* inhabiting Tambaraparani river fed system were studied by Jesu Arockia *et al.* (2004).Gut contents showed small fishes (12.6 per cent), cladocerans (11 per cent),mollusks (10.8 per cent),annelid worms (10.5 per cent),rotifers (9 per cent),insect larvae (7.8 per cent),copepods (7.6 per cent),detritus (7.6 per cent),crustaceans (7.6 per cent),fish scales (5 per cent),algae (5 per cent) and unknown items (5.5 per cent).Alimentary canal modification in fish suits for omnivorous type of feeding. The fish appears to be suited for pond culture as it feeds on food of lower trophic levels.

Moderate built jaws, thin canine and weak vomarine teeth, less and moderately long gill rackers and moderately long and few coils in intestine indicated that *M.montanus* is an omnivorous fish. It was categorized under bottom feeders due to the occurrence of detritus (including sand) and filamentous algae in the gut (Menon and Chacko, 1956).

Mystus cavasius

Fish can be categorized as eury-omnivorous as it feeds on wide range of diet including both the vegetable as well as animal diet but on the basis of biomass of food material, animal material contributes a major portion of the diet. Hence it may be pointed out as carnivore in its feeding habit (Chaturvedi and Parihar, 2014).

Xenentodon cancila

The gut contents of *X. cancila* consist of large portion of animal material in biomass and plant materials is in lesser proportion. Hence it comes in the category of carnivorous fish (Chaturvedi and Parihar, 2014). Empty stomach were observed in May and June (pre-spawning period) due to bigger size of the gonads which occupying larger space in the body cavity and allowed a little space for the food. This may be concluded that food and feeding habits of the fish is correlated with its natural habitat.

Notopterus notopterus

Food and feeding habits of a threatened freshwater 'featherback', *Notopterus notopterus* were studied by Jesu *et al.*,(2004).The gut contents were belonged to 10 different genera along with sand granules, detritus matter and some unidentified food items. The most preferable food items were insect larvae (23.24 per cent), algae (17.18 per cent), fish scale (16.25 per cent) and annelid worms (15.51 per cent) had a highest value of index of preponderance. Another important prey groups *viz.*, cladocera, crustacean, rotifer, aquatic insects, smaller own young ones, fry of *Puntius* sp. and juveniles of prawn. The cannibalistic nature was confirmed by Jesu *et al.* (2004).

A study on the feeding habits of fish from Hemavathy reservoir indicated that the major stomach contents of fish includes insects (42 per cent), semi-digested organic matter (24 per cent),fish remains (16 per cent), prawns (7 per cent), miscellaneous items (7 per cent) and plant tissues (4 per cent) (Sugunan, 1995).

Chanda nama

Copepods, mainly *Naupli, Cyclops, Diaptomus* (65 per cent), insects, mainly dipteran larvae, chironomids, mayfly nymphs (17.5 per cent),semi digested organic matter (15 per cent),fish remains (2.5 per cent) comprise the main food items in species recorded from Nagarjunasagar reservoir of Andhra Pradesh (Anon,1982).

Channa punctatus

It is commonly known as 'Spotted murrel' or 'Green snakehead'. The fish is distributed throughout the south-east Asian countries and is a hardy air-breathing fish. The fish breeds during south-west monsoon and north-east monsoon in flooded rivers, at streams and ponds. It is found in swamps and beels and widely cultured in paddy fields of Assam and West Bengal. It fetches a higher price (about Rs.175/kg).

Fish is carnivorous and changes its food habit with the change in seasons. Index of pre-ponderance of various food compositions in the gut of *C. punctatus* indicated that the fish was the most dominant food item in the gut, followed by the insects, crustaceans, plant matter mucks and unidentified materials, annelids and molluscs (Saikia *et al.*, 2012).In fry stages, most dominant food item was zooplankton followed by insects, annelids, crustaceans and plant matter. In juvenile insect was the most preferred food item, followed by trash fishes, crustaceans, annelids, plant matter and molluscs. In adult fish was found to be the most dominant item in the gut next food items in order of preference were insects, mucks and unidentified material, plant matter, crustaceans, annelids and mollusks. Maximum feeding intensity in females was from August through October and lowest value was recorded in winter (November –January).

Babu Rao (1990) studied effect of eutrophication on the biology of *C. punctatus* from Hussainsagar Lake of Hyderabad. Hussainsagar waters being polluted, the insects and insect larvae had a thick population, since *C.punctatus* being carnivorous fish, insects and insect larvae naturally dominated among the gut contents. The oligochete annelids formed the other important component of the food, while in fishes from Guntur (Reddy, 1980) and Aligarh (Qayyum and Qasim,1964) fishes formed the main food item.

Channa striatus

It is commonly known as 'Striped murrel', and is the commonest of all murrels found in India. The striped murrel, *C.striatus* is a native freshwater fish of tropical Africa and Asia (Ng and Lim, 1990).It is the most widely cultured murrel species in many southeast Asian countries (Ling,1977).It is highly predatory in habit and mainly feeds on small fishes, tadpoles, insects and live items in ponds (Selvaraj and Francis, 2005).Based on available literature, the natural feed of the post larvae of *C.striatus* consists of cladocerans, copepods, phytoplanktonic organisms, rotifers and small aquatic insects.

Osteobrama belangeri

It is an important food fish in Manipur and enjoys a high commercial status and is preferred by the local consumers.

Basudha and Vishwanath (1999) studied food and feeding of *Osteobrama belangeri* from Loktak lake and adjoining lakes of Manipur. Animal matters including zooplankton, insects and worms are preferred by juvenile. However, as the fish grows, it has a higher preference for the plant food items. The adult fish prefers plant matters (large amounts of leaves, stems and roots of aquatic plants).Other food items such as insects, worms, zooplankton and phytoplankton are also found in large amount in the gut. Therefore, *O.belangeri* is an omnivorous fish which mainly feeds on zooplankton and algae in the juvenile stage and macrovegetation in the adult stage.

Osteobrama bakeri

It is an endemic and vulnerable fish found in Kerala. This rare species is a valuable food fish and has great potential as an ornamental fish. Fish is omnivorous and mainly feeds on insects (22.4 per cent), microinvertebrates (29.48 per cent),algae (8.63 per cent),plant matter (25.4 per cent), animal matter (7.46 per cent) and miscellaneous items (7.04 per cent).There is a little variation in the gut content of male and female fishes (Biju, 2003).

Amblypharyngodon mola

It is found abundantly in the rivers, canals, ponds, lakes, floodplains, beels and ditches of West Bengal. Fish has high protein, vitamin and mineral contents. Zafri and Ahmed (1981) reported that *A.mola* contains 200IU of vitamin A per gram of edible protein.Taking 3 whole A.mola specimens per day can save a child from blindness due to shortage of vitamin A (BSS, 1988).Fish is planktivorous and mainly feeds on chlorophyceae (78.77 per cent), bacillariophyceae (11.85 per cent), and debris with mud (3.09 per cent).These food item clearly indicated that fish preferred phytoplanktonic food. The fish incidentally took the animal nature plankton foods of rotifera and crustacean in negligible amount (Mamun *et al.,* 2004).

Musa (2008) studied food and feeding habits of *A.mola* and observed algae as most dominant food item (37.97 per cent), followed by rotifers (13.66 per cent),aquatic insects (12.36 per cent), crustaceans (9.69 per cent), protozoans (8.65 per cent),aquatic plants (7.40 per cent) and unidentified food materials (10.27 per cent).The greatest percentage of occurrence of animal food *viz.,* algae, rotifers, aquatic insects, crustaceans, protozoans etc strongly indicated that *A.mola* is a planktivorous fish. During spawning period, it showed a reduced level of feeding (Musa, 2008).

Heteropneustes fossilis

Heteropneustes fossilis, commonly known as 'Singhi' or 'Kari', is a highly priced air breathing Indian catfish. The fish found in all types of freshwaters such as rivers, canals, swamps, ponds, ditches and derelict water areas. It adapts well to hypoxic conditions, withstands high stocking density and utilizes atmospheric oxygen where there is depletion in dissolved oxygen in the water body. The excellent flesh quality, high protein content, physiologically available iron, essential amino acids and low fat contents make *H.fossilis* commercially important. Its protein content is higher but fat content is lower than that of major carps. Because of high haemoglobin content, it is recommended for recuperation of patients.

Insect larvae, insects and insect eggs formed the major food items of this species collected from Hussainsagar lake (Babu Rao, 1990).Crustaceans, plant materials and algae, gastropods, bryozoans etc were the other groups of food items in order of abundance.

In specimens collected from Aligarh, the gut contents consisted mainly of crustaceans followed by molluscs, insect larvae, small fishes, plant materials etc (Bhatt,1968), and in specimens from Punjab the major food item was crustaceans followed by oligochaetes, fish, insects and insect larvae (Johal,1981).

Clarias batrachus

Popularly known as '*Magur*' or '*Walking catfish*', it is found across southern Asia including India, Pakistan, Sri Lanka, Bangladesh, Thailand, Indonesia, China and Philippines. It is a hardy and voracious feeder and become active at night. It is an important air-breathing cat fish with good markets especially in North-Eastern parts of India where it fetches a higher price than the carps. In some parts of Assam, the fish is sold at more than Rs. 300 per kg. The fish has medicinal value, better taste, rich protein content and of fewer spines. The fish is in great demand in the northeastern part of India particularly in West Bengal, Assam, Orissa and Bihar for its high nutritional value. It contains higher percentage of protein and iron as compared to other edible freshwater fish species. Its fat content is also very low and is therefore easily digestible so that it is very useful during convalescence.

Sakhare and Chalak (2014) analyzed gut content of *Clarias batrachus* collected from different water bodies around Ambajogai and revealed that the food consisted of insect larvae, small fish, shrimps and organic debris. In *Clarias batrachus* small fish and insect larvae were preferred as the primary food item in all the seasons. On average for all months of the study period, small fish dominated the list with a percentage of 30.27.The other food items in descending order are insect larvae (27.66 per cent), worms (20.27 per cent), shrimps (14.3 per cent) and organic debris (7.05 per cent).In the present investigation it is found that the feeding intensity in mature fishes was found to be very poor during August to September. This period of poor feeding activities in case mature fish coinicides with the peak spawning season.

Clarias gariepinus

The African sharp tooth catfish, *Clarias gariepinus* (generally referred as thai magur) is extremely hardy and withstands adverse environmental conditions and habitat instability. Furthermore, it can efficiently assimilate a wide variety of animal and plant protein.

It is highly carnivorous fish feeding on insects, crabs, snails and fish and can also eat live birds, rotting flesh, natural food organisms, plankton, plants and fruits (Fish Base, 2004).In culture they are fed on trash fish mixed with rice bran and boiled eggs, chicken and slaughterhouse waste (Singh and Mishra, 2001).

It is carnivorous in feeding habit and fish is the most important food item. It contributed 81.7 per cent of the food items of the juveniles and 86.8 per cent of the adults by volume (Elias, 2000). *Oreochromis niloticus* was the most utilized prey of *C.gariepinus. O.niloticus* accounted for 71 per cent of the food eaten by juvenile fish (16.3 to 35 cm TL) and 77.5 per cent of the food of adults by volume. Other food items found in the stomachs of *C.gariepinus* include insects, fish eggs, gastropods, pieces of macrophytes, detritus and zooplankton.

Many authors concluded that the African magur is an omnivorous slow moving predatory fish which feeds on a wide variety of food items from zooplankton to fishes of half of its own length (Janseen,1987).

Rita rita

It is a sluggish, bottom dwelling, carnivorous catfish and the bulk of its food primarily consists of molluscs.In addition, it feeds on small fishes, crustaceans, insects, as well as on decaying organic matter (Yashpal *et al.,* 2006).The surface architecture of the mouth cavity of *Rita rita* was also examined by Yashpal *et al.,*(2006) for better understanding its role in relation to the species' food and habitat preferences. In *R. rita* the mouth cavity is spacious and opens anteriorly through a wide transverse mouth, which is bordered by the upper and the lower lips. The roof of the mouth cavity comprised antero-posteriorly an upper jaw consisting of the premaxillae and the maxillae, the velum; the palatine regions bilaterally supported by the palatine bones; and the palate extending up to the pharynx.

Syeda *et al.* (2013) studied food and feeding habits of *R. rita* and observed that the diet of fish consisted of a broad spectrum of food types but crustaceans were dominant, with copepods constituting 20.73 per cent,other non-copepode crustaceans constituted 12.01 per cent.The next major food group was insect (15.97 per cent), followed by molluscs (14.76 per cent), teleosts (12.98 per cent) and fish eggs (8.608 per cent).Food items like teleosts, molluscs, insects and shrimps tended to occur in the stomachs in higher frequencies with an increase in *Rita rita* size (up to 30.5 to 40.5 cm),while fish eggs, copepods and non-copepode crustaceans tended to increase in stomachs at sizes between 10.5 to 20.5 cm. Analysis of monthly variations in stomach fullness indicated that feeding intensity fluctuated throughout the year with a low during June and August corresponding to the spawning period.

Garra lamta

Kanwal and Pathani (2012) recorded a total of 32 taxa of chlorophyceae, 30 taxa of bacillariophyceae, 2 genera of cyanophyceae and some miscellaneous items in the gut of *Garra lamta* from some tributaries of Suyal river of Uttarkhand. Algae formed the main item of the gut contents. The food items indicate the bottom grazing planktoherbivorous feeding habits of the fish.

The minimum quality of food items were observed during summer, especially in April-May (onset of spawning), while the maximum during winter, especially in November. During summer months (April and June) the variety of bacillariophyceae and chlorophyceae were lower in comparison to other season of the year.

Horabagrus brachysoma

Commonly known as 'Golden catfish', is widely distributed in backwaters of Kerala and sold in live condition. It enjoys a high commercial status and is preferred over carps due to its superior flesh quality and flavor. With its attractive and brilliant greenish-yellow colour, it is much sought after for aquarium keeping. It attains a size of 45 cm total length.

Fish is omnivorous, feeding mainly on filamentous algae (39 per cent), detritus (23 per cent) and fish offal (22 per cent). Macrovegetation (8 per cent) and crustaceans, mainly prawn larvae (6 per cent) and others (2 per cent) were also recorded as food items in smaller quantities (Padmakumar *et al.,* 2009).The percentage of empty stomachs were higher in January-February, July and October-November months. The intake of food was reduced significantly among females during pre-spawning and spawning months. In culture systems, the fish accept commercial pellets copiously (Padmakumar *et al.*, 2004).

Tenualosa ilisha

Commonly known as Hilsa, the Indian shad, *Tenualosa ilisha* has established itself as one of the most important commercial fishes of the Indo-Pacific region. It has a wide range of distribution and it occurs in marine, estuarine and riverine environments. The fish is found in the Bay of Bengal, Persian Gulf, Red Sea, Arabian Sea, Vietnam Sea and China Sea. The riverine habitat covers the Satil Arab, and the Tigris and Euphrates of Iran and Iraq, the Indus of Pakistan, the rivers of eastern and western India, namely the Ganga, Bhagirathi, Hooghly, Rupnarayan, Brahmaputra, Godavari, Narmada, Tapti and other coastal rivers, the Irrawaddy of Myanmar, and the Padma, Jamuna, Meghna, Karnafully and other coastal rivers of Bangladesh. The major portion of hilsa fishery (about 90 per cent) is captured in waters of Bangladesh, India and Myanmar (Bhaumik and Sharma, 2011).

Hilsa is mainly a plankton feeder. Hora (1938) recorded that hilsa fry (20-40mm) mainly fed on diatoms, copepods, daphnia and ostracods whereas, the younger hilsa (up to 100 mm) feeds on smaller crustaceans, insects and polyzoa. Jones and Sujansingani (1951) confirmed that hilsa are plankton feeders but they do not exhibit any selectivity in feeding. Pillay (1958) expressed that there is no appreciable change in feeding intensity during spawning period. Halder (1968) studied the food of young hilsa and concluded that the young of hilsa is dominantly plankton feeder and the species feed at all depths of water either in the freshwater zone or in the tidal zone of the estuary.

Pillay and Rao (1962) observed that from January to March feed was fairly intensive and from June to November feeding intensity was less as little amount of feed was available in the stomach of collected species and they found that the diatoms are the predominant food items in the river Godavari. Gut content analysis of different stages of hilsa revealed that copepods were the most important food items consumed by the fish of all sizes as observed by De and Datta (1990).De *et al.* (2013) confirmed that hilsa prefer copepods in their early stages and shift their preference towards diatoms when they grew beyond 50 g size.

References

Abbas, Ashraf. 2010. Food and feeding habits of freshwater catfish, *Eutropiichthys vacha* (Bleeker). *Indian J. Sci. Res.* 1(2): 83-86.

Anna Mercy, T. V., Raju Thomas, K. and Eapen Jacob. 2002. Food and feeding habits of *Puntius melanampyx* (Day)-An endemic ornamental fish of the Western Ghats. In: riverine and reservoir Fisheries of India (Eds. Boopendranath, M. R., Meenakumari, B., Jose Joseph, Sankar, T. V., Pravin, P. and Leela Edwin) pp. 172-175, Society of Fisheries Technologists (India), Cochin.

Anon, 1982. Final Report of All India Coordinated Project on Ecology and Fisheries of Freshwater reservoirs, Nagarjunasagar, Research Information Series, 3 March 1983, Central Inland Capture Fisheries Research Institute, Barrackpore, West Bengal, India, pp. 148.

Arthi, T., Najarajan, S. and Sivakumar, A. A. 2011. Food and feeding habits of two freshwater fishes, *Ompak bimaculatus* and *O. malabaricus* of river Amaravathy, Tamil Nadu. *The Bioscan*, 6(3): 417-420.

Babare, R. S., Chavan, S. P. and Kannewad, P. M. 2013. Gut content analysis of *Wallago attu* and *Mystus (Sperata) seenghala*, the common catfishes from Godavari river system in Maharashtra State. *Adv. Biores.* 4(2): 123-128.

Babu Rao, M. 1990. Effect of eutrophication on the biology of fishes in reservoirs-A case study at Hussainsagar. p. 48-52. In: Jhingran, A. G. and Unnithan, V. K. (eds.). Reservoir Fisheries in India. Proceedings of the National Workshop on Reservoir Fisheries, 3-4 January, 1990, Special Publication 3, Asian Fisheries Society, Indian Branch, Mangalore, India.

Badola, S. P. and Singh, H. R. 1980. Food and feeding habits of fishes of the Genera Tor, Puntius and Barilius. Proc. Indian Natn. Sci. Acad. B 46(1): 58-62.

Basudha, Ch. And Vishwanath, W. 1999. Food and feeding habits of an endemic carp, *Osteobrama belangeri* (Val.) in Manipur. *Indian J. Fish.* 46(1): 71-77.

Bhatt, V. S. 1968. Studies on the biology of some freshwater fishes. Part VII. *Heteropneustes fossilis* (Bloch). *Indian J. Fish.* 15 (1 and 2): 99-115.

Bhaumik, Utpal and Sharma, A. P. 2011. The fishery of Indian shad (*Tenualosa ilisha*) in the Bhagirathi-Hooghly river system. *Fishing Chimes*, 31(8): 20-27.

Bhuiyan, A. S., Afroz, S. and Zaman, T. 2006. Food and feeding habit of the juvenile and adult snake head, *Channa punctatus* (Bloch). *J. Life. Earth. Sci.* 1(2): 53-54.

BSS (Bangladesh Sanbad Sangstha). 1988. Ten lakhs people are blind in Bangladesh. The daily Ittefaq, January 15, 35(23).

Chakrabarty, N. M., Chakrabarty, P. P. and Mondal, S. C. 2007. On breeding of Tangra (*Mystus vittatus*) and production of its post larvae/fry. *Fishing Chimes*, 27(4): 16-18.

Chaturvedi, Jaya and Parihar, Deepak Singh. 2014. Food and feeding habits of two freshwater catfish, *Mystus cavasius* and *Xenentodon cancila* from Chambal river (MP). *International Journal of Science and Research,* 3(8): 639-642.

Choudhuri, Sagarika. 2010. Food and feeding strategy of an indigenous ornamental fish-*Puntius conchonius. Fishing Chimes*, 29(10): 72-74.

Das, S. M. and Moitra, S. K. 1955. Studies on the food of some common fishes of Uttar Pradesh, India. 1. The surface feeders, the mid feeders and bottom feeders. Proceedings of the National Academy of Sciences, India, 25: 1-6.

Das, S. M. and Moitra, S. K. 1963. Studies on the food and feeding habits of some freshwater fishes of India, *Ichthyologica*, 2(182): 107-116.

Das, Saon, Nandi Sudarshan, Majumdar sandip and Saikia Surjya Kumar. 2013. New characterization of feeding habits of *Puntius sophore* (Hamilton, 1822) through morphometry. *Journal of Fisheries Sciences. Com*, 7(3): 225-231.

Dasgupta, M. 1988. A study on the food and feeding habits of the copper mahseer, *Acrossochilus hexagonolepis* (McClld). *Indian J. Fish.* 35(2): 92-98.

David, A. and Rajagopal, K, V. 1975. Food and feeding relationships of some commercial fishes of the tungabhadra reservoir. *Proc. Ind. Nat. Sci. Acad.* B41: 61-74.

De, D. K. and data, N. C. 1990. Studies on certain aspects of the morphology of Indian shad hilsa, *Tenualosa ilisha* (Hamilton) in relation to food and feeding habits. *Indian J. Fish.* 37(3): 189-198.

De, Debasis, Shyne Anand, P. S., Sinha Subhasmita and Suresh, V. R. 2013. Study on preferred food items of Hilsa (*Tenualosa ilisha*). *International Journal of Agriculture and Food Science Technology*, 4(7): 647-658.

Debnath, C., Das, S. K. andDatta, M. 2011. Biology and culture potential of Butter catfish (*Ompak bimaculatus* Bloch, 1794). *Fishing Chimes.* 30 (10 and 11): 30-31.

Dube, Kiran. 2003. Biology, reproductive biology and embryonic development of carps. In: Carp and catfish breeding and culture (Ed. Langer, R. K., Kiran Dube and Reddy, A. K.), Central Institute of Fisheries Education, Mumbai, pp. 12-29.

Dutta, S. P. S. 1989, 1990. Food and feeding ecology of *Mastacembelus armatus* (Lecep.) from Gadigarh stream, Jammu. *Matsya*, 15 and 16: 66-69.

Elias, Dadebo. 2000. Reproductive biology and feeding habits of the catfish *Clarias gariepinus* (Burchell) (Pisces: Clariidae) in Lake Awassa, Ethiopia. *Ethiopian Journal of Science.* 23(2): 213-246.

Fish Base, 2004. Fish base relational Database CD Rom, ICLARM;205 Bloomingdak Bldg; Salceds St. legapsi Village, Makati city, Metro Manilla 1200, Philippines.

Geetha, S., Suryanarayanan, H. and and Balakrishnan Nair, N. 1990. On the food and feeding habits of *Puntius vittatus* (Day). *Proc. Indian natn. Sci. Acad.* B56(4): 327-334.

Gomos, A., Yilmaz, M. and Polat, N. 2002. Relative importance of food items in feeding of *Chondrostoma regium* Heckel, 1843, and its relation with the time of annulus formation. *Turk. J. Zool.* 26: 271-278.

Halder, D. D. 1968. Observations of the food of young *Hilsa ilisha* (Ham.) around Nabidwip in the Hooghly estuary, *J. Bombay Nat. Hist. Soc.*, 65(3): 796-798.

Hall, D. 1981. The feeding Biology of the common carp (*Cyprinus carpio* L.) in Lake Alexandria and the lower River Murray South Australia. Hons. thesis, Department of Zoology, University of Adelaide. 57p.

Hora, S. L. 1938. A preliminary note on the spawning grounds and bionomics of the so-called Indian shad, *Hilsa ilisha* (Ham.) in the river Ganges. *Rec. Indian. Mus.*, 40(2): 147-158.

Hora, S. L. and Pillay, T. V. R. 1962. Handbook on fish culture in the Indo-Pacific Region, FAO Fish. *Biol. Tech. Paper*, 14: 204.

Janseen, J. 1987. Hatchery management of the African clariid catfish *Clarias gariepinus* (Burechell, 1822). In: Selected Aspects of Warmwater Fish Culture, (Eds) A. Coche and D. Edward. FAO/UNDP, Rome, 1989, 181pp.

Jesu Arockia Raj, A., Haniffa, M. A., Seetharaman, S. and Singh, S. P. 2004. Food and feeding habits on endemic catfish *Mystus montanus* (Jerdon) in river Tambaraparani. *Indian J. Fish.*, 51(1): 107-109.

Jesu Arockia Raj, A., Seetharaman, S., Haniffa, M. A. and Singh, S. P. 2004. Food and feeding habits of a threatened featherback, *Notopterus notopterus. J. Aqua. Biol.* 19(1): 115-118.

Jhingran, V. G. 1975. Fish and Fisheries of India, Hindustan Publishing Corporation, New Delhi, India, 954pp.

Jhingran, V. G. and Khan, H. A. 1979. Synopsis of biological data on the *Cirrhinus mrigala* (Hamilton, 1822), FAO Fisheries Synopsis No. 120, Rome, FAO, 78pp.

Jhingran, V. G. and Pullin, R. S. V. 1985. A hatchery manual for the common, Chinese and Indian major carps, ICLARM Studies and reviews 11, 191pp, Asian Development Bank, Manila, Philippines and International Center for Living Aquatic Resources Management, Metro Manila, Philippines.

Johal, M. S. 1981. Food and feeding habits of some fishes of Punjab. Vest. Cs. Spolac. Zool. 45: 87-93.

Jones, S. and Sujansingani, K. H. 1951. The Hilsa fishery of the Chilka Lake, *J. Bombay Nat. Hist. Soc.* 50(2): 264-280.

Kant Sashi and Vohra, Shama. 1993. Role of algae as primary producer in fish production in Kashmir lakes. In: Advances in Limnology (Edited by H. R. Singh), Narendra Publishing House, Delhi, pp. 79-86.

Kanwal, B. P. S. and Pathani, S. S. 2012. Food and feeding habits of a hillstream fish, *Garra lamta* (Hamilton-Buchanan) in some tributaries of Suyal river, Kumaun Himalaya, Uttarakhand (India), *International Journal of Food and Nutrition Science*, 1(2): 16-22.

Khan, R. A. 1972. Studies on the biology of some important major carps. Ph. D. thesis, Aligarh Muslim University, Aligarh, India, 185pp.

Khumar, Faroogh and Siddiqui, M. S. 1989. Food and feeding habits of the carp *Labeo calbasu* Ham. in North Indian Waters. *Acta Ichthyologica Et. Piscatoria*, XIX(1): 33-48.

Kumar, Raj; Sharma, B. K. and Sharma, L. L. 2007. Food and feeding habits of *Catla catla* (Hamilton-Buchanan) from Daya reservoir, Udaipur, Rajasthan. *Indian J. Anim. Res.* 41(4): 266-269.

Lakra, W. S. and Singh, A. K. 2007. Exotic fish introduction in Indian waters-past experience and lesson for the future. *Fishing Chimes.* 27(1): 30-34.

Lammens, E. H. R. R. and Hoogenboezem, W. 1991. Diets and Feeding behavior. In cyprinid fishes: Systematics, Biology and Exploitation (Eds. I. J. Winfield and J. S. Nelson). pp. 353-376, Chapman and Hall, London.

Ling, S. W. 1977. Aquaculture in Southeast Asia: A historical overview. Seattle: University of Washington Press, pp. 108.

Mamun, A., Tareq, K. M. A. and Azadi, M. A. 2004. Food and feeding habits of *Amblypharyngodon mola* (Hamilton) from Kapti reservoir, Bangladesh. *Pakistan Journal of Biological Sciences,* 7(4): 584-588.

Menon, M. D. and Chacko, P. I. 1956. Food and feeding habits of freshwater fishes of Madras State. Proc. Indo-Pacific Fish. Coun. Sections II and III. IPFC.

Mondal, Debjit Kumar and Kaviraj, Anilava. 2010. Feeding and reproductive biology of Indian shad *Gudusia chapra* in two floodplain lakes of India. *Electronic Journal of Biology.* 6(4): 98-102.

Musa, A. S. M. 2008. Food and feeding habits of *Amblypharyngodon mola* (Hamilton-Buchanan). *Fishing Chimes,* 28(1): 124-125.

Ng, P. K. L. and Lim, K. K. P. 1990. Snakeheads (Pisces: Channidae): Biology and economic importance. In: C. L. Ming and P. K. L. Ng (Eds.). Essays in Zoology. Papers commemorating the 40th anniversary of the Department of Zoology, National University of Singapore, Singapore, 127-152.

Ni, D. and Wang, J. 1999. Biology and Diseases of Grass Carp. Ed. Ni D and Wang J., Science Press, Beijing, China, pp. 437.

Nikolskii, G. V. 1963. The ecology of fishes, translated from Russian by L. Birkett, Academic Press, London, New York, 352pp.

Nusrat Rasool, Ulfat Jan and G. Mustafa Shah. 2012. Feeding habits and diet composition of brown trout (*Salmo-trutta fario*) in the upper streams of Kashmir valley. *International Journal of Scientific and Research Publications,* 2(12): 1-8.

Padmakumar, K., Anuradha Krishnan, G., Bindu, L., Sreerekha, P. S. and Joseph, Nita. 2004. Captive breeding for conservation of endemic fishes of Western Ghats, India, NATP Publication, Kerala Agricultural University, India, 79 pp.

Padmakumar, K. G., Bindu, L., Sreerekha, P. S. and Joseph, Nitta. 2009. Food and feeding behaviour of the golden catfish, *Horabagrus brachysoma* (Gunther). *Indian J. Fish.* 56 (2): 139-142.

Pillay, S. R. and Rao, K. V. 1962. Observation on the biology and the fishery of hilsa, *Hilsa ilisha* (Ham.) of river Godavari. Proc. Indo. Pacif. Fish. Coun, 2: 37-62.

Pillay, T. V. R. 1958. Biology of the hilsa, *Hilsa ilisha* (Hamilton) of the river Hooghly. *Indian J. Fish.*, 5: 201-257.

Piska, Ravi Shankar. 1999. Fisheries and Aquaculture, Lahari Publications, Hyderabad, pp. 452.

Popma, T. and Masser, M. 1999. Tilapia: Life history and biology. Southern Regional Aquacultural Centre, 283pp.

Qayyum, A. and Qasim, S. Z. 1964. Studies on the biology of some freshwater fishes. Part 1, *Ophiocephalus punctatus* Bloch. *J. Bombay Nat. Hist. Soc.*, 61(1): 74-98.

Reddy, P. B. S. 1980. Food and feeding habits of *Channa punctata* (Bloch) from Guntur. *Indian J. Fish.* 27(1 and 2): 123-129.

Ross, S. T. 1986. Resource partitioning in fish assemblages: A review of fiele studies. *Copeia*, 2: 352-388.

Routray, P., Kumaraiah, P. and Chakraborty, N. M. 2001. Some aspects of fishery, biology, and conservation of a peninsular carp, *Puntius pulchellus* (Day). *Fishing Chimes*, 21(5): 53-55.

Saikia, A. K., Abujam, S. K. S. and Biswas, S. P. 2012. Food and feeding habit of *Channa punctatus* (Bloch) from the paddy field of Sivasagar district, Assam. *Bulletin of Environment, Pharmacology and Life Sciences*, 1(5): 10-15.

Saikia, S. K. and Das, D. N. 2008. Feeding ecology of common carp (*Cyprinus carpio L.*) in a rice-fish culture system of the Apatani Plateau (Arunachal Pradesh, India). *Aquat. Ecol.* 43(2): 559-568.

Sakhare, V. B. 2010. Food and feeding habits of common carp, *Cyprinus carpio* (Linn.). *Fishing Chimes*, 30(1): 180-182.

Sakhare, V. B. and Chalak, A. D. 2014. Food and feeding of *Catla catla* (Hamilton) from waterbodies around Ambajogai, Maharashtra. *Eco. Env. and Cons.* 20(2): 783-785.

Sakhare, V. B. and Chalak, A. D. 2014. Food and feeding habits of *Clarias batrachus* (Linnaeus, 1758) from Ambajogai, Maharashtra, India. *J. Fish.* 2(2): 148-150.

Sakhare, V. B. and Rawate, S. G. 2012. Indian freshwater Shark, *Wallago attu* (Schneider)-Its food and feeding habits. *Fishing Chimes*. 32(1): 124-125.

Sarkar, Uttam Kumar and Prashant Kumar Deepak. 2009. The diet of clown knife *Chitala chitala* (Hamilton-Buchanan): An endangered notopterid from different wild population (India). *Electronic Journal of Ichthyology*, 1: 11-20.

Sarma, Debajit. 2009. Chocolate mahseer (*Neolissochilus hexagonolepis*): A candidate fish for hill aquaculture. *Fishing Chimes*, 29(7): 8-11.

Scaperclause, W. 1933. US Department of Interior, Fish Wildl. Serv. Fish. leafl. No. 31, 261pp.

Selvaraj, C., Murugesan, V. K. and Unnithan, V. K, . 1997. Ecology-based fisheries management in Aliyar reservoir, Central Inland Fisheries Research Institute, Barrackpore (West Bengal), Special Publication No. 72.

Selvaraj, S. and Francis, T. 2005. Breeding and culture of murrels in India. *Fishing Chimes*. 25(9): 19-22.

Serajuddin, M., Khan, A. A. and Mutsafa, S. 1998. Food and feeding habits of the spiny eel, *Mastacembelus armatus, Asian Fisheries Science*, 11: 271-278.

Sharma, Jyoti, Alka Parashar, Deepti Dubey and Adarsh Kumar. 2013. Gut content analysis of mahseer (*Tor tor*), cultured in cages, ponds and Narmada river, *Biosci. Biotech. Res. Comm*. 6(1): 68-70.

Sibbing, F. A., Osse, J. M. W. and Terlouw, A. 1986. Food handling in the carp (*Cyprinus carpio*): Its movement patterns, mechanisms and limitations. *Journal of the Zoological Society of London*. 210: 161-203.

Singh, A. K. and Mishra Arvind. 2001. Environmental issue of exotic fish culture in Uttar Pradesh. *J. Environ. Biol*. 22(3): 205-208.

Siva Reddy, Y. and Babu Rao. 1987. A note on the food of *Mystus vittatus* (Bloch) from the highly polluted Hussainsagar Lake, Hyderabad. *Indian Journal of Fisheries*, 34(4): 484-487.

Sugunan, V. V. 1995. Reservoir Fisheries of India, FAO fisheries Technical Paper No. 345, Daya Publishing House, Delhi.

Sundararaj, V. and Srikrishnadhas, B. 2000. Cultivable aquatic organisms, Narendra Publishing House, Delhi, pp. 165.

Syed Mushahida-Al-Noor, Md. Abdus Samad and N. I. M. Abdus Salam Bhuiyan. 2013. Food and feeding habit of the critically endangered catfish *Rita rita* (Hamilton) from the Padda river in the North-Western region of Bangladesh. *International Journal of Advancements in Research and Technology*, 2(1): 1-12.

Talwar, P. K. and Jhingran, A. G. 1991. Inland fishes of India and adjacent countries (Volume 1), Oxford and IBH Publishing Co. Pvt. Ltd. New Delhi.

Verma, Rakesh. 2013. Feeding biology of *Labeo dyocheilus*: a vulnerable fish species of India. *International Journal of Research in Fisheries and Aquaculture*, 3(3): 85-88.

Vilizzi, L. and Walker, K. F. 1999b. The onset of the juvenile period in carp, *Cyprinus carpio*: a literature survey. *Environmental Biology of Fishes*. 56: 93-102.

Yashpal, Madhu, Usha Kumari, Swati Mittal and Ajay Kumar Mittal. 2006. Surface architecture of the mouth cavity of a carnivorous fish *Rita rita* (Hamilton, 1822) (Siluriformes, Bagridae), *Belg. J. Zool*. 136(2): 155-162.

Yeragi, S. S. and Yeragi, S. G. 2014. Food and feeding habit of *Mystus seenghala* (Sykes) the common catfish of Mithbav estuary of south Konkan, Sindhudurg District, Maharashtra, India. *Int. Res. J. of Science and Engineering*, 2(2): 71-73.

Zafri, A. and Ahmed, K. 1981. Studies on vitamin A content of freshwater fishes: Content and distribution of vitamin A in mola (*A. mola*) and Dhela (*Rohtee cotio*). *Bangladesh J. Biol. Sci*. 10: 47-53.

Chapter 2

An Overview of Giant Freshwater Prawn, *Macrobrachium rosenbergii* (de Man)

☆ *Ansuman Das, Jayalaxmi Mohapatra*
and A. Anusikha

Introduction

The giant freshwater prawn, *Macrobrachium rosenbergii* (de-Man) is the largest natantian of the world. The largest specimen reported so far is about 654gm but there are unconfirmed data which reports even still larger sizes. It has become the largest target species for large scale farming in freshwater ecosystems because of its fast growth, attractive size, better meat quality and omnivorous feeding habit; in addition to good market demand both in domestic as well in overseas trade. Its farming is well developed and attained commercial level in several Asian countries like Thailand, Taiwan and also in western countries like Hawaii, Mexico, Brazil, etc. This paper presents an overview on the biological aspects of this prawn relevant to the hatchery, grow-out phase management including its marketing and transportation.

Taxonomy

Phylum: Arthropoda

Class: Crustacea

Sub-class: Malacostraca

Division: Eucardia

Order: Decapoda

Sub-order: Natantia

Infra-order: Caridea

Family: Palaemonidae

Genus: *Macrobrachium*

Species: *rosenbergii*

Identifying Features

The key identifying features of *Macrobrachium rosenbergii* are:

1. Overlapping of the pleura of second abdominal segment over those of first and third segment
2. Presence of large second pair of thoracic legs in males.
3. Rostrum long and bent at middle and upturned distally.
4. Tooth formula is 12-13/11-13.
5. Presence of distinct black bands at the dorsal side at the junction o all the abdominal segments.
6. Its juveniles can be identified by the presence of several horizontal black bands on the lateral side of the carapace.

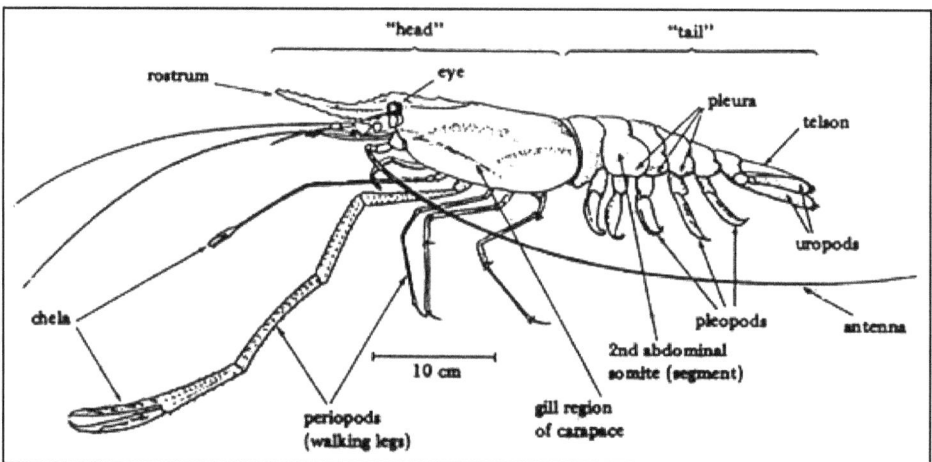

Figure 2.1: *Macrobrachium rosenbergii.*

General Biology

Distribution

It is a tropical species, widely distributed in the Indo-pacific region, but its distribution is limited to the estuarine and freshwater zones of river mouths and backwaters with a temperature usually ranging from 25-34°C and salinity between 0-

20ppt. In India, this species is distributed in the lower stretches of most of the river systems of both the coasts. It accounts for considerable fishery in Hooghly, Godavari, Narmada rivers and also in Kolleru lake and Kerala backwaters.

Habits

This species is slow moving and sluggish by nature and hides under shades in the shallow areas of rivers, lakes or ponds during daytime to avoid direct sunlight and is very active during night time. It has strong agnostic, pugnacious and territorial instincts where no other prawns were allowed.

Food and Feeding Habits

This species is a bottom feeder and an omnivore, feeding on a variety of food starting from grains to flesh of different crustaceans, mollusk, and fishes including its own molted skin.

It locates its food by help of its feelers and capture food with its first pair of chelipeds supported by the second pair.

Food provided in the culture ponds should be broadcasted in the entire pond or provided at corners in the form of pellets having high water stability and attractability. The feed should have 25-30 per cent protein for obtaining optimal growth.

Growth

Unlike other crustaceans, growth is not continuous in this prawn due to the presence of hard exoskeleton. Growth occurs only during molting. Frequency of molting depends on a number of factors like age, food and environmental parameters. Growth rate in this prawn is rapid but shows sexually dimorphic pattern, *i.e.*, both male and female have similar growth pattern till they attain sexual maturity, after which the males out grow the females due to ovarian development in females. Under culture conditions, it attains a weight of 40-50g in five months period.

Reproduction

The sexes are separate in this prawn showing distinct sexual dimorphism. Males are proportionately larger in size with large head, their second periopods are longer, robust with spinous claws and have narrow abdomen, whereas females are smaller in size with small head and broad abdominal space to serve as brood chamber. Males also possesses appendix masculine on the endopodite of the second pleopod.

Maturation, Breeding and Incubation

Both males and females of this species attain sexual maturity at about 150mm in total length at an average weight of 40gm in 4-7 months in different ecological regions. The orange coloured ovary can be clearly seen through the carapace of the ripe female prawn which occupies most of the cephalothorax.

Breeding season differs in different river systems, usually between march to september.Under controlled conditions, they breed when the water temperature is between 28° to 32°C. Mating occurs in freshwater. Before mating the ripe female prawn undergoes pre-mating molt. Then the male deposit sperm as a gelatinous mass on the ventral thoracic region of female between the walking legs. Egg laying occurs in about 5-6 hours after mating in brackishwater. The fertilized eggs adhere to the ovigerous setae of the first four pairs of pleopods. The number of eggs carried by a female varies between 20,000 to 1,60,000 depending on the size.

Incubation occurs in brackishwater. The female continuously incubates the eggs by constantly fanning the pleopods to provide sufficient aeration for the development of the embryo. Incubation period varies between 15-21 days depending on the ambient temperature and water quality parameters. The larvae hatches out mostly during night time and their survival and growth takes place only in brackishwater environment at a salinity of 18-20ppt. After they metamorphose into post larvae after passing through eleven larval stages within 21 -25 days, they migrate back to freshwater region in the rivers where they grow into adults.

Life Cycle of *Macrobrachium rosenbergii*

The giant freshwater prawn has four distinct phases in its life cycle – egg, larva (zoea), post larva and adult. Detailed information on the life history of the species has been given by Ling (1969) and Uno and Soo (1969).

Eggs

The fertilized eggs of this species are slightly elliptical in shape, measures 0.6-0.7mm on long axis and are bright orange in colour. As the incubation period proceeds, the colour of the eggs gradually changes to grey and at the time of hatching they appears as black spots. Normally hatching takes place in 19 days with water temperature of 26-28°C.

Embryonic Development

Fertilization is external and takes place soon after the eggs are extruded. First division occurs after four hours of fertilization and cleavage is completed by the end of 1st day. Ventral plate is formed at the end of 2nd day. By 3rd day body rudiments of the embryo are formed and buds of appendages appear on 4th day. Development of optic vesicles and eye pigmentation are observed by 8th day. The functioning of the heart started by 10th day and the embryo is well developed by 12th day. Further development of the embryo continues till hatching.

Hatching takes place through continuous vibration of the mouth parts of the larvae, followed by stretching of the body outwards that breaks the egg shell. The emerging larvae are inactive and settle at the bottom immediately. Gradually the larvae become active and come to the surface.

Larval Development

There are eleven distinct larval stages (Zoeal stages) in the larval cycle of this species lasting 23-32 days depending on the temperature, food and water quality.

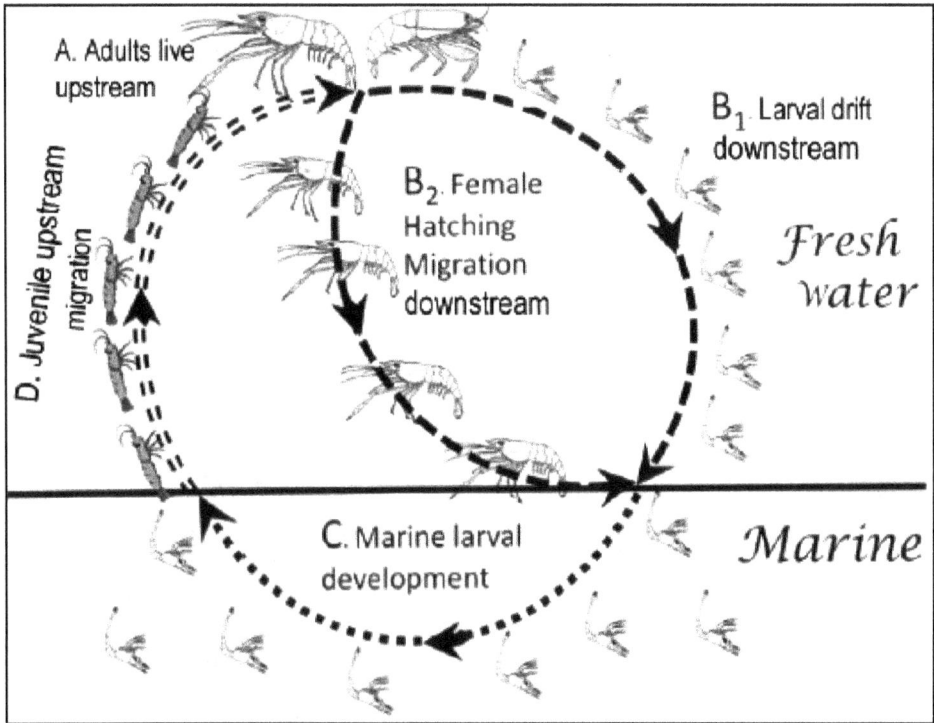

Figure 2.2: *Macrobrachium rosenbergii* **(Life cycle).**

Larval Stages	Age (Days)	Total Length (mm)	Recognized Characters
I	1	1.92	Sessile eyes
II	2	1.99	Stalked eyes
III	3-4	2.14	Uropod appeared
IV	4-6	2.50	Two dorsal rostral teeth, uropod biramous with setae
V	5-8	2.80	Telson narrow and elongated
VI	7-10	3.75	Pleopod buds appeared
VII	11-17	4.06	Pleopod biramous and bare
VIII	14-19	4.68	Pleopods with setae
IX	15-22	6.07	Endopods of pleopods with appendices interna
X	17-24	7.05	Three or four dorsal rostral teeth
XI	19-26	7.73	Teeth on half of the dorsal margin of rostrum
PL	23-32	7.69	Teeth on both dorsal and ventral side of rostrum, adult behavior

Larval Behaviour

Larvae of this species are free swimming and planktonic in habit. They are photo tactic but avoid direct and bright light. The larvae swim with their tail up and head down on their dorsal side down at an oblique angle. The early larval stages, up to stage V, shows gregarious habit which slowly disappears with further development. The larvae actively feed on both live and prepared feed of suitable particle size supplied in suspension.

Post Larvae

Post larvae are the miniature prawns that resemble juveniles except for their underdeveloped body parts. They are benthic in habit. They slowly migrate to freshwater where they develop into juveniles in a few months.

Juveniles

They are bottom crawlers or cling to submerged objects. There are six to eight horizontal black bands prominently situated on the carapace. These bands disappear as the juveniles metamorphose into adults.

Hatchery Technology of *Macrobrachium rosenbergii*

Owing to the good market demand of this species both in the domestic and export trade and also being an ideal species for culture, this species has been incorporated in commercial aquaculture ventures. This species being a fast growing species can be easily introduced either as a component in carp culture or in monoculture ponds to obtain export oriented products. The first successful rearing of all its larval stages were carried out by Ling and Merican, 1961; Ling, 1969 and Uno and Soo, 1969 after which many countries were successful in controlled breeding and seed production of this species because success of its culture depends on its viable seeds.

Hatchery Equipments and Facilities

Storage Tanks

Required to store seawater, freshwater and mixed water. These can be of cement or ferro-cement material. Their size and number varies with the size or capacity of the hatchery.

Larval Rearing Tanks

These are of two types required for two phase larval rearing system.

   a) Cylindro-conical tanks: Fiber glass tanks of 500-1000 L capacity.
   b) Rectangular tanks: Cement tanks with bottom U-shaped and of 5000 - 10,000 L capacity with epoxy coating inside the tank.

Conical Fiber Glass Tanks

100 to 200 L capacity for Artemia hatching.

Aeration and Water Supply System

Two air blowers of 5 to 10HP. Aeration distribution system should be provided throughout the hatchery with PVC pipes fitted with regulators.

Electricity

A dependable three phase power supply with a stand by generator is essential (10-25 KVA).

Brood Stock Ponds

Number and area of ponds are related to the hatchery size. A dependable freshwater source is highly essential.

Water Analysis Kit

It is required to analyze different water quality parameters in hatchery and the rearing ponds.

Brood Stock Management

Matured prawns or brood stock are an important component of hatchery. Brood stocks are maintained in freshwater ponds adjacent to the hatchery to have an uninterrupted supply of berried prawns for hatchery operation.

For raising brood stock, early juveniles of the species are stocked in the rearing ponds at the density of $3/m^2$ and the adults are maintained at $2/m^2$ in the ratio of one male to four females. During this period, prawns are feed with pelletized diet having 40 per cent protein @ 3-5 per cent of their body weight per day.

The juveniles attain sexual maturity within 4-7 months depending on water temperature, food and water quality parameters. This species have different breeding period under different climatic conditions but can be made to breed throughout the year under optimum climatic temperature of 28-32°C.

For hatchery operation, grey egg bearing berried females is directly collected from the rearing ponds, disinfected with 0.3ppm Copper sulphate solution and released into hatching tank containing 5ppt brackish water till hatching. Normally hatching completed within 2-3 days with the hatching fecundity of about 500 larvae per gram body weight of the berried prawn.

Water Intake and Treatment System

Water intake and treatment system is an important management aspect for optimal larval survival and seed production. Sea water and freshwater should be drawn preferably through bore wells near the sea shore and should be treated with 10ppm Sodium Hypochloride and allowed to settle. The saline water should be diluted to 12ppt and passed through pressurized sand filters before letting it into the larval rearing tanks.

Methods of Larval Rearing

Larval rearing is a critical phase in hatchery operation which depends on three factors *i.e.*, water quality, water temperature and feed. Hence, suitable regime of these factors need be followed for best result.

Based on the availability of sea water for hatchery operation, two major systems of larval rearing are followed:

Open System

The open system of larval rearing is generally practiced in coastal based hatcheries where there is no dearth of seawater and freshwater. Under this system, about 50 per cent of the medium is exchanged per day with freshly prepared medium. This system utilizes huge volume of water.

Closed System

The closed system of larval rearing is suitable in non-coastal hatcheries where sea water availability is the major constraint. Under this system, sea water is transported, stored and diluted to 12ppt larval medium. The used water is either recycled or used recirculated through a biological filter.

Technique of Larval Rearing and Seed Production

A two phase larval rearing system is usually followed to increase larval survival seed output.

Stocking

In the first phase, the first stage zoea larvae are stocked in conical tanks at a very high density of 500-1000 larvae per litre for 10-15 days. 75 per cent of the medium is usually exchanged daily with fresh medium of identical salinity.

In the second phase, the advanced larvae are stocked in large cement tanks with a greater surface area @ 50-80 per litre and reared till metamorphosis. About 50 per cent of the medium is exchanged daily in open water system management or connected to a biofilter unit by means of an air-lift.

Feeding

Availability of adequate and right kind of feed in the growing environment is essential for the survival and growth of larvae and post larvae in the hatchery. In the natural environment, the larvae feed on plankton available in the environment. In a hatchery system, such planktons are isolated and stock culture is maintained to fed the larvae. Apart from planktons, several live feeds along with compounded feeds and micro- encapsulated diets are provided in specific feeding regimes for the optimal survival and growth of prawn larvae. Further, these larvae are also specific to the size and type of feed which need to be taken care off.

Types of Feed

a) Live Feeds

Among the live feeds the nauplii of *Artemia salina* is recognized as the most suitable food for crustacean larvae. Artemia cysts are freshly hatched in the hatchery as per the hatching instructions of the supplier at the rate of 100 - 200 cyst per litre and then feed to the larvae at the rate of 2-10 per ml. Before introducing the nauplii into the tank, they should be disinfected with Copper Sulphate solution and thoroughly washed with water.

Apart from *Artemia nauplii* a zooplankton named Moina is also predominantly used as a live feed for prawn larvae. This species is cultured in the hatchery either in mass culture or stock culture methods, then harvested and introduced into the larval rearing tank after conditioning for an hour to excrete the fecal matter. Moina being a freshwater species survive in the larvae culture tank for 10- 15 minutes; hence, the dead Moina should be siphoned out from the tank to prevent pollution of water.

b) Inert Diets

Tubifex worms and freshwater mussels (*Lamellidens* sps.) also form excellent larval diet. Whole Tubifex worms and gonad and foot portion of freshwater mussels are collected, sliced thoroughly, disinfected with few drops of Copper Sulphate solution, washed with water and sieved to get required particle size of 300 µ and feed to different sizes of zoea (Zoea VI to Post larvae). The feed is dispersed throughout the larval rearing tanks 3-5 times a day.

c) Mixed Diets/Prepared Diets

A variety of feed ingredients of vegetable and animal origin are used to prepare combination diet for prawn larvae. Some of ingredients that are commonly used are-wheat flour, soya flour, corn flour, fish flesh, mussel, squid hen's egg, skimmed milk powder, vitamin – mineral pre mixes etc.

The selected feed ingredients are mixed together into dough, steamed, cooled, and refrigerated. This feed with 50 per cent protein can be used for 2-4 days after screening through different sieves to get appropriate particle size of feed suitable for different larval stages (stage IV to post larvae).

d) Microencapsulated Diets

Microencapsulated diets made from egg, gelatin and agar when feed to zoea larvae from Zoea-II onwards in combination with Artemia nauplii has shown encouraging result only on laboratory trials but further through investigation is necessary under commercial hatchery conditions.

Mode and Frequency of Feeding

Feeding of prawn larvae is initiated on the second or third day of hatching (stage-II or III). For first ten days the larvae should be fed with Artemia only twice a day and after that it is supplemented with prepared diets. Prepared diets are given 4-5 times a day during day time and Artemia during night time. Depending on the acceptance of the feed by the prawn larvae and larval progression, the diets are classified into primary or secondary diets. For example, Artemia nauplii is a primary diet whereas soya products is a secondary diet.

In practice the exact quantity of prepared feed to be given each time cannot be prescribed. It depends on the utilization of feed by the larvae and has to be adjusted by visual observation and experience. However, it has been observed that different stages of prawn larvae require 5-50 Artemia nauplii and 50-150µgm of prepared feed per larvae per day. It should be noted that there should not be excess feed or inadequate feed which leads to deterioration of larval medium or cannibalism and delayed metamorphosis respectively.

Proximate Composition of Selected Diets for Larvae of *Macrobrachium rosenbergii*

Diet	Protein Per cent	Carbohydrate Per cent	Fat Per cent	Ash Per cent
Artemia sps.	55.60	—	15.28	15.25
Moina sps.	56.69	13.47	23.73	0.611
Mixed zooplankton	46.00	6.00	23.00	25.00
Tubifex sps.	65.00	15.00	14.00	6.0
Acetes sps.	54.46	—	3.74	15.20
Lamellidens sps.	40.20	—	7.00	13.17
Prepared feed	41.80	20.40	11.50	4.58
Mushroom	32.30	—	2.20	6.90

Cleaning

Water quality in the larval tank deteriorates due to accumulation of metabolites, unutilized food and molted skin in the tank. So, cleaning of larval rearing tanks is very essential to maintain good water quality in the medium. It can be done by siphoning method preferably during evening hours before introducing live feed into the tank. Aeration should be stopped before cleaning.

Water Quality Management

In aquaculture water quality is usually defined as the suitability of water for survival and growth of the aquatic animals and it is normally governed by only a few variables. Regulation or maintenance of the optimum levels of these variables for better growth, survival and optimum production is termed as water quality management.

Success in hatchery production of *M. rosenbergii* seeds largely depends on the water quality management. The deterioration of water quality in a hatchery is mainly due to:

- ☆ Accumulation of metabolic wastes of living biomass.
- ☆ Decomposition of unutilized feed.
- ☆ Decay of biotic materials.

In a *M. rosenbergii* hatchery water, pH, totally alkalinity, temperature, dissolved oxygen, NH_3-N, NO_2-N and salinity can be considered as gross and primary water quality parameters. These factors if controlled within the safe range would result in the maintenance of many other parameters within the safe level.

Water Temperature

This is considered as a critical ecological parameter with influence growth, molting and also regulate the seed production cycle. 29- 31°C is suitable temperature for optimum production of giant freshwater prawn seed.

pH

It is a key indicator of water quality in larval rearing. It influences the acid base condition of the body fluid thus it affect metabolism and physiological process of larvae. pH slightly greater than 7.5 is considered good for larvae because certain salts like Bicarbonate are in the water at this pH which is essential for good growth and other physiological activities.

Dissolved Oxygen

It is the most important environmental parameter that regulates the hatchery output. It is not only an important factor for respiration of the larvae but also to maintain favorable chemical and hygienic environment for the water body. Dissolved oxygen level of 5 mg- saturation level is the best condition for good larval growth.

Salinity

It is one of the most important factor for successful rearing of all the larval stages of *M. rosenbergii*. Though the larvae can withstand a wide range of salinity (8-18 per cent) optimum result have been obtained at 12 per cent. Salinity variation of ± 2 per cent is not detrimental to the larvae. However, sudden wide variations in salinity must be avoided while exchanging water in the larval rearing tanks.

Total Alkalinity

It refers to the total concentration of bases in water and is expressed in mg/l of equivalent $CaCO_3$. In most waters bicarbonate (HCO_3) and carbonate (CO_3) or both are predominant bases. Alkalinity primarily determines the fluctuation of pH of water. Water with low alkalinity has low buffering capacity which results in wide fluctuation in pH value. Total alkalinity of *M. rosenbergii* larval rearing tank water should be 550 ± 50 ppm.

Ammonia

It is the second most important water parameter to limit larval production after oxygen. In a system ammonia originates through excretion (protein metabolism) or as a product of microbial decomposition.

The toxicity of ammonia is attributed to unionized ammonia. As the pH and temperature rises, the concentration of unionized ammonia increases relatively. As ammonia level the water medium increase, excretion of ammonia by the organism decreases. As a result the ammonia level in the body adversely affects enzyme catalyzed reaction and membrane stability. Ammonia increases oxygen consumption by tissue, damage gills and reduce the ability of the body fluid to transport oxygen. The tolerance level is 0.05 ppm.

Ammonia level in the larval rearing medium can be managed by maintaining sufficient level of dissolved oxygen in the medium, providing suitable slope to the tank bottom to facilitate collection and removal of wastes and periodic partial removal of cyano-bacteria and also algal bloom.

Nitrite

Nitrite (NO_2) is an intermediate product in the bacterial oxidation of ammonia to nitrate (NO_3), a process called nitrification. Nitrite toxicity affects oxygen carrying capacity causing hypoxia. It increases with increasing pH and decreases with increasing calcium and chloride concentration.

It can be effectively managed in the larval rearing tank by effective removal of organic wastes, frequent water exchange and adequate aeration.

A brief description of the mixed water for the larval rearing is given below:

Parameters	Concentrations
pH	7.8 – 8.3
Dissolved Oxygen	5 – 7 ppm
Water temperature	25 – 31 ppm
Total alkalinity	550 ± 50 ppm
Unionized ammonia (NH_3-N)	Trace – 0.02 ppm
Nitrite (NO_2-N)	Trace – 0.2 ppm
Salinity	8 – 10 per cent
Total Iron	0.01 ppm

Growth and Metamorphosis

Larval growth and metamorphosis occurs through molting which are related to ambient temperature, water quality and food that need to be critically monitored daily to obtain optimal growth. Larvae normally take 1-4 days to pass from one larval stage to the next. Larvae exhibit uniform growth up to VIth zoeal stage, thereafter two to five larval stages are can be observed in the same larval tank due to differential growth among them. Healthy larvae swim at the tank surface and feed actively, whereas unhealthy larvae settle at the bottom of the tank. In Indian conditions, first post larvae is observed within 22-32 days of hatching under variable agro-climatic conditions and 90 per cent of larvae metamorphose within next 10 days. The post larval production ranges between 25-35 per litre and the cycle lasts for 35 days.

The post larvae are characterized by radical change in appearance and behavior. They resemble miniature adults and swim freely. They crawl or cling to the tank surface.

Shelters

In larval rearing of this species, pre and post-metamorphosis stages are critical and mortalities due to cannibalism can be minimized by providing a variety of shelter or substrata. Shelters can be provided in the from of molluscan shells, plastic or asbestos materials fabricated into multitier system and placed in larval rearing tanks for the post larvae to settle, since they prefer shades after their molting.

Acclimatization

The post larvae are gradually acclimatized to freshwater under continuous aeration system. Once they are acclimatized to freshwater, they are weaned from larval diet to variety of feed like small shrimps, worms, broken rice etc. after a fortnight of acclimatization they are suitable for rearing in grow-out ponds. If they are not released into the culture ponds, then their survival rate decreases due to cannibalism under crowded conditions in rearing tanks.

Seed Transporatation

The prawn seed or post larvae exhibits cannibalistic tendency which is more pronounced under crowded conditions hence suitable methods are to be followed for packaging and transportation of these seeds from hatchery to grow-out ponds for successful aquaculture operations. Further, hatcheries are normally located in coastal regions whereas farming areas are located throughout the country, hence suitable methods of packaging and transportation are a pre-requisite for supply of healthy seeds to the farms which involves 24-40 hours of duration, depending upon the distance to be covered.

Estimation of Prawn Seed for Packing

Maintaining the accuracy in estimation of prawn seed is a time consuming and difficult task. Estimation method for prawn seeds varies with the volume of the seeds.

a) For a less number of prawn seed, individual counting is possible.

b) For seeds exceeding 5000 numbers, individual counting method is not viable. Hence, methods like volumetric estimation or sampling method is preferred.

Methods followed or estimation of prawn seed should be such that time factor should be less and handing period should be minimal so as to avoid stress to the prawn seeds.

Packaging Material

Standard quality and economically viable packing materials should be used during packing and transportation of prawn seed, especially during long distance journey. The materials that are recommended for packing of prawn seeds are:

a) Polythene bags

b) Filtered freshwater that is sufficiently aerated

c) Plastic strips as substrates

d) Ice cubes sealed in small polythene packets for long distance transportation

e) Thermocol sheets of 10mm thickness of variable sizes to fit into inner lining of cardboard boxes

f) Cardboard boxes of suitable sizes

g) Jute thread for tying polythene mouths

h) Sealing tapes

i) Oxygen cylinders

Cost of packing should also include the labor charges apart from the material cost.

Packing Methods

The standard method for packing involves certain basic steps as follows:

☆ The bottom end of the polythene bags should be tied and hot sealed to prevent leakage of oxygen and water. It should be checked before packing.

☆ Insert plastic strips into the polythene bags which will provide shelter to the prawn seeds.

☆ Estimate the prawn density depending on the distance to be transported and release into the polythene bags with 5 litre of water (1/3rd of the bag)

☆ Do not introduce any feed into the bags. Seeds should be preferably feed one hour before packing.

☆ Remove air from the bags and introduce oxygen into it (2/3rd of the bag) to provide sufficient dissolved oxygen.

☆ Tie the bags with jute and place into the cardboard boxes lined with thermocol and seal it with tape.

☆ Stocking densities in bags varies with the size of the prawn seed (15-20mm) and the distance to be transported.

Transportation of Prawn Seed

Normally afternoon is more convenient for harvesting, estimation, packing and transportation of prawn seed. For short distance transportation up to 4hrs, any convenient vehicle can be used but for long distance transportation above 24hrs, insulated vehicles are preferred. They can be transported either by road, rail or air.

Nursery Rearing and Management of *Macrobrachium rosenbergii*

Nursery is an intermediate phase between hatchery and grow-out phase. During this phase the newly metamorphosed post-larval prawns are reared to larger juveniles for stocking in various grow-out systems. Nursery system helps in accountability of seed, predictability of yield, efficient utilization of feed and efficient utilization of grow-out facilities. Based on the stocking density and the management practices of rearing ponds, nursery rearing facilities are of two types:

Low Density Nursery Rearing

In low density rearing the rearing system usually are the earthen ponds, which are small scale versions of grow-out ponds. Size of the pond varies with the farm area, usually 0.2ha or less in size. The most frequently used are flat bottomed rectangular ponds having a gentle slope towards the outflow which can be emptied rapidly and completely. The pond soil should be clay-or clay-loam. The pond should be prepared in advance. Eradication of all predators and competitors is very important. Liming should be done @ of 200- 1000kg/ha to correct the pH as well as to maintain the pond hygiene. Following liming, manuring should be done @ 1000 kg cattle dung/ha, 500 kg poultry manure/ha and 100 kg super phosphate/ha to encourage the development of pond fauna on which the juveniles feed. It also produces a plankton cover which in turn will prevent the growth of rooted vegetation.

Stocking density is generally about 50-100 post larvae/m^2 of pond surface area and the duration of the nursery phase is usually 4-8 weeks. Juvenile prawns in nursery ponds are generally fed the same prepared diet used in grow-out ponds. The crude protein content of these feeds usually does not exceed 30 per cent. Once a day feeding is sufficient and is given in the evening. The daily ration at optimum temperature is 8 per cent of the biomass at the start of the nursery stage and 4 per cent at the end. The prawns also feed on the natural food- small benthic organisms available in the ponds.

Management procedures for nursery ponds are essentially the same as for growout ponds. Proper maintenance of water quality is one of the important aspects in the management of nursery system. Some of the important water quality parameters for successful rearing are as follows:

Parameters	Optimum Range
1. Temperature	28–32ºC
2. pH	7.0–8.5
3. Dissolved oxygen	>5 ppm
4. NH_4^+	<1 ppm
5. NO_2	<0.1 ppm
6. Hardness	40–100 ppm
7. Transparency	20–50 cm

Diseases have always been found to be linked to poor rearing conditions like over-feeding, water shortages, poor water quality, etc. and poor rearing practices which need to be managed judiciously for optimal growth. Growth depends on density, feed and environmental conditions. Usually, under good management practices, the post larvae grows to a mean size of 1-2 g in 4-8 weeks of nursery period with 90 per cent survival rate.

Harvesting can be carried out by dewatering/drying the pond or by selective removal of the largest individual with a seine net of suitable sized mesh. Harvesting must be rapid and efficient to minimize handling and transfer stress.

High Density Nursery Rearing

In high density rearing, the rearing system currently in use are tanks fabricated of concrete, fibre glass, plastic liners or other materials (10 – 100 m²). The stocking density varies from 2000- 5000 post larvae/m². Usually the duration will be shorter (15 -30 days). The higher densities are made possible through intensive management quality and the concentrated use of artificial habitats and feeding.

Preferably the prawns should be fed several times daily. A formulated feed is usually used supplemented with natural food. Sometimes large aquatic plants such as water hyacinth and elodea are placed in nursery tanks to provide additional habitat area, supplemental feed and water purification. Water quality has to be carefully monitored and if needed 30- 50 per cent of water should be exchanged daily. Continuous aeration through blower is an absolute necessity.

Growth is usually slower at the higher densities and the prawn generally attain a mean weight of 0.1- 0.5 g after 2-4 weeks of rearing. At the end of the nursery period, the prawns are harvested by tank drainage with 75 per cent survival rate.

Common Diseases of *Macrobrachium rosenbergii*

Disease has been defined as a definite morbid process having a characteristic strain or syndrome, it may affect the whole body or any of its parts and the etiology, pathology and prognosis may be known or unknown. Disease has been recognized as one of the several biological factors which can limit and hinder the development of aquaculture by causing tremendous economic and production loss.

Both host and pathogen maintain equilibrium during their existence in the environment, disease condition outbreaks when pathogen gets upper hand. If the environment is more congenial for the pathogen, then it cause harm to the organism. Chronic diseases in fishes may remain sub-clinical for a prolonged period whereas acute cases can be detected easily by clinical signs. Some of the common diseases of *M.rosenbergii* which are frequently encountered during different stages of its life history are given as follows:

Based on the causative factors, diseases are classified into two major categories-

Infectious Diseases

Bacterial Disease

A) Black Spot Disease

Causative agent: Chitinoclastic bacteria, *i.e.*, *Benkea* sps., *Pseudomonas* sps., and *Aeromonas* sps.

Life stage affected: Juveniles and adults.

Gross signs: Progressive erosion of exoskeleton beginning as small black lesions.

Most affected areas: Gill filaments, ventral abdominal muscles, telson and walking legs.

Diagnosis: By visual and microscopic observation.

Effect on host: Mechanical damage of the exoskeleton

Preventive measures: Minimizing stocking density, maintaining good environmental conditions in ponds, avoiding injuries to prawn exoskeleton, management of organic load in pond and providing adequate diet.

Treatments: i) Daily water exchange of 70-80 per cent, ii) Static treatment with Malachite green (0.6 ppm dose) for one day, iii) Dip treatment with Furance (1 ppm dose).

B) Filamentous Bacterial Disease

Causative agent: *Leucothrix* sps., *Benkea* sps., *Pseudomonas* sps., and *Aeromonas* sps.

Life stage affected: Larvae and Juveniles

Gross signs: Filamentous growth on the body surface of the affected individuals.

Most affected areas: Gills, pleopods and uropods of the affected individuals.

Diagnosis: By microscopic observation.

Effect on host: Mechanical damage of the exoskeleton.

Preventive measures: Maintaining good quality with optimum dissolved oxygen level and management of low organic load in pond.

Treatments: i) Potassium permanganate treatment with 10 ppm for 1 hour or 2.5 – 5 ppm for 3-4 hours ii) Static treatment with Malachite green (0.0075 ppm dose) for one day, iii) Dip treatment with Furance (1 ppm dose),iv) Terramycin   treatment at 2 ppm or Erythromycin at 4-6 ppm.

C) Bacterial Necrosis

Causative agent: Unknown but probably involves several different genera of bacteria.

Life stages affected: Post-larvae and adult.

Gross sign: Localized necrosis or discoloration on any appendages.

Most affected areas: Antennae and abdominal appendages.

Diagnosis: By visual and microscopic observation.

Effect on host: Damages to one or more appendages and rapid killing of larvae.

Preventive measures: Avoiding overcrowding and over feeding, maintaining good water quality in ponds and adequate water exchange.

Treatment: (i) Penicillin/streptomycin treatment @ 2 g/m^3 helps in regeneration of necrosed appendages. (ii) Static treatment with Erythromycin (65 ppm dose), iii) Dip treatment with Furance (7 ppm dose).

D) Black Nodule Diseases

Causative agent: It is a bacterium hither to unidentified.

Life stages affected: Larvae, post-larvae and adult prawns.

Gross sign: Formation of black nodules in the hypodermis.

Most affected areas: Thoracic region.

Diagnosis: Microscopic observation.

Effect on host: The affected larvae becomes very weak, lose their appetite and larvae comes to a moribund stage.

Preventive measures: Maintaining good water quality and providing sufficient feed.

Treatment: Dip treatment with furance @ 0.09 mg/litre.

Viral Disease

Larval mid-cycle disease (MCD)

Causative agent: Still unknown (Entero-bactereerogenes, a virus or an unidentified toxin. Johnson, 1878; Brock, 1988).

Life stages affected: Early larval stages between zoea IV to PL.

Gross sign: Similar to bacterial necrosis. The affected larvae are bluish grey in color and swim weakly often in a cork screw manner.

Most affected areas: Appendages.

Diagnosis: Microscopic observation.

Effect on host: Affected larvae lose their appetite and the moribund larvae are eaten up by the healthy ones and causes mass mortalities.

Preventive measures: Maintaining proper sanitation and good environmental condition, avoiding over stocking and providing adequate nutrition.

Treatment: Not yet known. The whole batch should be discarded.

Parasitic Disease

A) Isopod Parasitic Disease

Causative agent: Bopyrid isopod, *Probopyrus*.

Stages affected: Juveniles (25-55 mm)

Gross sign: Melanisation in the affected areas, swelling on lateral side of the carapace at bronchial stegial region.

Diagnosis: Microscopic observation.

Effect on the host: The parasite gets itself lodged on the gills below branchial stagite. The presence of the parasite is clearly visible by the conspicuous swelling of the lateral side of the carapace. Due to attachment of the parasite the gills become highly compressed. The area of infection shows severe melanization. The infected prawns are very thin and emaciated.

Preventive measure: Avoid overcrowding and maintain good water quality.

Treatment: (i) Formalin @ 30 ppm bath for one hour, (ii) Copper sulphate 1 ppm static treatment.

B) Protozoan Parasitic Diseases or Ciliate Infection

Causative agent: Several protozoan parasites - *Zoothamnium, Vorticella* sp., *Acineta* sp. and *Epistylis* sp. (Ciliate).

Stages affected: Larvae and juvenile.

Gross sign: Slight opaqueness of the body color in case of protozoan infection and fuzzy mat like growth on the general body surface.

Most affected areas: Whole body, mostly appendages.

Diagnosis: Microscopic observation.

Effect on host: In mild infections, the protozoans are easily shed during molting but in heavy infections they obstruct molting, suppress growth resulting in mortality of the affected prawns.

Preventive measures: Maintain good water quality and avoid high organic load, siltation, turbidity and low oxygen levels.

Treatment: (i) Formalin @ 20-30 ppm for 2 hours static bath, (ii) Malachite green @ 0.0075 ppm in static condition, (iii) Acetic acid @ 2.0 per cent $_0$ for one minute dip is the best treatment and (iv) Sulfa-quinine dip treatment.

Non-infectious Disease

A) Muscle Opacity and Necrosis

Causative agent: Environmental stress, high salinity, temperature, overcrowding.

Stages affected: Post-larvae and juveniles.

Gross sign: Usually a small opaque whitish patches occur first at the base of appendages and in tail of the larvae and then spread throughout the entire body followed by necrosis.

Most affected areas: Appendages and then the whole body.

Diagnosis: By visual observation.

Effect on host: Produce sporadic heavy larval mortalities.

Preventive measures: Infected larvae should be removed, tanks and other equipments should be cleaned and disinfected and overcrowding should be avoided.

Treatment: Includes environmental manipulation such as- reduce stress, maintain optional culture condition and if heavily infected it is better to sacrifice the whole batch.

B) Black Gill Disease

Causative agent: Chemical contaminants or precipitating chemicals and nitrogenous waste products increasing levels of ammonia and nitrite in rearing ponds.

Stages affected: Larvae, post-larvae, juveniles and adults.

Gross signs: The gills show reddish, brownish to black discoloration and atrophy at the tip of the filaments. In advanced stages most of the filaments are affected and the gills become black. Physical deformities are also found.

Most affected areas: Gill filaments.

Diagnosis: By visual and also by microscopic observation.

Effect on host: Blackening of the gills due to melanisation or due to the heavy deposition of pigment at sites of heavy haemocyte activity (inflammation), respiratory disturbances and secondary infections by bacteria, fungi and protozoans (via) the dying cells of the gills. Lastly it leads to death.

Preventive measures: Avoid over feeding, change water frequently, avoid heavy metal discharge into rearing system and routine monitoring of nitrogenous compound levels.

Treatment: (i) Adequate water exchange, (ii) Methylene blue 8-10 ppm, (iii) Prefuran 1 ppm and (iv) Malachite green 0.0075 ppm in static conditions.

C) Exuviae Entrapment Disease (EED) or Metamorphosis Molt Mortality Syndrome

Causative agent: Poor water quality, inadequate nutrition.

Stages affected: Late larval stage and early post-larvae.

Gross sign: The larvae become lethargic and settle at the bottom, unable to feed.

Most affected area: Appendages, eye and rostrum.

Diagnosis: By visual observation and microscopic observation.

Effect on host: Affected larvae are unable to free their appendages, eyes or rostrum from the exuviate during and after molting and become entrapped. The larvae even after complete molting generally appear to have malformed appendages and die shortly after molting.

Preventive measures: Adequate water exchange, maintenance of good water quality and proper feeding.

Chapter 3

Method of Data Collection for the Stranded Marine Mammals

☆ *V. Ravi*

Introduction

Marine mammals are important group of vertebrates as they influence the marine food webs and structure and functions of marine ecosystem. They are classified into three Orders: Cetacea (whales, dolphins and porpoises), Sirenia (manatees and dugong) and Carnivora (sea otters, polar bears and pinnipeds). The Indian seas greatly support a variety of marine mammals including toothed whales, baleen whales, dolphins, porpoise and dugong.

Despite their large body size and predatory in nature, most often marine mammals die or become weakened at sea may be brought passively to shore through wind and wave action. At the same time, the natural and unnatural causes of death and disablement leading to single stranding are many: low sea temperature or ice entrapment, parasites, disease, biotoxin, entanglement associated with fisheries, starvation due to decreased food supply, collision with vessels, contaminants, oil spills and death or direct injury inflicted by predators, other marine mammals, or at the hands of humans (Geraci and Lounsbury, 1993). In addition, several causes for stranding include areas with broad tidal flats, strong or unusual currents, or extreme tidal volume leading to errors of navigation/judgment or impaired echolocation (Geraci and Lounsbury, 1993). In fact, the learning of diversity and distribution of various marine mammals of the world are mainly because of their stranding event. For example, a mass stranding offers a population sample, opening to view parameters such as sex ratio, age structure, pregnancy rate, lactation rate and relatedness within a group. Most of the records of the whales along the Indian coast are on their stranding only (Kasinathan, 2002; Melinmani, 2004; Ravi and Murugan, 2010). Therefore, every stranding event should be considered a potentially unique opportunity to learn something that cannot be learned any other way (Perrin and Geraci, 2002). The method of collecting data from the stranded marine mammals is described in this article.

Methods

Identification of Marine Mammals

Immediately after reaching the location of stranded marine mammal, the investigator should take necessary images with the digital camera especially total animal, head, dentition, dorsal and ventral view, sex organ, blow hole, dorsal fin, flukes, wounded part, etc., These images will be helpful for correct identification of species of marine mammal. Then, measurement for morphometric data should be gathered with a helper using long measuring tap and scale. The identification of marine mammals is followed as described by Geraci and Lounsbury (2005) and Vivekanandan and Jeyabaskaran (2012) and the list of stranded marine mammals in India is given in Table 3.1.

Table 3.1: List of Identified and Stranded Marine Mammals in India

Sl.No.	Common Name	Species Name
1.	Blue whale	*Balaenoptera musculus* (Linnaeus, 1758)
2.	Fin whale	*Balaenoptera physalus* (Linnaeus, 1758)
3.	Bryde's whale	*Balaenoptera edenis* Anderson, 1878
4.	Common Minke whale	*Balaenoptera acutorostrata* Lacepede, 1804
5.	Humpback whale	*Megaptera novaeangliae* (Borowski,1781)
6.	Sperm whale	*Physeter macrocephalus* Linnaeus, 1758
7.	Pygmy sperm whale	*Kogia breviceps* Blainville, 1838
8.	Dwarf sperm whale	*Kogia sima* (Owen, 1866)
9.	Cuvier's beaked whale	*Ziphius cavirostris* G.Cuvier,1823
10.	Indo-Pacific beaked whale	*Indopacteus pacificus* (Longman, 1926)
11.	Short-finned pilot whale	*Globicephala macrorhynchus* Gray, 1846
12.	Killer whale	*Orcinus orca* (Linnaeus, 1758)
13.	False Killer whale	*Pseudorca crassidens* (Owen,1846
14.	Pygmy killer whale	*Feresa attenuata* Gray, 1875
15.	Melon-headed whale	*Peponocephala electra* (Gray, 1846)
16.	Irrawady dolphin	*Orcaella brevirostris* (Owen in Gray, 1866)
17.	Indo-Pacific humpbacked dolphin	*Sousa chinensis* (Osbeck, 1765)
18.	Rough-toothed dolphin	*Steno bredanensis* (G.Cuvier in Lesson, 1833)
19.	Risso's dolphin	*Grampus griseus* (G.Cuvier,1812)
20.	Bottlenose dolphin	*Tursiops aduncus* (Ehrenberg, 1833)
21.	Pan tropical spotted dolphin	*Stenella attenuata* (Gray, 1846)

Contd...

Table 3.1–*Contd...*

Sl.No.	Common Name	Species Name
22.	Spinner dolphin	*Stenella longirostris* (Gray, 1828)
23.	Striped dolphin	*Stenella coeruleoalba* (Meyen, 1829)
24.	Long beaked common dolphin	*Delphius capensis* Gray, 1828
25.	Finless porpoise	*Neophocaena phocaenoides* (G. Cuvier,1829)
26.	South Asian River Dolphin	*Platanista gangetica* (Roxburgh, 1801)
27.	Sea cow	*Dugong dugon* (Muller, 17776)

Data Collection and Analysis

The condition of stranded mammal (Figure 3.1) determines much about what can be collected from it and should be specified in the field notes. Standard condition codes (Geraci and Lounbury, 1993) are: (1) alive, (2) freshly dead (3) decomposed, but organs basically intact, (4) advanced decomposition (*i.e.,* organs not recognizable, carcass intact) and (5) mummified or skeletal remains only.

Geraci and Lounsbury (1993) described three levels of collection of data for the stranded marine mammal. They are as follows:

Level A Data: Basic Minimum Data
1. Name and institutional address of investigator
2. Reporting source
3. Species (including preliminary identification and voucher material in the form of photographs in several views, teeth, skulls, and other specimens) - Scientific and common name
4. Field number
5. Number of animals, including total and subgroups
6. Location (preliminary description, plus longitude and latitude)
7. Date and time of discovery and specimen recovery
8. Length (and girth and weight if possible)
9. Sex

Level B Data: Supplementary On: Site Information and Samples
1. Weather and tide condition
2. Offshore human/predator activity
3. Presence of prey species
4. Behaviour before and during stranding
5. Samples collected for life history studies (teeth, earplugs, or bone for age determination, reproductive tracts, stomachs)
6. Samples collected for blood studies
7. Samples collected for genetic studies

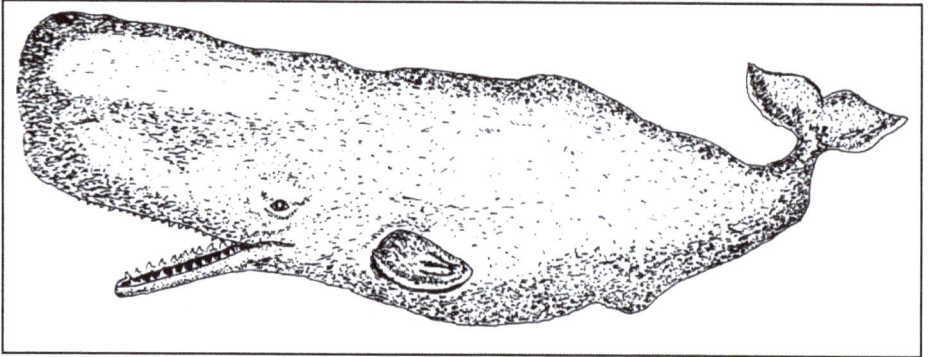

Sperm whale

Bottlenose Dolphin

Sea cow

Figure 3.1: Marine Mammals.

8. External measurements (Figure 3.2 and Table 3.2)
9. Disposition of carcass

Table 3.2: Datasheet for Morphometric Analysis of a Cetacean

Sl.No.	Parameters	Measurement (cm)
1.	Snout to melon	
2.	Snout to angle of mouth	
3.	Snout to blowhole	
4.	Snout to centre of eye	
5.	Snout to anterior insertion of dorsal fin	
6.	Snout to tip of dorsal fin	
7.	Snout to fluke notch	
8.	Snout to anterior insertion of flipper	
9.	Snout to caudal end of ventral grooves (when present)	
10.	Snout to centre of genital aperture	
11.	Snout to centre of anus	
12.	Flipper length	
13.	Flipper width (maximum)	
14.	Fluke width	
15.	Dorsal fin height	
16.	Girth: axillary	
17.	Girth: maximum	
18.	Girth: at level of anus	
19.	Blubber thickness: dorsal (anterior and lateral to dorsal fin)	
20.	Blubber thickness: lateral at mid- length	
21.	Blubber thickness: ventral at mid- length	

Level C Data: Necropsy Examination and parasite collection
1. Collection of tissues for toxicology, microbiology and gross histopathology
2. Collection of parasites

Marine mammals: Wild Life Protection Act (1972)

The Indian Parliament passed the Wild Life (Protection) Act in the year 1972 in order to protect the wild life from destruction Wild life is nature's gift and its decline has an adverse effect of ecology and hence there is an urgent need to protect them.

The Act prohibits hunting of wild animals. No person shall hunt any wild animals as specified in the Schedules. The State govt. by notification, may declare any area within the reserved forest or territorial waters as a sanctuary if it considers fit the area for protection and conservation of wild life.

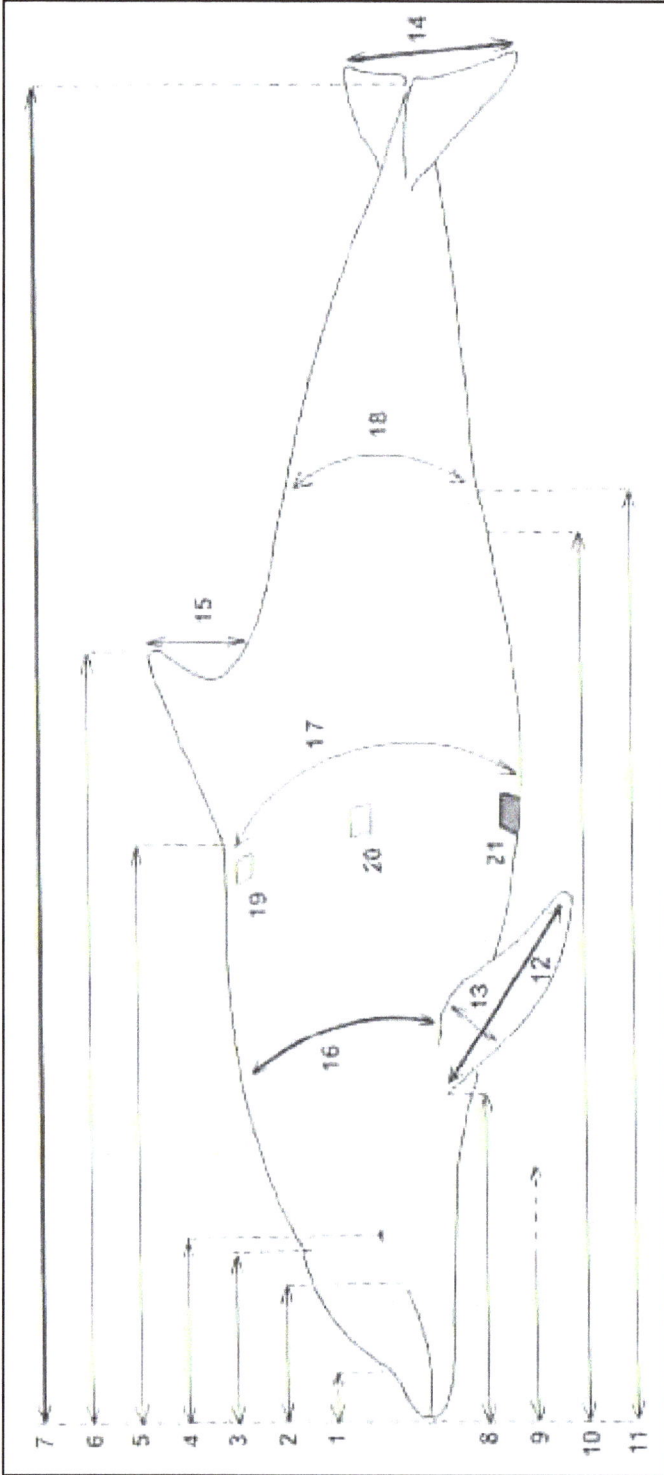

Figure 3.1: Morphometrics for a Cetacean (Geraci and Lounsbury, 2005)

1. Snout to melon. 2. Snout to angle of mouth. 3. Snout to blowhole. 4. Snout to centre of eye. 5. Snout to anterior insertion of dorsal fin. 6. Snout to tip of dorsal fin. 7. Snout to fluke notch. 8. Snout to anterior insertion of flipper. 9. Snout to caudal end of ventral grooves (when present). 10. Snout to centre of genital aperture. 11. Snout to centre of anus. 12. Flipper length. 13. Flipper width (maximum). 14. Fluke width. 15. Dorsal fin height. 16. Girth: axillary. 17. Girth: maximum18. Girth: at level of anus. 19. Blubber thickness: dorsal (anterior and lateral to dorsal fin). 20. Blubber thickness: lateral at mid- length. 21. Blubber thickness: ventral at mid- length.

Some of the important endangered marine animals have been accorded special protection under the Wild Life (Protection) Act, 1972 (WPA) wherever they occur in Indian territories. These include all species of Cetaceans such as whales and dolphins, dugongs, reptiles such as Salt water or Estuarine crocodile and sea turtles such as Green Sea, Hawksbill, Leatherback, Olive Ridley turtles, hard corals, some fishes specially shark species, sea cucumbers, certain molluscs etc. Other marine animals get full protection within the Protected Areas (PAs), such as National Parks and Sanctuaries, which are constituted under the WPA.

References

Geraci, J. R. and V. J. Lounsbury. 1993. Marine Mammals Ashore: A Field Guide for Strandings. Texas A and M University Sea Grant College Program, Galveston, Texas.

Geraci, J. R. and V. J. Lounsbury, 2005. Marine Mammals Ashore: A Field Guide for Strandings, Second Edition. National Aquarium in Baltimore, Baltimore, MD.

Kasinathan, C. 2002. On the landing of Black Porpoise *Neophocaena phocaenoides* at Sangumal (Palk Bay) near Rameswaram. Mar. Fish. Infor. Serv., T and E Ser., No. 173 (CMFRI Publications, India), 3.

Melinmani, B. S. 2004. Stranding of a whale, *Balaenoptera* sp. near Vijaydurg landing centre of Maharashtra coast. Mar. Fish. Infor. Serv., T and E Ser., No. 182 (CMFRI Publications, India), 14.

Perrin, W. F. and J. R. Geraci, 2002. Stranding. *In*. Encyclopedia of Marine Mammals-Eds.

Perrin, W. F., B. Wursig and J. G. M. Thewissen, p. 1192- 1197. Academic Press.

Ravi, V. and S. Murugan, 2010. On the Pygmy Sperm Whale *Kogia breviceps* (Blainville, 1838) Stranded at Cuddalore Silver Beach, Tamil Nadu, Southeast Coast of India, Proenvironment 3 92010): 400- 401.

Vivekanandan, E. and R. Jeyabaskaran, 2012. Marine mammal species of India. Central Marine Fisheries Research Institute, Kochi, 228p.

Wild Life (Protection) Act, 1972. www. moef. nic. in.

Chapter 4

Trends and Issues in Animal Science Research

☆ *Chhaya Panse*

Introduction

Animal experimentation has played a very important role in the life sciences, particularly in health science, which is directly concerned with the survival and health of humankind. It goes without saying that the information gleaned from animal experimentation has contributed a great deal to research into fundamentally important human medicine and veterinary medicine, which of course helps to cure and prevent diseases in both humans and animals, and also contributes to education and training in healthcare technology (Hiromi Omoe, October 2006). Animal Science is described as "studying the biology of animals that are under the control of mankind". (Wikipidea).

Degrees in Animal Science are offered at a number of colleges and universities. Typically, the Animal Science curriculum not only provides a strong science background, but also hands-on experience working with animals in the laboratory.

People for the Ethical Treatment of Animals (PETA) is an American animal rights organization based in Norfolk, Virginia, and led by Ingrid Newkirk, its international president. A non-profit corporation with 300 employees, it claims to have three million members and supporters and to be the largest animal rights group in the world. Its slogan is "animals are not ours to eat, wear, experiment on, use for entertainment or abuse in any way."

According to PETA countless monkeys, dogs, rats and other animals are burned, blinded, cut open, poisoned, starved and drugged behind closed laboratory doors every year for convenience, for economic reasons and because of old habits. Not only are animal tests extremely cruel, they are also completely inaccurate because of the vast physiological variations between species.

Vivisection

Vivisection is the practice of experimenting on live animals. Many vivisectors come to India because, in their own countries, they cannot get away with doing the type of animal testing they can here. Every year, research facilities across India – including the Animal Research Centre, the Patel Chest Institute, the National Institute of Nutrition (NIN) and the All India Institute of Medical Sciences (AIIMS), just to name a few – squander valuable time and resources as well as millions of rupees conducting experiments on monkeys, dogs, cats, rabbits, rats, mice and other animals.

AAVS (American Vivisection Society) also believes that animals have the right to not be exploited for science. New technologies, alternatives and clinical and epidemiology studies in human can provide better, more relevant answers without causing animal suffering.

Opponents to any kind of animal research—including both animal-rights extremists and anti-vivisectionist groups—believe that animal experimentation is cruel and unnecessary, regardless of its purpose or benefit. There is no middle ground for these groups; they want the immediate and total abolition of all animal research. If they succeed, it would have enormous and severe consequences for scientific research. The UK has gone further than any other country to write such an ethical framework into law by implementing the Animals (Scientific Procedures) Act 1986 (Simon and Robin, 2007).

The aims of this additional review process are: to provide independent ethical advice, particularly with respect to applications for project licences, and standards of animal care and welfare; to provide support to licensees regarding animal welfare and ethical issues; and to promote ethical analysis to increase awareness of animal welfare issues and to develop initiatives for the widest possible application of the 3Rs—replacement, reduction and refinement of the use of animals in research (Russell and Burch, 1959).

Animal-rights groups also disagree with the 3Rs, since these principles still allow for the use of animals in research; they are only interested in replacement.

Committee for the Purpose of Control and Supervision of Experimentation on Animals (CPCSEA) – which was created under the provisions of the Prevention of Cruelty to Animals Act 1960 – is supposed to help implement good laboratory practices and ensure that animal testing is carried out under proper conditions, animal research in India is notoriously riddled with problems.

Many pharmaceutical companies do not employ full-time veterinarians to take care of animals on a day-to-day basis or caretakers to look after the animals at night. Most of the procedures are performed by students. Housing conditions are bleak because many laboratories do not provide animals with air conditioning, proper lighting, or hygienic water bottles, cages and food.

PETA US has conducted many undercover investigations in laboratories. Every time it does, physical abuse and neglect are documented. Animals are yelled at, hit, left to suffer after surgery without any painkillers, crammed into small cages, denied veterinary care and more.

A few years ago, PETA and the CPCSEA rescued a monkey named Paro and 36 others from Pune's National Institute of Virology (NIV) after uncovering horrid conditions. Unable to provide even one record for any of the animals it used, NIV had confined most of its monkeys to tiny cages for more than a decade, and some had been disfigured or paralysed from confinement and abuse. Some monkeys were missing fingers and teeth, while others – who had gone insane from years of intensive confinement – spun in circles around their cages.

Medical Research

The most significant trend in modern research in recent years has been the recognition that animals are rarely good models for the human body. Studies have shown time and again that researchers often waste lives – both animal and human – and precious resources by trying to infect animals with diseases which they would not normally contract.

According to PETA in many cases, not only does animal testing hurt animals and waste money, it also harms and kills humans. For example, thalidomide, Zomax and DES were all tested on animals and judged safe, but they had devastating consequences for the people who used them.

Unsurprisingly, medical general practitioners (GPs) are even more aware of the contribution that animal research has made and continues to make to human health. In 2006, a survey by GP Net showed that 96 per cent of GPs agreed that animal research has made important contributions to many medical advances (RDS News, 2006).

Product Animal Testing is Animal Cruelty

As reported by Massachusetts Animal Rights Coalition, Animal testing uses animals to determine the toxicity of the chemicals in the products that we use in our homes. Companies poison, burn, and blind hundred of thousands of animals every year to unnecessarily test soaps, toothpaste, mouthwash, oven cleaners and more.

Few of the tests conducted are as follows:

Eye Irritancy Testing

This test forces chemicals into the eyes of fully conscious, restrained rabbits. Anesthetics are not required to be used. The extreme pain often causes them to struggle so severely that they break their own backs, dying in agony.

Skin Irritancy Testing

This test places corrosive chemicals onto the shaved/raw skin of rabbits and guinea pigs. The caustic nature of these substances causes severe injuries to the animals. Gaping wounds and bleeding are common.

Oral Toxicity Testing (LD50)

This test force-feeds chemicals to fully conscious animals for 14-28 days till death. In 1983, David Rall, Director of the National Toxicity Program, called this test "an anachronism" and said, "I do not think this test provides much useful information about the health hazards to humans."

Good Alternatives

With so many sophisticated non-animal product tests now available, companies have no excuse for continuing to torment animals. Instead of measuring how long it takes a chemical to burn away the cornea of a rabbit's eye, manufacturers can now drop that chemical onto donated human corneas Alternatives to animals include use of cell cultures, corneal and skin tissue cultures, corneas from eye banks.

In addition, companies can use computer and mathematical models. They can also choose to use ingredients and chemicals which have already been proved to be harmless and are known to be safe.

Dissection

Many schools still use dissection to teach courses in biology, anatomy and other life sciences. Every year, millions of frogs, cats, dogs, pigs, worms, mice, rats, rabbits and fish are killed for use in classroom dissections.

Animals used by the dissection industry suffer terribly before they even reach the classroom. CPCSEA While working undercover inside one biological supply house, PETA US investigators documented that employees were embalming cats and rats while they were still alive. Frogs were held by their hind legs and slammed headfirst onto a hard surface to kill them.

Danger to Students

Animals used for dissection are often embalmed with formaldehyde or a chemical derived from formaldehyde – a preservative linked to cancer of the throat, lungs and nasal passages as well as a variety of other health problems.

Danger to the Environment

Frogs are the most commonly dissected animals. Frogs feed on lot of insects every day, so removing frogs from the ecosystem disrupts nature's delicate balance, which results in increased crop destruction and the spread of diseases such as malaria. Several species of frogs are now endangered in India because so many were taken from the wild.

Cutting Dissections Out of the Curriculum

While there was a time when dissection and the use of live animals in the classroom went unchallenged, PETA has campaigned to ban dissection by writing letters to the University Grants Commission (UGC) and to all the universities in India. PETA has urged these institutions to replace dissection in zoology courses with humane, non-animal methods which are both more effective and readily available. Recently, PETA achieved a partial but crucial victory when the expert committee set up by the UGC to consider banning dissections recommended doing away with dissections by students at the undergraduate level.

UGC Guidelines for Discontinuation of Dissection and Animal Experimentation in Zoology/Life Sciences in a Phased Manner

Animal dissection as an aspect of Zoology curriculum is about ninety year old. Over the years there has been a tremendous expansion of knowledge content of Zoology in the light of emergence of newer branches such as biodiversity, biochemistry, biophysics, molecular biology, etc. Thus, in the contemporary scenario, there is over-emphasis of learning of anatomy as laboratory exercises. It has been felt that the curriculum must be revamped to accommodate the latest developments where in there is pertinent need to underplay animal dissections. Further, when there were fewer higher learning institutions and fewer students, fewer animals were used in dissections. Now the number of such institutions has become manifold and more than a million students take to programs requiring animals for dissections. Most of these animals are caught from the wild, and their indiscriminate removal from the natural habitats disrupts the biodiversity and ecological balance. Thus, use of animals in dissections has come to be a factor compounding with habitat loss, pollution and climate changes in depletion of animal populations. It is a fact that the demand for dissection specimens increases pressure on threatened species. The case of frogs, the population of which has declined to alarming levels in the recent times, is often cited as the example. Also, it has been noticed that laws/regulations/guidelines about animals and their welfare are not taken to cognizance while prescribing animal use in the curriculum.

Following are the Recommendations

Recommendation 1

All Institutions of Higher Education to strictly adhere to the Wild Life Protection Act, 1972 and the Prevention of Cruelty to Animals Act, 1960. The Wildlife Protection Act, 1972, amended from time to time, has all Elasmobranchii (sharks and rays) included in the Schedule I, and all frogs belonging to genus Rana included under Schedule IV. Further, "Animal Ethics" should be included as a chapter in an appropriate course of study. Therefore, all educational institutions coming under the purview of UGC shall prescribe laboratory curriculum involving animals in such a way to be compassionate with the animals, avoid experiments on animals, wherever possible, and use alternatives in their place, experiments on animals are not performed merely for the purpose of acquiring manual skill, and not to use animals protected under the Wildlife Protection Act 1972, particularly frogs belonging to genus Rana and any elasmobranch fish, in laboratory exercises.

Recommendation 2

All Institutions of Higher Education are required to constitute "Dissection Monitoring Committees" (DMC) to look into the use of animals and UGC to provide guidelines for the same.

The tenure of DMC shall be 2 years, and on expiry of a term, the DMC should be reconstituted wherein only the Convener and Chairperson (the Head of the Department) may continue for two or more terms if he/she happens to continue to be the Head of the Department.

A vacancy arising during the tenure of a DMC shall be filled with a faculty belonging to the respective category.

The DMC shall be convened by giving one week written notice to the members records thus maintained.

The quorum for the meeting shall be 3 out of 6 where in at least one member from the neighboring institution must be present.

The DMC shall meet at least once each semester/half year and approve/review use of animals in dissections/experiments for laboratory exercises, within the purview of the Guidelines here in.

It shall be the responsibility of DMC to ensure that animals that are permitted to be used for dissections/experiments in the Guidelines herein are procured from ethical sources, and not removed from the wild for these purposes, and transported to the laboratory without stress or strain to the animals if alive and anesthetized appropriately if they are to be used in dissections.

The Institution shall maintain appropriate records of procurement of animals, their transport if alive, number of animals used, use of anesthesia/euthanasia if applicable, etc. The DMC shall scrutinize the records thus maintained.

Recommendation 3

For both UG and PG programs, there shall be reduction in the number of animals for dissection and experimentation as well as in the number of species with all ethical considerations. Preference shall be given to laboratory bred animal models. Removal of animals from their natural habitats should be best avoided. Removal of animals from their natural habitats should be best avoided.

Recommendation 4

For UG: 'Only one species' to be adopted for 'demonstration only' by the faculty and 'students should not do any dissection'. In lieu of this, curriculum must be developed to encourage students to take up field work.

Recommendation 5

For PG: Students shall have the option to perform dissection of 'selected species' as per the curriculum or to have a project related to biodiversity/biosystematics, etc.

Long Term Actions

Human Resource Development through training programs towards adopting alternative modalities for animal dissection.

Software development for alternative modalities for animal dissection, experimentation and dissemination.

Empowering Zoology/Life Sciences departments with appropriate information communication technology (ICT) for implementing the above recommendations.

Curriculum related to invertebrates, vertebrates, etc., to be enriched with bio-systematics, population dynamics, evolution and bio-diversity, etc

References

Animal Diversity. Cleveland P. Hickman, Larry S Roberts, Susan L. Keen, Allan Larson, David Eisenhour. McGraw-Hill Higher Education, 2008.

Animal Diversity. Diana R. Kershaw. University Tutorial Press, 1984.

Animal Diversity: A Textbook of Invertebrate Zoology. Eylers. Mosby, Incorporated, 1991.

Digital Zoology: Version 2. 0 CD-ROMand Student Workbook. Jon G. Houseman. McGraw-Hill, 2003.

General Zoology Laboratory Guide. Charles F. Lytle, John R. Meyer. McGrawHill Higher Education, 2008.

Glencoe Science Modules: Life Science, Animal Diversity, Student Edition. Lucy Daniel, Dinah Zike. McGraw-Hill, 2007.

Hiromi Omoe, Life science Research unit Ouatrely Review No. 21/October 2006.

Invertebrate Zoology: A Functional Evolutionary Approach. Edward E. Ruppert, Richard S. Fox, Robert D. Barnes. Thomson-Brooks/Cole, 2004.

Invertebrate Zoology: A Laboratory Manual. Robel1 L. Wallace, Walter Kingsley Taylor. Prentice Hall, 2002.

Laboratory Studies in Animal Diversity. Cleveland P. Hickman, Lee B. Kats, William C. Ober. McGraw-Hill, 2006.

Laboratory Studies in Animal Diversity. Cleveland P. Hickman, Lee B. Kats. McGraw-Hili Higher Education, 2008.

RDS News. 2006. GPs Back Animal Research. London, UK: Research Defence Society.

Russell WMS, Burch RL (1959). The Principles of Humane Experimental Technique. London, UK: Methuen.

Simon Festing1 and Robin Wilkinson1Journal List EMBO Rep v. 8(6); Jun 2007.

Vertebrate Zoology: An Experimental Field Approach. Nelson G. Hairston Cambridge University Press, 1994.

http: //www. aavs. org/research_problem

http: //www. peta. org/issues/animals-used-for-experimentation/animal-testing-bad-science/

Chapter 5

Global Climate Change and the Activities of Marine Microbes: A Review

☆ *S.S. Navami and Ayona Jayadev*

ABSTRACT

Climate change is a serious issue which contributes various threats to biological diversity. The day to day increase in CO_2 and the resultant global warming represents the clearest and best documented signal of human alteration of the earth system. This will affect various ecosystems in many ways. Coastal marine systems are the most ecologically and socio-economically vital on our planet. The climate change seems to be increasingly stressful for many marine organisms including the microbial population. The aim of this review paper is to present about the consequences of climate change on organisms, particularly the marine microbes.

Keywords: *Climate change, Global warming, Marine ecosystem, Marine Microbes.*

Introduction

Human induced alterations on various systems on earth result in further alterations to the normal functioning of the earth system; most notably by driving global climatic change and causing irreversible losses of biological diversity (Inter-governmental Panel on Climate Change, 1995., United Nations Environment Program, 1995). The natural green house effect is the mechanism by which the earth's climate is being controlled. This mechanism is operated as a result of the heat trapping mechanism of various gases like water vapour, carbon dioxide, ozone, methane, and nitrous oxide, which absorb heat radiated from the earth's surface and lower atmosphere and then radiate much of the energy back toward the surface. Without this effect the earth would be colder. Thus warming of earth's climate normally has made the conditions helpful for the origin and development of life on earth. The climate is changing all over the world both due to natural and anthropogenic activities. The modern increase in CO_2 represents the clearest and best documented signal of human alteration of the earth system. Increase in the concentration of CO_2 is a major cause of the increase in the global temperature beyond normal rate over the last 50 years. Anthropogenic emissions of carbon dioxide (CO_2) have increased to nearly 394 ppm in 2012 (NOAA Earth System Research Laboratory, 2012). The best known postulated consequence of an increasing atmospheric CO_2 concentration is increased global warming rate, which may, among other things, lead to sea level changes, promote ocean stratification, and alter the sea-ice extent and patterns of ocean circulation (Doney *et al.*, 2012).

Global average surface air temperature has increased substantially since 1970. Over the past three decades, human influences on climate have become increasingly obvious, and global temperatures have risen sharply. Warming leads to the melting of glaciers and ice sheets, which raises sea level by adding water to the oceans. The natural productivity of oceans will be affected by the changes in the temperature. There is strong correlation between climate induced changes in the marine environment –including rising sea surface temperatures, increases in UV radiation, and even increased deposition of iron rich dust arising from dust storms – and the emergence of bio-toxin based human diseases. Emissions of heat-trapping gases will cause further warming in the future. Human pressures, including climate change, are having profound and diverse consequences for marine ecosystems. Atmospheric concentrations of ozone-depleting substances, increased greenhouse gas concentrations have the potential to affect the spatial distribution of ozone and will influence the ultra violet radiation reaching the Earth's surface (UNEP, 2012). The aim of this review paper is to present the global climate change scenario and its effect on marine microbial community and consequences of interactions between ocean acidification, increased UVR (Ultra Violet Radiation), anthropogenic pollutants, and marine microbial communities.

Effect of Climate-Change on Life on Earth

Climate change is identified as an average weather condition of an area characterized by its own internal dynamics and by changing in external factors that affect climate (Trewartha *et al.*, 1980). It has imparted significant effect on the existence of biota all over the world. It is generally agreed that climatic regimes influence species' distributions, often through species-specific physiological thresholds of temperature and precipitation tolerance (Hoffman and Parsons, 1997; Woodward, 1987). Climate change responses have significant role in the phenology of species (Parmesan, 2006). Studies by Parmesan and Yohe (2003) estimated that more than half (59 per cent) of 1598 species exhibited measurable changes in their phonologies (the timing of seasonal activities of animals and plants) and/or distributions over the past 20 to 140 years. Species responding to recent, relatively mild climate change (global average warming of 0.6°C) is in very high proportion. The responses vary according to the species with differential magnitudes of climate change experienced. The warming trend that is apparent in the temperature records is confirmed by other independent observations, such as the melting of Arctic sea ice, the retreat of mountain glaciers on every continent, reductions in the extent of snow cover, earlier blooming of plants in spring, and increased melting of the Greenland and Antarctic ice sheets. Human altered changes are expected to alter the precipitation pattern all over the globe. After at least 2,000 years of a little change, sea level rose by roughly 8 inches over the past century. Satellite data available over the past 15 years show sea level raising at a rate roughly double the rate observed over the past century. A given change in climate is expected to have the largest proportional effect on biodiversity in those biomes characteristic of extreme climates, although biodiversity in all biomes likely will be sensitive to climate.

Invasions of exotic species are promoted by human disturbance and changes in climate variability (Sala *et al.,* 2000). Walther *et al.* (2002) reported that with climate change, non-native species from adjacent areas may cross frontiers and become new elements of the biota. A permanent establishment at the new locality may not be possible without changes in local conditions. Climate-linked invasions might also involve the immigration of unwanted neighbours such as epidemic diseases (Epstein, 1998) and also involve alterations in trophic interactions (Brander, 2010). Many polar fishes and invertebrates are adapted to contemporary cold conditions and have limited tolerance to seemingly small increases in water temperature (Somero,2012). Species that have adapted to life at the sea ice edge, such as crustaceans, may also experience population declines, thereby decreasing the available food sources for penguins, seals, polar cod and narwhals (IPCC, 2007).

Marine Ecosystem

Coastal marine systems are among the most ecologically and socio-economically vital on the planet. Much of deep biodiversity is exclusively of marine origin. Marine communities are biological networks in which the success of species is linked directly or indirectly through various biological interactions (*e.g.,* predator-prey relationships, competition, facilitation, mutualism) to the performance of other species in the community (Doney *et al.,* 2012). Our knowledge of marine diversity in the present is poor compared to our knowledge for terrestrial organisms. Marine biodiversity is the variety of life in the sea, encompassing variation at levels of complexity from within species to across ecosystems (Sala and Knowlton, 2006). Millions of people around the world depend upon the oceans for a diverse array of resources and services. The sustainable provision of these materials is determined by a diversity of complex ecological interactions between a variety of different living and non -living factors. Human alterations of marine ecosystems are more difficult to quantify than those of terrestrial ecosystems, but several kinds of information suggest that they are substantial (Vitousek *et al.,* 1997).

Climate Change and Marine Ecosystem

Marine biodiversity provides most services we obtain from the sea, including food security, protection against coastal erosion, recycling of pollutants, climate regulation, and recreation. Species depletions due to the climate change can change ecological processes that are vital to the persistence of marine communities (Stachowicz *et al.,* 2002). Global warming acts in synergy with other anthropogenic factors, such as pollution and over exploitation, negatively impact upon the complex marine ecosystem driving changes in the marine environment with important socio-economic consequences. The direct components of climate change affecting marine organisms over the next century are (i) temperature increase, (ii) sea level increase and subsequent changes in ocean circulation, and (iii) decrease in salinity (Marcais *et al.,* 1998). Projected declines in ocean oxygen levels reflect the combined effects of reduced oxygen solubility from warming and reduced ventilation from stratification and circulation changes (Keeling *et al.,* 2010). This condition is increasingly stressful for many marine organisms. Decades of ecological and physiological research document that climatic variables are primary drivers of distributions and dynamics of marine plankton and fish (Hays *et al.,* 2005, Roessig *et al.,* 2004). Owing to thermal expansion and melting of land fast ice (glaciers and ice caps and sheets), warming is causing sea level to rise, with a current rate of approximately 3 mm per year (Cazenave *et al.,* 2008). Climate change has resulted in rising sea temperatures and levels, changes in ocean circulation, pH and salinity, and has exposed the world's oceans to increasing levels of ultraviolet radiation. These physical and chemical changes influence the prevalence and potency of marine pathogens and bio-toxins, with serious ecological and socio-economic ramifications including the loss of species. Marine biodiversity naturally changes locally at scales of years to centuries in what has been called ecological succession. During a natural successional sequence and in the absence of further disturbance, biodiversity tends to slowly increase over time in a self-organization process that is a consequence of the activities of the organisms themselves (Sousa, 1979). This process will be affected by the changes in the normal physico-chemical conditions of marine ecosystem.

As the ocean absorbs carbon dioxide from the atmosphere, seawater is becoming less alkaline (its pH is decreasing) through a process generally referred to as ocean acidification. This is a normal condition. But increased activities of industries and increased exhausts from vehicles releases oxides of gases like nitrogen, sulphur etc. This will get dissolved in rain water and falls on the surface of land as well as water as nitric, nitrous, sulphuric and sulphurous acids and causes the acidity of these places. The pH of seawater has decreased significantly since 1750. This ocean acidification is irreversible over a time scale of centuries. This affects the calcification process of molluscs and some plankton by which the formation of shells takes place (Guldber, 1999) and also affects the bio-geo-chemical cycles. Immersion in more acidic waters can also disturb the internal acid-base balance of organisms, which in turn can affect a wide variety of metabolic processes (Portner, 2010). Humans have directly caused the global extinction of more than 20 described marine species, including seabirds, marine mammals, fishes, invertebrates, and algae (Carlton *et al.,* 1999; Dulvy *et al.,* 2003).

The most obvious consequence of sea level rise will be an upward shift in species distributions. Most species are expected to be able to keep pace with predicted rates of sea level rise, with the exception of some slow-growing, long-lived species such as many corals (Knowlton 2001). Marine systems, which are often dominated by organisms with planktonic life history stages, are also sensitive to alteration in coastal oceanographic patterns. As a result of ozone layer depletion will likely result in increased ultraviolet radiation at the earth's surface, which would in turn have negative effects on invertebrate larvae and algae (Hoffman *et al.*, 2003; Peachey 2005).

Marine Microbes

A simple definition says marine microbes are–any microorganisms found in marine systems and can reproduce in the marine environment. These organisms hold a position of unique importance in the biosphere and are being studied for a couple of decades. Study of marine microbial biodiversity is of vital importance to the understanding of the different processes of the ocean, which may present potent novel microorganisms for screening of bioactive compounds. Microbial communities play a central role in the global recycling of pollutants. For example, the oil-catabolic versatility of microbes, particularly bacteria, ensures that oceans are not completely covered with an oil film (Head *et al.*, 2006). They also play more direct roles in the health of corals and other marine organisms. As the microbial communities have a complex ecosystem process, biodiversity study explores the distribution and roles in the habitat (Das *et al.*, 2006).

Marine microbes carry out many steps of the biogeochemical cycles that other organisms are unable to complete, which are very important for the stability of earth system. The metabolic rate of the marine microbial community is very high and the marine phytoplanktons contribute 45 per cent of the total supply of oxygen to the atmosphere. Beneficial microbial symbioses have enabled many invertebrate species to take advantage of habitats that would otherwise be unavailable to them (Cevera *et al.*, 2006). Marine microbes are adapted to survive in the very extreme climatic condition in their ecosystem. The metabolic capabilities of these organisms are very significant as they are efficient sources of many bioactive products and energy. Marine microbes may also be used to manufacture compounds called "nutriceuticals," natural products used as diet supplements to promote health. Current evidence indicates that most marine microbes are not cosmopolitan, but, instead, are restricted to specific habitat types or geographic locations; however there are a few examples of truly cosmopolitan organisms. Free-living marine microbes may be more cosmopolitan than symbionts, biofilm-associated microbes, and others.

Microbial habitats in the oceans are influenced by an almost innumerable array of forces and factors, including salinity, currents, terrestrial inputs, and climate. With more than a billion micro-organisms living in one cubic litre of seawater, the biodiversity of these mighty midgets is unparalleled. Parasites, including bacteria and viruses, and bio-toxins compromise the health of marine organisms and therefore are key regulators of marine populations and habitats in marine ecosystems. Venter *et al.*(2004) using genome shotgun sequencing, found 1800 distinct microbial genomic species in only 1500 litres of surface seawater in the Sargasso Sea. This technique did not allow them to record most of the rare species; they estimated that a more in-depth coverage would have revealed approximately 48,000 microbial species. Nearly every measurable physical, chemical, and biotic variable in the marine environment has been found to increase, decrease, or otherwise alter microbial diversity.

Marine Microbial Community and Climate Change

Climate change should be considered a major top-down controller of microbial communities. The microbial life on our planet plays a central role in either accentuating or mitigating the effects of climate change. Microbes are central to global systems such as the cycling of carbon, nitrogen, and other gases through the environment. Climate change is likely to have significant impacts on marine microbes, potentially altering their diversity, function and community dynamics. Temperature, precipitation, and wind (including windborne particulate matter) can each impact marine communities in a number of ways. Globally distributed planktonic records show strong shifts of phytoplankton and zooplankton communities in concert with regional oceanic climate regime shifts, as well as expected poleward range shifts and changes in timing of peak biomass (Beaugrand *et al.,* 2002, deYoung *et al.,* 2004, Hays *et al.,* 2005, Richardson and Schoeman 2004).

The impacts of global temperature changes resulting from climate changes will be felt by microbial enzyme systems. Different enzyme systems have different temperature optima, so micro-ecological processes in a given area may be depressed or stimulated depending on the nature of the local temperature change (up or down) and the optima of the enzyme systems involved. Temperature changes in the oceans may also disturb the delicate balance between the numbers of bacteria and phages. Changes in water temperature and ultraviolet radiation (UV), two factors known to be impacted by human activities, are known to disturb the relative numbers of bacteria and viruses in the oceans, with possibly disastrous results for human health. Atmospheric changes that increase the amount of light in these wavelengths that reaches the oceans can upset the balance between viruses and their hosts, possibly leading to uncontrolled epidemics in fish, invertebrates, or humans (Cevera *et al.,* 2005). Microbial communities in the oceans maintain the balances that can keep harmful algal blooms in check. The nutrient contamination can disturb coastal marine microbes, triggering these blooms. Altering microbially- mediated equilibria in one part of the ocean will often have impacts on adjacent areas and far-flung regions.

Planktons are particularly good indicators of climate change in the marine environment for several reasons. First, unlike other marine species, such as fish and many intertidal organisms, few species of plankton are commercially exploited; therefore, any long-term changes can be attributed to climate change. Second, most species are short lived and so population size is less influenced by the persistence of individuals from previous years. Plankton can show dramatic changes in distribution because they are free floating and can respond easily to changes in temperature and oceanic current systems by expanding and contracting their ranges. Finally, recent evidence suggests that plankton are more sensitive indicators of change than are even environmental variables themselves, because the nonlinear responses of biological communities can amplify subtle environmental perturbations. The rate of N_2 fixation by ocean planktons may vary in intensity with changes in climate. These changes may include increased upper-ocean stratification, which could enhance N_2 fixation and shift organic matter export from being mostly particulate to mostly dissolve. Consequently, changes in N_2 fixation can markedly alter the N inventory of the ocean and, hence, the stoichiometric balance between C, N and P available as nutrients (Karl *et al.,* 2002).

Key ecosystem processes like bacterial production, respiration, growth efficiency and bacterial–grazer trophic interactions are likely to change in a warmer ocean (Sarmento *et al.,* 2010). Climate change and acidification will undoubtedly affect microbial food webs as well, although there are many uncertainties (Sarmento *et al.,* 2010, Joint *et al.,* 2011). Santos *et al.,* 2012 reported that ocean acidification and increased UVR have the potential to affect microbial assemblages. Phytoplanktons have differing sensitivities to CO_2 concentration and have a variety of mechanisms for carbon utilization. Thus, an increase in seawater CO_2 concentration will not only change the activity of individual phytoplankton species, but will also tend to favour some species over others (Hays *et al.,* 2005), since it will affect the higher trophic level communities. The changes in the pH and CO_2 concentration can have both positive and negative effect on the growth of marine plankton with a corresponding impact on their role as a net source or sink of CO_2 to the atmosphere (Riebesell, 2004). High concentration of CO_2 during calcification creates a negative feedback in the normal physiology of planktons, because increasing CO_2 levels and the resultant decrease in pH inhibit calcification (Riebesell, 2000). The activity of bacteria (which produce CO_2) and the zooplankton (which consume phytoplankton) might also be affected by pH.

Ocean acidification affects the metal bio-availability. Iron is a limiting nutrient for marine phytoplankton in large oceanic regions (Sunda 2010) which are involved in the growth and metabolism of microbes (Gadd, 2010). It is also an important factor in the detoxification of hydrocarbons. Reduced oceanic pH has the potential to affect the adsorption of metals by organic particles. Organic particles are negatively charged and, as pH declines, surface sites become less available to adsorb positive ions like metals. Small deviations in the concentration of elements such as Cu and Cd can have a serious effect on the health of marine organisms (Millero *et al.,* 2009).

The amount of UVR that reaches the Earth's surface has important consequences for aquatic ecosystems. UVR is the most photo-chemically reactive waveband of incident solar radiation and can have genotoxic, cytotoxic, and ontogenetic effects on aquatic organisms (Bancroft *et al.,* 2007). The biological effects of UV-A are usually considered indirect, resulting from intracellular generation of reactive oxygen species (ROS), which cause oxidative damage to lipids, proteins, and DNA (Pattison and Davies 2006). UV-C wavelengths are generally not deemed to be environmentally relevant, given that they are almost completely screened out of the atmosphere by oxygen and ozone. UV-B is the highest energy wavelength of solar radiation that reaches the Earth's surface and the UV wavelength that is mostly affected by shifts in the ozone layer (Andersen and Sarma 2002). Garcia-Pichel, 1994 reported that UVR represents an important stressor for bacteria in aquatic ecosystems, as their simple haploid genomes provide little or no functional redundancy. The differential sensitivity to UVR exhibited by the most abundant bacterial groups present in the bacterio-plankton is of great importance for the biogeochemical impact of enhanced UVR on ecosystems. Photo-enhanced toxicity of PAH (Poly Aromatic Hydrocarbons) due to UVR exposure has already been observed in a variety of organisms (Peachey, 2005).

According to Coelho *et al.* (2013) the use of microcosms (simplified systems that are constructed to mimic natural environments under controlled conditions) is combined with recent advances in microbe characterization technologies can provide an important framework to start unraveling how climate change and pollution may interact to affect several levels of biological organization. The recent developments in molecular technologies have the capacity to provide relevant information about how the recent climate change scenario affects different types of microbes in various spheres of earth. Larsen *et al.* (2012) used an artificial neural network to develop a model that predicts the abundance of microbial taxa as a function of environmental conditions and biological interactions.

Conclusion

Climate change is a serious issue which is a very relevant topic today. All climate models projects that human-caused emissions of heat-trapping gases will cause further warming in the future. Human activities are without doubt now the strongest driver of change in marine biodiversity at all levels of organization; hence, future trends will depend largely on human-related threats. A long- term ocean biological data sets are essential for understanding climate change impacts on marine ecosystems. All ecosystems are affected by the global climate change. Rising CO_2 and climate change may modify overall ecosystem properties such as trophic structure, food-web dynamics, and aggregated functioning such as energy and material flows and biogeochemical cycles, eventually impacting the ecosystem services upon which people and societies depend. Climate change will exacerbate the stress on living resources already impacted by pollution, over fishing and other anthropogenic activities. This phenomenon, which will be felt by marine microbial communities as changes in ocean temperatures, will undoubtedly alter the diversity of communities in unforeseen ways. Understanding the interactive effects of climate change and anthropogenic pollutants on microbial communities is a complex task. Further advanced studies are needed to assess the impact of climate change on the entire microbial community in ocean. The data available today suggests that the global warming and ultra violet radiation significantly affect the biological functioning of microbes.

References

Andersen, S. O., and Sarma, K. M. 2002, Protecting the ozone layer: the United Nations history, Earthscan, London.

Bancroft, B. A., N. J. Baker, and A. R. Blaustein. 2007, Effects of UVB radiation on marine and freshwater organisms: a synthesis through meta-analysis. *Ecol. Lett.* 10: 332–345.

Beaugrand, G, Reid PC, Ibanez F, Lindley J. A. and M. Edwards. 2002. Reorganization of North Atlantic marine copepod biodiversity and climate. *Science,* 296: 1692–1694.

Brander, K. 2010, Impacts of climate change on fisheries. *J. Mar. Syst.* 79: 389–402

Carlton, J. T., Geller, J. B., Reaka-Kudla, M. L., Norse, E. A. 1999. Historical extinctions in the sea. *Annu. Rev. Ecol. Syst.* 30: 515–538.

Cazenave, A., Lombard, A. and Llovel, W. 2008, Present-day sea level rise: a synthesis. C. R. *Geosci.* 340: 761–70.

Cevera, J. H, Karl, D. and Buckley, M. 2006, Marine Microbial Diversity: The key to earth's habitability, A report from the American Academy of Microbiology.

Coelho Francisco J. R. C., Santos, A. L, Coimbra, J, Almeida, A, Cunha A, Cleary Daniel F. R., Calado, R. Newton and C. M. Gomes. 2013, Interactive effects of global climate change and pollution on marine microbes: the way ahead. *Ecology and Evolution, 3*: 1808–1818.

Das, S, Lyla, P. S. and Khan, S. 2006, A Marine microbial diversity and ecology: importance and future perspectives, *Current Science.* 90(10).

DeYoung B, Harris R, Alheit J, Beaugrand G, Mantua N. and Shannon L. 2004., Detecting regime shifts in the ocean: data considerations. *Prog. Oceanogr.* 60: 143–164.

Doney, S. C, Ruckelshaus, M, Duffy, J. E Barry, J. P, Chan, F, English, C. A, Galindo, H. M, Grebmeier, J. M Hollowed, A. B, Knowlton, N, Polovina, J, Rabalais, N. N Sydeman, W, J and Talley, L. D. 2012, Climate Change Impactson Marine Ecosystems, *Annu. Rev. Mar. Sci.* 4: 11–37.

Doney, S. C., Ruckelshaus, M, Emmett Duffy, J, Barry, J. P., Chan, F. and English, C. A., 2012. Climate change impacts on marine ecosystems, . *Annu. Rev. Mar. Sci.* 4: 11–37.

Dulvy, N. K, Sadovy, Y. and Reynolds, J. D. 2003, Extinction vulnerability in marine populations, *Fish Fish.* 4: 25–64.

Epstein, P. R. *et al.,* 1998, Biological and physical signs of climate change: focus on mosquito-borne diseases. *Bull. Am. Meteorol. Soc.* 79: 409-417.

Gadd, G. M. 2010., Metals, minerals and microbes: geo-microbiology and bioremediation., *Microbiology.,* 156: 609–643.

Guldberg, O. H. 1999, Climate change, coral bleaching and the future of the world's coral reefs., *Mar. Freshw. Res.* 50: 839–66.

Hays, G. C., Richardson, A. J. and Robinson, C. 2005, Climate change and marine plankton, *Trends in Ecology and Evolution*, 20: 337-344.

Head, I. M., Jones, D. M. and Rcoling, W. F. M. 2006. Marine microorganisms make a meal of oil. *Nat. Rev. Microbiol.* 4: 173–182.

Hoffman, A. A. and Parsons, P. A. 1997. Extreme Environmental Change and Evolution Cambridge Univ. Press, Cambridge.

Hoffman, J. R., Hansen, L. J. and Klinger, T. 2003. Interactions between UV radiation and temperature limit inferences from single-factor experiments. *J. Phycol.*, 39: 268–272.

Intergovernmental Panel on Climate Change, Climate Change. 1995. Cambridge Univ. Press, Cambridge, pp. 9–49.

IPCC. Intergovernmental Panel on Climate Change. 2007.

Joint, I., Doney, S. C. and Karl, D. M. 2011. Will ocean acidification affect marine microbes? *ISME J.* 5: 1–7.

Karl, D. *et al.,* 2002, Dinitrogen fixation in the world's oceans. *Biogeochemistry* 57: 47–98.

Keeling, R. F, Kortzinge, A. and Gruber, N. 2010, Ocean deoxygenation in a warming world. *Annu. Rev. Mar. Sci.* 2: 199–229.

Knowlton, N. 2001, The future of coral reefs. Proc. Natl. Acad. Sci., 98: 5419–5425.

Larsen, P. E., Field, D. and Gilbert, J. A. 2012., Predicting bacterial community assemblages using an artificial neural network approach., *Nat. Methods.,* 9: 621–625.

Marcais, B, Dupuis, F. and Desprez, L. M. L. 1996, Annales des sciences forestiere., 53- 369.

Millero, F. J., Woosley, R. Ditrolio, B. and Water, J. 2009, Effect of ocean acidification on the speciation of metals in seawater., *Oceanography,* 22: 72–85.

NOAA Earth System Research Laboratory. 2012. Available at http: //www. esrl. noaa. gov/gmd/ccgg/trends.

Parmesan, C. 2006, Ecological and EvolutionaryResponses to Recent Climate Change, *Annu. Rev. Ecol. Evol. Syst.* 37: 637–669.

Peachey, R. B. J. 2005, The synergism between hydrocarbon pollutants and UV radiation: a potential link between coastal pollution and larval mortality., *J. Exp. Mar. Biol. Ecol.* 315: 103–114.

Portner, H. O. 2010, Oxygen and capacity limitation of thermal tolerance: a matrix for integrating climate related stressors in marine ecosystems. *J. Exp. Biol.* 213: 881–893.

Richardson, A. J. and Schoeman, D. S. 2004., Climate impact on plankton ecosystems in the Northeast Atlantic. *Science.* 305: 1609–1612

Riebesell, U. 2004, Effects of CO_2 enrichment on marine phytoplankton., *J. Oceanogr.* 60: 719–729.

Riebesell, U. *et al.,* 2000, Reduced calcification in marine plankton in response to increased atmospheric CO_2. *Nature,* 407: 634–637.

Roessig, J. M, Woodley, C. M, Cech, J. J. and Hansen L. J. 2004. Effects of global climate change on marine and estuarine fishes. *Rev. Fish Biol. Fish.* 14: 215–75.

Sala, E and Knowlton, N. 2006, Globalmarine Biodiversity Trends, *Annu. Rev. Environ. Resour.* 31: 93–122.

Sala, O. E, Chapin, F. S, Armesto, J. J, Berlow, E, Bloom, J, Dirzo, R Sanwald, E. H, Huenneke, L. F, Jackson, R, B, Kinzig, A, Leemans, R, Lodge, D. M Mooney, H. A,

Poff, N. L. R, Sykes, M. T, Walker, B. H, Walker, M, Wall and D. H Wall. 2000, Global Biodiversity Scenarios for the Year 2100., *Science*, 287: 1770-1774.

Santos, A. L., V. Oliveira, I. Baptista, I. Henriques, N. C. Gomes, A. Almeida, *et al.*, 2012. Effects of UV-B radiation on the structural and physiological diversity of bacterio neuston and bacterioplankton., *Appl. Environ. Microbiol.* 78: 2066–2069.

Sarmento, H, Montoya, J, Vazquez-Domingues, E, Vaque, D. and Gasol, J. 2010, Warming effects on microbial food web processes: How far can we go when it comes to predictions? *Philos. Trans. R. Soc. B Biol. Sci.* 365: 2137–49.

Somero, G. 2012. The physiology of global change: linking patterns to mechanisms. *Annu. Rev. Mar. Sci.* 4: 39–61.

Sousa, W. P. 1979, Experimental investigations of disturbance and ecological succession in a rocky intertidal algal community., *Ecol. Monogr.* 49: 227–254.

Stachowicz, J. J, Fried H, Osman, R. W. and Whitlatch, R. B. 2002, Biodiversity, invasion resistance, and marine ecosystem function: reconciling pattern and process. *Ecology* 83: 2575–2590.

Statistics from World Health Organisation website, http: / /www. who. int/topics/ cholera/en.

Sunda, W. G. 2010., Iron and the carbon pump. *Science*, 327: 654–655.

Tretwartha, T. G. and Horn, H. L. 1980, An introduction to climate. Mc Graw Hill, Aucland.

UNEP. 2012. Environmental effects of ozone depletion and its interactions with climate change: progress report, 2011. *Photochem. Photobiol. Sci.* 11: 13–27.

United Nations Environment Program, Global Biodiversity Assessment, V. H. Heywood, Ed. (Cambridge Univ. Press, Cambridge, 1995).

Venter, J. C, Remington, K, Heidelberg, J. F, Halpern, A. L. and Rusch, D. 2004, Environmental genome shotgun sequencing of the Sargasso Sea. *Science* 304: 66–74.

Vitousek, P. M, Mooney, H. A, Lubchenco, J. and Melillo, J M. 1997, Human Domination of Earth's Ecosystems., *Science*, 277: 494-499.

Walther, G. R, Post, E, Convey, P, Menzel, A, Parmesank, C, Beebee, T. J. C, Fromentin, J. M, Guldberg, O. H. and Bairlein, F. 2002, Ecological responses to recent climate change, *Nature.*, 416: 389-395.

Woodward, F. I. 1987 Climate and Plant Distribution., Cambridge Univ. Press, Cambridge.

Chapter 6

Sunfish (*Mola mola*) at Veraval Landing Centre, Gujarat

☆ *H.L. Parmar, P.V. Parmar, J.B. Solanki
and A.R. Dodia*

ABSTRACT

Sunfish was recorded from Veraval landing center, Gujarat, India. A single specimen of *Mola mola* (99.0cm total length and weighing 53.0kg) was collected from Veraval fish landing centre, on the north-west coast of India during April 2013. Morphometric and meristic character of *Mola mola* are presented in this paper.

Keywords: Sunfish, Mola mola, Veraval, Gujarat.

Introduction

This report is about capture of sunfish (*Mola mola*) from Veraval fish landing center, Gujarat. Sun fishes belong to the family Molidae. They are large, and are primarily pelagic members of the Tetraodontiforms. They are commonly referred as ocean sunfish and known as 'Popicho' locally. Fishes of Molidae have a distinctive laterally compressed shape and "chopped off" appearance (Fraser-Brunner, 1951; Smith and Heemstra, 1986).

In Gujarat, Saurashtra coastal belt is replete with fisheries activity related to marine fish capture (Solanki *et al.,* 2011).The present report is based on a single specimen of sunfish (*Mola mola*) measuring 990 mm in total length, caught on 02.04.2013 in trawl net (in Gujarat called "Jaal") from Veraval - Gujarat, west coast of India (Figure 6.1). The morphometric measurements of the specimen are listed in Table 6.1.

Figure 6.1: Sun Fish (*Mola mola*).

The sunfishes (or head fishes) widely occur in warmer oceanic regions and inhabit open waters upto 200-300 m depth. These fishes occur either solitarily or in small schools and are known for their peculiar drifting nature (Manoj Kumar *et al.,* 1998).

The diet of the ocean sunfish consists primarily of various jellyfish. It also consumes salps, squid, crustaceans, small fish, fish larvae, and eel grass. This range of food items indicates that the sunfish feeds at many levels, from the surface to deep water, and occasionally down to the seafloor in some areas. The diet is nutritionally poor, forcing the sunfish to consume a large amount of food to maintain its size (Powell, 2001).

Four male sun fishes (weighed: 40, 43, 46 and 49 kg) were caught by trawl nets off veraval at 25-50 m depth in march 1997 (Manoj Kumar *et al.,* 1998).

Table 6.1: Morphometric Measurements (in mm) of *Mola mola* Caught at Veraval

Particulars	
Total length (mm)	990
Weight (kg)	53
Breadth at the middle region	720
Length of dorsal fin	490
Length of anal fin (mm)	470
Shout to anal fin insertion	810
Snout to dorsal fin insertion	610
Eye diameter-horizontal	55
Eye diameter-vertical	45
Pectoral fin base breadth	70
Dorsal fin base breadth	265
Anal fin base breadth	220
Snout to Insertion of pectoral	310
Length of gill slit	70
Inter-orbital distance (mm)	300
Dorsal fin tip to anal fin tip	1500
Pectoral fin length	150
Eyeball to snout	150
Sex	Male
Liver weight (g)	1600
Gut length (mm)	2400
Gut length to body length ratio	2.42:1

There are incidences of sun fish capture by trawl nets but due to low magnitude of catch and uncertainty of market fisherman generally do not sell it. The catch remains completely unreported and neglected. Hence, scientific and reliable information about the fish landing is sparsely available. Proper steps toward processing and preservation of Sun fish should be taken to utilize the new resource and to earn vital foreign exchange.

Acknowledgments

The authors wish to express their gratitude to Mr. Manoj Solanki and Mr. Harji Kotavadiya (Progressive fisherman, Veraval) for collection of sunfish sample.

References

Franser-Brunner, A. 1951. The ocean sunfishes (family molidae). Bulletin of British Museum, 1: 89-121.

Manoj Kumar, Kozhakudan, J. K., Thomas, S. and Dineshbabu, A. P. 1998. A record of sunfish *Mola mola* from coastal waters at Veraval. Marine fisheries information service, No. 157, Article No. 877, pp. 21-23.

Powell, David C. 2001. Pelagic Fishes". A Fascination for Fish: Adventures of an Underwater Pioneer. Berkeley: University of California Press, Monterey Bay Aquarium. pp. 270–275.

Smith, M. M. and Heemstra, P. C. 1986. Tetraodontidae, In: Smith's Sea fishes. Smith, M. M. and Heemstra, P. C. (eds.). Springer-Verlag, Berlin, pp. 894-903.

Solanki, J. B., Kotiya A. S., Jetani K. L., Dodia, A. R. and Parmar, H. L. 2011. Traditional storage method of dried fishes by sea sand for consumption along Saurashtra coast, Gujarat. *Fishing Chimes* 31(2): 50-52.

Chapter 7

Traditional Food Product TCR from Crab *Barytelphusa cunicularis* (Westwood, 1836): Routine Diet of Fisher Tribes along Godavari River Coast of Maharashtra

☆ *S.P. Chavan, Pandurang Kannewad*
and Shivaji Poul

ABSTRACT

Traditional knowledge of fisher tribes Bhoi, Zinga-bhoi, Koli living in Godavari river and its tributaries in Marathwada Region of Maharashtra was recorded through the interviews on preparation of their routine food recipe named 'Khekda Curry'or'Crab Curry' called TRC. *Barytelphusa cunicularis* (Westwood, 1836) was best species of crab they select to prepare this recipe. It was medicinal and nutritional dish in the routine diet of tribes in this region. Locally available spices but with specific method of preparation of TCR is a speciality of this documentation important to share with researchers and common people.

Keywords*: Traditional food, Crab, Tribes, Maharashtra.*

Introduction

Food is one of the prime need of every creature in any ecosystem. Aquatic ecosystem is one of the most complex ecosystem on the earth. River as an ecosystem, there is a dependence of animals on each other to fulfil the food requirement forming in to simple and complex food chains. Godavari river is named as 'Dakshin Ganga' similar to the river Ganges in Northern India; it has 1,465 km total length from its origin to merge in to Bay of Bengal of which 692 km distance it cross in Maharashtra. The river originates in the mountain ranges of Sahyadri (Western Ghats) near Trembak - Panchavati in Nashik district of Maharashtra State at about 1067 m. height. About 10,000 fisher tribes found living in different villages in the Godavari river basin in Marathwada region where the study was conducted. The fisher tribes were still depending on the fisheries resources of Godavari river as their full time fishing business. After spending many years in fishing activities in the river the fishermen have detail knowledge about fish species *like Catla catla, Labeo rohita, Cirrihina mrigala, Labeo calbasu, Cirrihina reba, Chela phulo, Chela bacaila, Puntius ticto, Puntius sarana* etc. and they also know about other species from cyprinidae and siluridae, bagarids, murrels, mugils; the tribes have similar detail knowledge on crab species in this region like *Barytelphusa cunicularis* (Westwood, 1836) (Black Crab) and *Barytelphusa guirini* (Milne-Edwards, 1853) (Whitish-Yellow small crab/rice field crab).

The main aim behind the preparation of this manuscript is to share the traditional knowledge of fisher tribes to all those who are interested to use the crabs in their food but do not know the procedure to prepare a tasty and nutritious recipe from it. The procedure, the steps in the preparation process, the ingredients to be added and the quantity, the time taken for the cooking process are the great concern in the formation of any food recipe having a specific taste, fragrance and deliciousness; same is the case in the formation of crab and prawn based food products the tribes apply to prepare the recipes which is their routine diet.

It is our opinion that, the existing generation of fisher tribes in the age group of 50 and above both men and women are having their traditional tricks, tactics and specific methods and procedures they have developed on their own as need base activities to prepare specific spices and foods (Mohammad, 2011). All this tribal life treasure may get lost forever and there are heavy chances of loss of this knowledge treasure with the death of this age group; because new generation of tribes are not interested to learn these activities from their parents and grandparents due to inclination of young generation towards the modern life style to use the readymade, junk food, packed and processed food. The young generation may not have the same kind of knowledge therefore the documentation of the tribal knowledge is most essential (Pandey, 2003; Roose, 2002; Rathakrishnan, 2009). The present investigation and record is also a part of this kind of approach.

Freshwater and marine crabs are used as an important natural protein source since ancient time as natural aqua-food by human being in many parts of the world (Ng, 1988), (Darren, *et al.,* 2008). From more than 6,700 globally distributed species of Brachyuran crabs there are 1,476 species of freshwater crabs in total, in 14 families. Of these 1,476 species over 1,300 species are true freshwater species and remaining are adapted for terrestrial or semi-terrestrial mode of life (Darren *et al.,* 2008), Martha (2008). In the context of global distribution Oriental Region (OR) which includes India and neighbouring countries, China, Japan, Malaysia, Indonesia, Korea etc. has

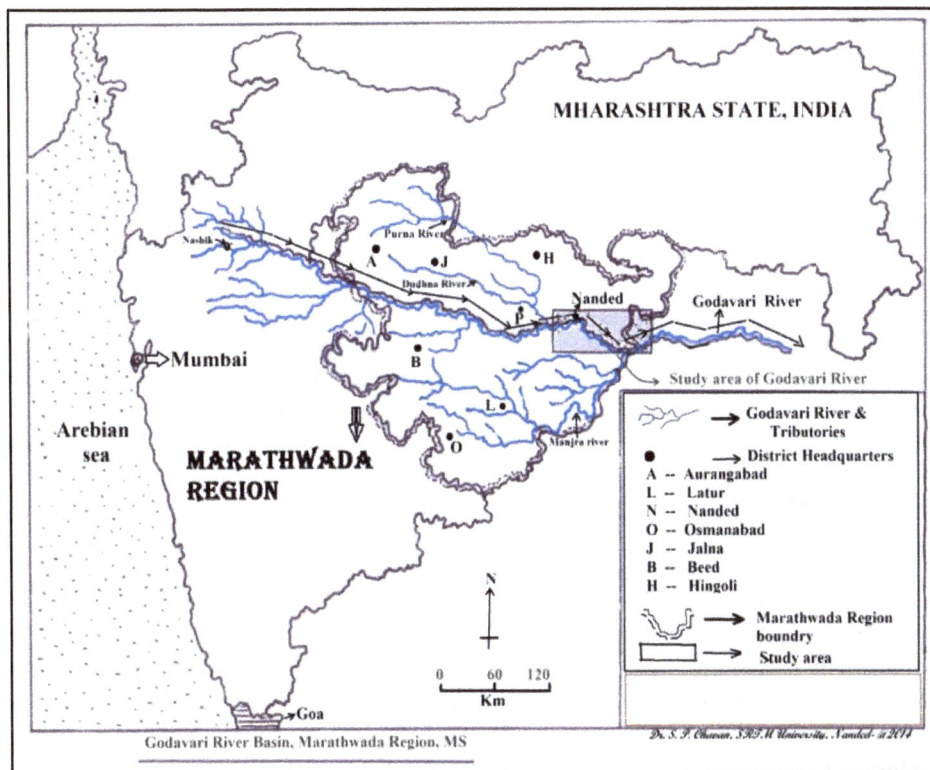

Figure 7.1: Study Area-Godavari River Basin, Marathwada Region, Maharashtra for the Sampling of Crab Species *Barytelphusa cunicularis.*

highest species diversity of crabs in the world with total 818 crab species of which 139 are true freshwater crabs (Cox, 2001). The Crabs comes under Phylum- Arthropoda (Latreille, 1829) Subphylum- Crustacea (Brilnnich, 1722), Class Malacostraca (Latreille, 1802), Order – Decapoda (Latreille, 1802), Suborder- Pleocyemata (Burkenroad, 1663), Infraorder- Brachyura (Linnaeus, 1758), Superfamily- Gecarcinucoidea (Rathbun, 1904). The crab species *Barytelphusa cunicularis* (Westwood, 1836) found in Godavari river basin was used by fisher tribes to prepare the traditional recipe TCR. *B. cunicularis* was one of the common crab species of genus *Barytelphusa*. The fisher tribes prepare traditional recipe named as 'Crab Rassa' (CR) which may be called as Traditional Crab Rassa (TCR). Other species of this genus are *Barytelphusa guerini* (H. Milne-Edwards, 1853), *Barytelphusa jacquemontii* (Rathbun, 1905) and *Barytelphusa pulvinata* (Alcock, 1909).

Materials and Methods

To determine the crab based food products prepared by fisher tribes living along the coast of river Godavari in Maharashtra State, 53 different villages with fisher tribe population and 74 different families of fisher tribes living in the temporary huts along Godavari river coast during fishing season were visited during June 2010 to April 2014. The interviews of individual or a group of fishermen and fisherwomen (age group 15-60 years) were conducted with the participation of 85 women and 30 men. The members of fisher population interviewed during this study were the tribes mainly from Bhoi, Zinga-Bhoi and Koli tribal communities which are common fisher tribes of this region (Chavan *et al.*, 2014). After repeated interviews of the same person after a gap of one month the earlier information given by the same person called informant from the selected tribal group was cross tallied for similarity confirmation and considered valid. It was observed that, about 10-15 per cent men were also involved practically in the preparation of crab and fish based food products at the fishing sites away from their home along with fisherwomen who play important role in daily cooking of meal at home. The information received from the fisher tribes about the typical and specific methodology of fish food preparation and the ingredients they use as additives during the preparation of foods were recorded in the form of videos and the photographs and stored as a traditional knowledge electronic data base in the Laboratory of Aquatic Parasitology and Fisheries Research, Department of Zoology, Swami Ramanand Teerth Marathwada University, Nanded, Maharashtra. A free and informed consent form was read out and made available to those who participated in the study. The interviews were recorded using digital recorders and later transcribed for analysis. The crab specimens collected were processed and identified by referring Alcock (1910) to the lowest possible taxonomic level and afterwards stored in the Aquatic Parasitology and Fisheries Research laboratory, Department of Zoology, School of Life sciences, Swami Ramanand Teerth Marathwada University, Nanded. The following crab based food product the tribes use to prepare is explained in detail with the method of preparation and ingredients used during preparation.

Crab rassa (TCR) from *Barytelphusa cunicularis* (Westwood, 1836)

Barytelphusa cunicularis (Westwood, 1836) (Figure 7.3) was the identified crab species by using standard literature on crab taxonomy (Alcock, 1910) from the tribes fishing samples which they use to prepare their specific recipe locally named as 'Khekda rassa' or 'Khekda curry'. It is a kind of curry (A kind of soup with spices) prepared mainly from *B. cunicularis* crab species commonly found in the Godavari river and its tributaries (Figure 7.1). It may be named as Traditional Crab Rassa (TCR) because it is a traditional knowledge based preparation of fisher tribes.

After discussion with the tribal informants it was found that, instead of using *B. guerini* they use *B. cunicularis* crab species for the preparation of this recipe because *B. cunicularis* species has a pair of large chelate legs with considerably good musculature, larger body size and more taste of formed TCR.

To prepare this recipe the fisher tribes apply the following two steps

1. Preparation of crab body extract
2. Treatment to body extract using spices.

Crab Collection and Preparation of Crab Body Extract/Filtrate (CBF)

Crab species found in river Godavari and its tributaries *Barytelphusa cunicularis* (Westwood, 1836) is one of the commonly occurring crab species in Marathwada region of Maharashtra. By handpicking, using nylon nets (5-7 cm mesh size), using light trap at night are the major common methods of catching this crab species. Usually night time is selected to catch this crab species by the fisher tribes because coastal area of river outside the river water is the feeding ground of this crab species. During day time it was collected from the rock crevices and hiding places under the rock pieces in the short, shallow burrows in mud (Figure 7.2) by digging. *B. cunicularis* of total body weight 100-400 g is the commonly occurring crab size in the river Godavari used to prepare the crab rassa.

Habitat of freshwater crab species *Barytelphusa cunicularis* (Westwood, 1836)
Locality- Godavari River, Near Jarikot, Dharmabad, Dist- Nanded, MS, India. Date- June, 2014
Photo by- Prof. S. P. Chavan, Deptt. of Zoology, School of Life Sciences, SRTM Univ. Nanded, MS,

Figure 7.2:Typical Habitat of Crab Species *Barytelphusa cunicularis* (Westwood, 1836) Rock Crevices and Muddy Coastal Area of River Godavari, Nanded District.

TCR is prepared by external cleaning of the crab body in clean cold water to remove the mud and other slimy materials. The flattened, reduced abdomen flexed under the thoracic sternum region in the form of bent flap of T-shape in male and rounded shape in female crab was removed to remove the faecal waste and abdominal food content. The crab was washed thoroughly in a container with water. Care was taken that the removal of chelate legs are essential before cleaning to prevent the cutting injury from crab but the walking legs (total 08) should not be removed to prevent the loss of nutritious body fluid (Proteins, fats, vitamins and minerals) from crab. Freshwater is boiled in a container; the quantity of water in the container depends on the quantity of crab to be boiled. The crab body and chelate legs should submerge in to the water is the essential sufficient quantity of water. After complete boiling of water the whole body of crab with walking legs and the separated chelate legs of crab are placed in the boiling water. After adding the crab in boiling water it was observed that the boiling activity stops, but it need to continue the heating process the container

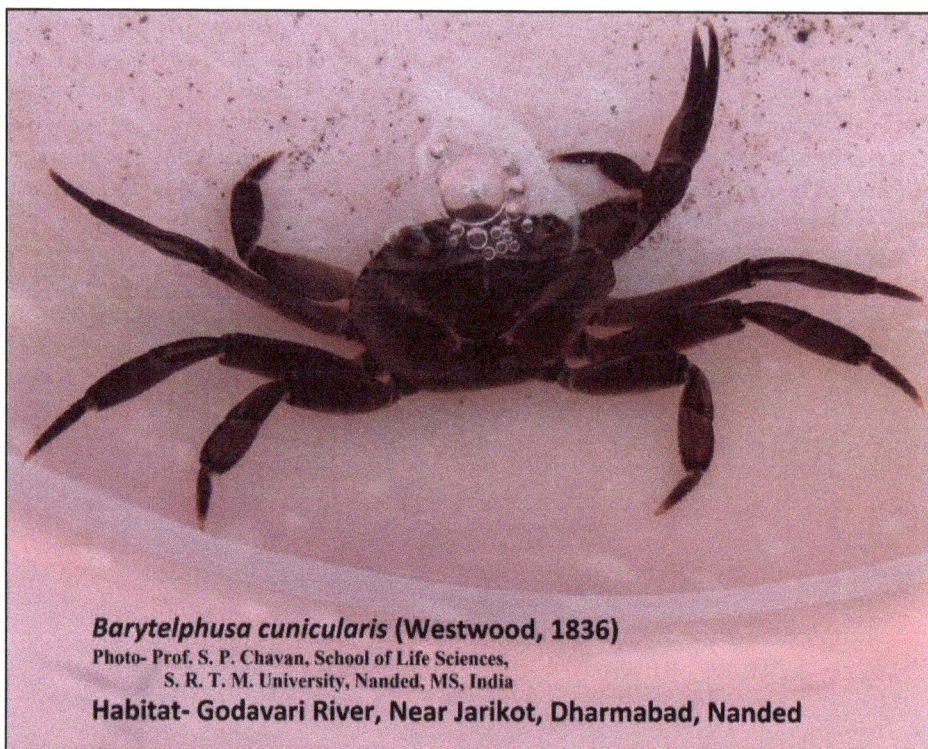

Barytelphusa cunicularis (Westwood, 1836)
Photo- Prof. S. P. Chavan, School of Life Sciences,
S. R. T. M. University, Nanded, MS, India
Habitat- Godavari River, Near Jarikot, Dharmabad, Nanded

Figure 7.3: External Morphology of Freshwater Crab *Barytelphusa cunicularis* (Westwood, 1836), Commonly Used to Prepare the Traditional Recipe TCR.

till boiling again starts. Boiling was continued till the water colour turns to reddish brown. It was taken care that, no any kind of lid was covered on the container during the boiling process. No need to add more water in the container during boiling. After 5-10 min the water content in the boiling pot reduced to nearly half of its whole amount. After 10 min. of boiling the whole content was cooled to room temperature then the upper carapace from crab body was removed and thrown as useless part. Remaining part of crab body, chelate legs, walking legs and all other body parts of crab in the container were removed and crushed in the mortar of stone and pestle of wood (Locally named as Khal-Batta in Marathi language) or it can be crushed by using electric grinder and mixer. During grinding process little quantity of edible salt was added for easy detachment of muscles from the crab body. Little quantity of extract formed during crab boiling process was added during crushing process. After complete crushing of crab as a paste the content was filtered by using common sieve which is commonly used for tea filtration. If there are some remains of crab muscles in the remaining residue then little quantity of cold water can be used to filter the content.

The Crab body filtrate (CBF) obtained in a container is the basic requirement for the preparation of crab rassa (TCR).

Treatment to Crab Body Extract (CBF) using Spices

Following was the procedure applied to prepare the Crab Rassa from 1 Lit. Quantity of Crab Body Extract or Filtrate (CBF) as prepared in earlier step a). As shown in Table 7.1 various ingredients were added in a specific quantity in to CBF and the content was mixed well with a spoon.

Table 7.1: Various Ingredients Used by Fisher Tribes to Prepare the Traditional Recipe TCR

Sl.No.	Name of Ingradient (Marathi)	Quantity
1	Edible salt (Namak or meeth)	1 Teaspoon
2.	Turmeric Powder (Haldi)	½ Teaspoon
3.	Cinnamom (Dalchini)	10 gm.
4.	Poppy Seeds (Khash Khash)	20 gm.
5.	Dry and Roasted coconut Powder/Crush (Khobra)	50 gm.
6.	Coriander seed powder (Dhaniya Powder)	1 teaspoon
7.	Nutmeg Powder (Jaiphal)	5 gm.
8.	Baked Gram flour (Harbhara Dal)	50 gm
9.	Roasted Rock flower Powder (Dagadful)	10 gm.
10	Aquaous extract of Tamarind	1 tea-cup
11.	Red Chilly Powder (Lal Mirch)	1 teaspoon
12.	Baked Bay leaf powder (Tez patta)	4 leaves
13.	Clove Powder (Lavang/Loong)	10 Cloves
14.	Black paper powder (Kalimirch/Mire)	10 papers

The words used for the spices in parenthesis are either vernacular or local.

The CBF with the addition of all spices mentioned in table - 1 above was the final mixture may be called as CBF-I to prepare the final product called Crab Rassa (CR/TCR). The CBF-I was further processed to prepare the CR. With the help of mortar and pestle a fine paste was prepared by using 30 gm. wet and fresh ginger, 50 gm. garlic, 5 g caraway seeds (Shah-jira). 100 ml. of edible oil (Sunflower seed oil) was heated well in a pan and 10-15 gm. of cumin seeds were added in the heated edible oil in a pan then the homogeneous paste of garlic and ginger was added in the oil and thoroughly mixed in the oil, then the heat was slowed down and the content in the pan was covered by a lid and cover lid was removed when a specific cracking sound comes due to water drops formed in evaporation process during heating falling in the oil content in the pan, any way the lid was removed after 1-2 min. time.

In the next step the crab extract (CBF-I) was poured slowly in the pan and boiled on slow heat till boiling start. Depending on what thickness of the CR needed, the water can be added in the CBF- ! container. To enhance the flavour of crab rassa (TCR) finely chopped green coriander leaves with cardamom (Velchi) powder were added after boiling completes and stop of heating. This will form a traditional most valuable, delicious, tasty and nutritional, medicinal food product of fisher tribes called as Crab Rassa (TCR) or Crab curry which they use routinely in their diet. To make this TCR more thicker one can add the powder of roasted gram (Dalwa/ Phutana).

As per the opinion of tribal people the TCR is useful to get the relief from cold, cough, headaque, Malaria, body pain, joint pain, to block the leprosy deformities, to avoid cancer problems and to recover from all the diseases and disorders of digestive system, it also improve the functioning of digestive system and liver.

The tribal informants reported that, after consumption of TCR when they rest or take a nip, the sweat ooze out from all over the body which removes all oxidants and disease forming wastes from the body and a person become very fresh and energetic; Therefore it is may have antioxidant property. The informant tribal person (90 per cent) told that those who use TCR routinely in their diet will never suffer suffer from cancer, Tuberculosis or Leprosy.

Discussion

Many food products with fish or crab as a basic material have been prepared by various communities in all coastal and inland states of India and the countries in oriental region (Ng *et al.,* 1988) but this is the first record of traditional knowledge of fisher tribes living in Godavari river basin on preparation of TCR. This will help common people to prepare the Crab curry just by referring this manuscript carefully and referring each step in the TCR preparation process for getting the tasty TCR. The tribes in the study area found do not waste the money in purchase of medicines to treat very common health problems like casual fever, headache, cold and flu etc., instead they consume TCR as a natural food material by traditional way of cooking. There are many reports on the use of plants, animals and forest produce by tribes in the world to cure about 36 different health problems (A. L. Sajem Betul, 2013) like Arthritis, Burns, Chronic Malaria, High blood pressure, Gall bladder stones, as health tonic to the aged, to prevent cancer and tuberculosis etc. but preparation of TCR and its medicinal use as a recipe has not been explained.

Acknowledgement

The authors are thankful to fisher tribes living in Godavari river basin to share their knowledge on preparation of traditional recipe and permission to publish this data. Thanks to University Grants Commission, New Delhi for providing the financial assistance under the Major Research Project F. No. 41-65/(SR) 2012- 11/07/2012. It was an additional work during parasite collection from different fish hosts from Godavari river basin.

References

Albert Lalduhawma Sajem Betul. 2013. Indigenous knowledge of zootherapeutic use among the Biate tribe of Dima Hasao District, Assam, Northeastern India. *Journal of Ethnobiology and Ethnomedicine*, 9: 56.

Alcock, A. 1910. Brachyura I. Fasc. II. *The Indian Freshwater crabs- Potamonidae. Catalogue of the Indian Decapod Crustacea in the collection of the Indian Museum.* Calcutta, pp. 1-135.

Cox C. B. (2001): The Biogeographic regions reconsidered, *Journal of Biogeography* 28: 511-523.

Chavan Shivaji, Kannewad Pandurang and Babare Rupali. 2014. Traditional Knowledge of Fish and Prawn Feeding Behaviour and Its Application by Fishermen for Successful Fishing in River Godavari in South Central India. *Int. Jour. Curr. Res. Acad. Rev.* 2 (7): 101-112.

Darren C. J. Yeo, Peter K. L. Ng, Neil Cumberlidge, Cello Magalhaes, Savel R. Daniels and Marth R. Campos. 2008. Global diversity of crabs (Crustacea: Decapoda: Brachyura) in freshwater. *Hydrobiologia* 595: 275-286.

Mohammed Mohiuddin and M. Khairul Alam. 2011. Opportunities of traditional knowledge in natural recourse management experience from the Chittagong hill tracts, Bangladesh, *Indian J. Tradit Knowle*, 10(3) 474-480.

Ng P. K. L. 1988. The freshwater crabs of Peninsular Malysia and Singapore. Department of Zoology National University of Singapore. Shinglee Press, Singapore. I-viii, 4 color plates 1-63.

Pandey, D. N. 2003. Cultural resources for conservation science, *Conserv. Biol.,* 17 (2): 633-635.

Rathakrishnan, T., Ramasubramanian M., Anandraja N, Suganthi N. and Anitha, S. 2009. Traditional fishing practices followed by fisher folks of Tamil Nadu, *Indian J Tradit Knowle,* 8(4) 543-547.

Roos N, Thilsted S H and Wahaly MdA. 2002. Culture of small indigenenous fish species in seasonal ponds in Bangladesh: The potential for production and impact on food and nutrition security, In: *Rural Aquaculture,* edited by Edwards P, Little D C and Denaine H, (CABI Publishing), 245-252.

Chapter 8

Limnological Aspects of Riverine, Estuarine and Marine Environment in Thrissur District, Kerala

☆ *P. Nimisha and S. Sheeba*

ABSTRACT

The aim of this study was to compare the physicochemical parameters and plankton population in the three environments such as river, estuary and sea. Water samples were collected monthly from the three different Stations-Kecheryriver, Chettuva estuary and Sea near Vadanapilly area at Thrissur district, Kerala during the period from January 2006 to August 2006. The hydrographic parameters such as temperature, pH, CO_2, ammonia, nitrate and phosphate showed slight changes between freshwater, estuary and marine water. But the concentration of salinity, alkalinity, calcium, chloride, TDS, TS and TSS showed very high fluctuation in the three environments. The algal population in the study area consisted of the members of bacillariophyceae, chlorophyceae, cyanophyceae and dinophyceae. Zooplanktons consisted of the members of rotifera, copepoda, protozoa and ostracoda.

Keywords: *Physico-chemical parameters, Phytoplankton, Zooplankton, River, Estuary, Sea.*

Introduction

Liquid water either as an ocean or as freshwater covers about three quarters of the earth's surface. Virtually all these water contains life in one form or other. River is an example of lotic ecosystem. River varies considerably over its length, as it changes from a mountain brook to a large river. It forms the connecting corridor through the landscape and its physical and biological features tend to vary in a predictable way from its headwater to the river mouth zones. The most distinctive feature of moving water ecosystems is the rate of flow and stream velocity. Another important factor of the biotic community is the turbulence or irregularity of the motion of the particular water.An estuary is a transitional zone between river and sea, which is either permanently or periodically open to the sea and within which there is a measurable variation of salinity due to mixture of seawater with freshwater derived from land drainage (Menon*et al.*, 2000). Estuaries differ in size, shape and volume of water flow, all influenced by the geology of the region in which they occur. Salinity also fluctuates during summer and monsoon season. Physico-chemical factors of this environment influence the abundance and diversity of biotic communities and also play an important role in productivity potential. The deep salty sea, unlike land and freshwater habitat is continuous and covers approximately 70 per cent of the earth's surface. The sea offers unusual stability in temperature, salinity and gaseous content. Life in sea is abundant and diverse and extends to all its depths. In all aquatic habitats planktons are the primary producers. Plankton community is a heterogeneous group of tiny plants - (phytoplankton) and animals (zooplanktons) added to suspension in the sea and freshwater. Phytoplanktons are important producers and have a major role in maintaining normal food chain and form the food for most of the pelagic fishes (Neal *et al.,* 2002). Fishery resources of aquatic environment are supported mainly by primary and secondary producers, which in turn are influenced by various physico-chemical and biological parameters of the environment. Also some plankton species are often used as indicators of water quality including pollution. The present study focuses on the comparison of hydrographic parameters and plankton population in the three different environments such as river, estuary and sea.

Materials and Methods

The present investigation was carried out in the three aquatic environments. Freshwaters, estuarine and marine environment all three are in Thrissur district. The Keecheririver was selected for freshwater environment. The study site was located near the river Pattikaraand the area was narrow and rocky.Water flow was slow in summer, stagnant between rocks, but during monsoon it was in floods. People use this area for domestic purposes. The second site is located in Chettuva backwater, and has a narrow passage of water with free communication with Arabian sea. Chettuva estuary is situated in Engandiyoor Village, Chavakkad Thaluk. Chettuva backwater lies at a latitude of $10°15^1$-$10°35^1$N and longitude $76°4^1$-$76°10^1$E. Two rivers namely the Kecheri and the Karuvannur drain into the backwater. The Kecheririver joins near the Chettuva estuary on the north portion. The Chettuva backwater is significant from the point of view of development of kolelands and also from the point of view of inland navigation, coir retting and fish culture. Mangrovevegetation is also seen in this area and it is one of the tourists resting spots facilitating boating. The estuary is connected to sea at Banglakadavu, a fishing harbour. The coastal area near Vadanapilly was selected as marine environment. The study area is situated 3Km away from Vadanapilly. For the present study water samples were collected monthly from the three stations – the Kechery river, Chettuva estuary and the sea near Vadanapilly at Thrissur district, Kerala during the period from January to August

2006.Physicochemical parameters were analyzed as per methods suggested by Trivedy and Goel(1986). Quantitative analysis of plankton was done by using standard references (Battish, 1992).

Result and Discussion

Physico-chemical parameters of the Kecheryriver, Chettuva estuary and marine water are presented in Tables 8.1–8.3 respectively. Water temperature is one of the most important factors in an aquatic environment. The maximum temperature (32°C) was in January and March, and the minimum temperature (27°C) in July in the riverine environment.Water temperature in the estuarine environment ranged from 27°C (February and July) to 32°C (May). Temperature variations of the estuarine environment clearly shared the investigations of Anilkumar and Abdul Aziz (1995) in Akathumuri estuary.Water temperature in the marine environment ranged from 26°C (July and August) to 31°C (March). Observations of water temperature at all the three stations were low during monsoon months may be due to the influence of south west monsoon.Hydrogen ion concentration is an important parameter that determines the quality of water in an aquatic environment. pH variations in the riverine environment ranged from 6.6 to 7. Hydrogen ion concentration in the estuarine environment varied from 6.5 to 7. Fluctuations of pH in estuarine value may be due to the riverine discharge during rainfall and exchange of water from the sea (Nair *et al.,* 1984). Observations of pH in the marine environment ranged from 6.5 to 7.Observation of salinity in the riverine sample ranged from 0.34°/$_{00}$ to 1.2°/$_{00}$. The highest salinitywas observed in March and the lowest salinity fluctuated in all months. Salinity in the estuarine environment ranged between 0.34°/$_{00}$ (August) to 32.8°/$_{00}$ (March). Maximum salinity in March may be due to high saline ingression from sea. Similar observations were made by Madhukumar and Anirudhan (1996) in Paravurbackwater. Salinity in the marine environment ranged between 28°/$_{00}$ (June) to 34.6°/$_{00}$ (May).

Table 8.1: Physico-chemical Parameters of Keecheri River from January to August 2006

Parameters	JAN	FEB	MAR	APR	MAY	JUN	JUL	AUG
Water temperature (°C)	32	29	32	28	30	30	27	29
pH	6.5	6.5	6.5	6.5	7	6.5	6.5	6.5
Salinity (°/$_{00}$)	0.342	0.654	1.2	0.36	0.34	0.34	0.34	0.34
Total Alkalinity (mg CaCO$_3$/l)	30	20	110	90	60	30	50	50
CO$_2$ (mg/l)	2.2	12.1	22	1.1	18.7	3.3	11	25
Dissolved oxygen (mg/l)	5.1	5.1	5	5	5.6	5.6	5.7	5.6
Ca (mg/l)	8.016	4.008	0	0	0	0	4.008	4.008
Chloride (mg/l)	0.1775	0.355	0.710	0.1775	0.1775	0.177	0.1775	0.1775
Ammonia (mg/l)	0.28	0.11	0.25	0.035	0.18	0.11	0.07	0.04
Nitrate-Nitrogen (mg/l)	0.6	0	0.4	0.1	1.8	1	1.4	0.4
Phosphate - Phosphorus (mg/l)	1.3	1.3	1.1	0.6	0.5	0.5	0.5	0.7
Total suspended solids (g/l)	0.019	0.019	0.017	0.019	0.015	0.016	0.015	0.014
Total dissolved solids (g/l)	0.027	0.028	0.027	0.026	0.021	0.022	0.022	0.022
Total solids (g/l)	0.04	0.045	0.044	0.045	0.036	0.038	0.037	0.036

Table 8.2: Physico-chemical Parameters of
Chettuva Estuary from January to August 2006

Parameters	JAN	FEB	MAR	APR	MAY	JUN	JUL	AUG
Water temperature(°C)	31	27	30	31	32	28	27	28
pH	7	6.5	6.5	7	7	6.5	6.5	6.5
Salinity (°/$_{oo}$)	28.7	28.13	32.8	30.6	5.9	1.2	0.966	0.34
Total Alkalinity (mg CaCO$_3$/l)	120	120	110	140	20	30	40	50
CO$_2$ (mg/l)	0.8	14.3	3.3	3.3	4.4	4.4	3.3	14.3
Dissolved oxygen (mg/l)	4.1	4.3	4	4	4.5	4.7	4.7	4.7
Ca (mg/l)	360.7	400.8	320.6	488.9	24.04	80.16	16.03	36.07
Chloride (mg/l)	1.633	1.597	1.863	1.739	3.372	0.710	0.532	0.177
Ammonia (mg/l)	0.105	0.07	0.245	0.32	0.11	0.11	0.07	0.04
Nitrate-Nitrogen (mg/l)	0.4	0	0.8	0	0	0.6	1	0.4
Phosphate - Phosphorus (mg/l)	0.1	1.3	0.9	0.6	0.5	0.4	0.5	0.8
Total suspended solid (g/l)	0.56	0.64	0.56	0.44	0.46	0.4	0.02	0.024
Total dissolved solid (g/l)	26	20	20	28	2.4	1.2	4	6
Total solids (g/l)	26.56	20.64	20.56	28.44	2.86	1.6	4.02	6.024

Total alkalinity in the riverine sample ranged between 20 mgCaCO$_3$/l(February) to 110 mgCaCO$_3$/l(March). Similarstudies were made by Gupta (1998) in the rivers of Dakshina Kannada.Total alkalinity in the estuarine sample ranged between 20mgCaCO$_3$/l(May) to 140 mgCaCO$_3$/l(April). In the marine environment alkalinity ranged between 110mgCaCO$_3$/l(May) and 170mg CaCO$_3$/l (January and April).

Table 8.3: Physico-chemical Parameters of Marine Environment
Sample from January to August 2006

Parameters	JAN	FEB	MAR	APR	MAY	JUN	JUL	AUG
Water temperature (°C)	30	29	31	30	28	28	26	26
pH	6.5	6.5	7	7	7	7	7	7
Salinity (°/$_{oo}$)	30.3	28.13	31.25	33.4	44.6	28.13	33.7	31.9
Total Alkalinity (Mg/CaCO$_3$/l)	170	120	120	170	110	120	120	150
CO$_2$ (mg/l)	4.4	16.5	1.5	17.6	1.5	4.4	7.7	23
Dissolved oxygen (mg/l)	4.1	4	3.9	3.8	4.1	4.2	4.25	4.3
Ca (mg/l)	440.8	400.8	521.04	416.8	416.8	384.7	160.3	180.36
Chloride (mg/l)	17.217	15.975	17.75	18.99	25.38	15.97	19.17	18.105
Ammonia (mg/l)	0.14	0.14	0.07	0.28	0.28	0.25	0.175	0.14
Nitrate-Nitrogen (mg/l)	0.4	0.2	0.4	0.1		1.2	2.4	0.6
Phosphate - Phosphorus (mg/l)	0.1	0.8	0.7	0.7	0.9	0.4	0.5	0.3
Total suspended solids (g/l)	200	240	400	240	380	1360	900	832
Total dissolved solids (g/l)	24	28	24	28	24	33.6	30	36
Total solids (g/l)	24.2	28.24	24.4	28.24	24.38	34.96	30.9	36.832

Phytoplanktons require CO_2 directly diffused from the airfor photosynthetic activity, respiration and decomposition. Free Carbon dioxide in the riverine environment ranged from 1.1mg/l (April) to 25mg/l (August).Free Carbon dioxide in the estuarine environment ranged from 0.8 mg/l (January) to 14.3mg/l (February and August). Free Carbon dioxide in marine environment ranged from 1.5mg/l (March and May) to 23mg/l (August). Dissolved Oxygen is a major constituent of interest among the water quality parameters of standard categories. Optimum concentration of dissolved oxygen is essential for maintaining aesthetic qualities of water as well as for supporting aquatic life.Concentration of dissolved oxygen in the riverine environment ranged between 5mg/l and 5.7mg/l. Maximum value was observed in July and August and minimum dissolved oxygen concentration was in a fluctuating manner at all the months. Concentration of dissolved oxygen in the estuarine environment ranged between 4mg/l (June, July and August) to 4.7mg/l (March and April) Observation of dissolved oxygen in the three environments revealed low level in pre monsoon months and high concentration in monsoon months. Mahukumar and Anirudhan(1996) also support these findings. Different factorssuch as velocity of water, biota and pollution from various sources influence the dissolved oxygen content, (Rajavaidhya and Markandey, 1998).Chloride occurs naturally in all types of water. However the concentration of chloride will be quite low in natural unpolluted freshwaters. Underlying alkaline rocks can also impart chloride to freshwater bodies. The other sources of chlorides were atmospheric fallouts, city drainage and ingression of saltwater from the sea.Concentration of chloride in the riverine sample ranged between 0.177mg/l to 0.71mg/l. Maximum concentration was observed in March and minimum chloride content in all the other months. Concentration of chloride in the estuarine environment ranged between 0.177mg/l (August) to 18.637mg/l (March). The high concentration may be due to high evaporation of water and saline ingression.Concentration of chloride in marine environment ranged between 15.975g/l (February) and 25.382mg/l (May). Calcium and Magnesium are major elements which impart hardness of water.In river calcium content varied between 0 and 8mg/l. Maximum value was observed at January and absences of calcium was observed in March, April, May and June. In the estuarine environment calcium concentration ranged between 16mg/l (July) to 488mg/l (April).In the marine environment calcium content ranged between 160mg/l (July) to 521mg/l (March). Calcium concentration increased from freshwater to marine area may be due to the increase in salinity or due to presence of molluscs with $CaCO_3$.Nitrogen reaches aquatic environment through diverse sources such as domestic/urban sewages, chemical industries and agricultural loads. Concentration of nitrogen compounds above permissible limit can lead to fish diseases or even fish mortality.Ammonia-Nitrogen concentration in the riverine environment ranged between 35mg/l (April) and 280mg/l (March). NH_3-N concentration in the estuarine sample ranged between 40mg/l (August) and 245mg/l (March). NH_3-N concentration in the marine environment ranged between 70mg/l (March) and 280mg/l (April and May).Nitrogen forms the major constituent of atmosphere (about 80 per cent) and is found in small amounts in surface water because of its low solubility. Most of the nitrogen comes in the form of organic nitrogen and ammonia cal nitrogen. Oxidation of ammonia to nitrates and subsequently to nitrogen which occurs under aerobic condition.Nitrate

– Nitrogen content in the riverine environment ranged from below detection level to 180mg/l (February). In the estuarine environment nitrate-nitrogen concentration ranged between BDL and 100mg/l (July). In the marine environment nitrate-nitrogen concentration ranged between BDL and 240mg/l (July). In the three ecosystems comparatively low concentration of nitrate-nitrogen was observed in monsoon months, may be, due to leaching of nitrate from the surroundings. Phosphorous is present in very low concentration in all types of natural water. Domestic and industrial effluents and agricultural drains are the major sources of phosphorus which occur both in inorganic and organic forms.In the riverine environment concentration of Phosphate – Phosphorous ranged between 50mg/l (May) to 130mg/l (January and February). In the estuarine environment phosphate-phosphorous concentration ranged between 10mg/l (January) and 130mg/l (February).In the marine environment phosphate-phosphorous concentration ranged between 10mg/l (January) and 90mg/l (May). In the riverine environment TSS (Total suspended solid) was ranged between 0.014g/l (August) and 0.019g/l (January, February and March).Suspended solids will always be present in natural water at different concentration. High amount of suspended solids can adversely affect the biological process of the ecosystem and make the water quality deteriorate. Total Suspended solids in the riverine environment ranged between 0.008g/l (April) and 0.32g/l (June). Maximum TSS was noticed in monsoon months, may be, due to turbulent and flooded water flowing very fast in the riverine environment. In the estuarine environment TSS concentration ranged between 0.02g/l (July) and 0.05g/l (March).Total suspended solids in marine environment ranged between 0.2g/l (January) and 1.3g/l (June). In the riverine environment TDS (Total Dissolved Solid) ranged between 0.021g/l (May) and 0.028 mg/l (February).The concentration of TDS in the estuarine environment ranged between 1.2g/l (June) and 28g/l (April). TDS content in the marine environment varied between 24g/l (January) and 36g/l (August).In the riverine environment TS (Total Solid) ranged between 0.036g/l (May and August) and 0.045g/l (February and April). The content of TS in the estuarine environment ranged between 1.6g/l (June) and 28.4g/l (April). The concentration of TS in the marine environment ranged between 24g/l (January) and 36g/l (August).

Phytoplanktons and Zooplankton

Regular and periodical changes in the climate synchronizing with seasons are ultimately reflected in the environmental parameters also, which in turn have a direct or indirect control over the Planktonic population. The algal (phytoplankton) component of the study areas consisted of the members of *Bacillariophyceae, Chlorophyceae, Cyanophyceae*and *Dinophyceae.*Zooplanktons consisted of the members of Rotifera, Copepoda, Protozoa and Ostracoda (Tables 8.4 and 8.5).

Riverine Environment

In the riverine environment phytoplanktons consisted of bacillariophyta, chlorophyta, and cyanophyta.In the bacillariophyceae *Cyclotella* sp., *Gomphonema* sp., *Navicula* sp., *Synedra* sp.and *Melosira* sp. were identified. *Melosira* sp.was the most predominant form. In the chlorophyceae *Cosmarium* sp.,*Scenedesmus* sp., *Chlorella* sp.,*Mougeotia* sp., *Chlamydomonas*sp., *Oedogonium* sp., *Pediastrum duplex, Kirchnerilla*sp.,

Table 8.4: Phytoplanktons in Riverine, Estuarine and Marine Environment during the Period January to August 2006 at Thrissur District

Riverine Environment	Estuarine Environment	Marine Environment
Bacillariophyta	**Bacillariophyta**	**Bacillariophyta**
Melosira sp.	Cyclotella sp.	Asterionella sp.
Navicula sp.	Fragellaria sp.	Biddulphia sp.
Cyclotella sp.	Gomphonema sp.	Chaetocerus sp.
Synedra sp.	Melosira sp.	Nitzshia sp.
Gomphonema sp.	Navicula sp.	Pleurosigma sp.
Chlorophyta	Nitzshia sp.	Tallasiosira sp.
Cosmarium reniformae	Pinnularia sp.	**Chlorophyta**
C. contractum	Synedra sp.	
Chlamydomonas sp.	**Chlorophyta**	Actinastrum sp.
Chlorella sp.	Closterium sp.	Diplosteron sp.
Kirchnerilla sp.	Cosmarium decoratum	
Mougoetia sp.	Eustrum sp.	**Dinophyta**
Oedogonium sp.	Micrasterias sp.	Ceratium sp.
Oocystis sp.	Mougoetia sp	
Pediastnum duplex	Pediatrum simplex	
Staurastrum arctiscon	P. boryanum	
S. armatus	P. duplex	
S. lefevrii	Staurastrum paradoxum	
Scenedesmus quandricula	Staurodesmus glaber	
Sphaerozosma sp	S. eboracence	
Volvox sp.	S. longibrachiatum	
Cyanophyta	Scenedesmus lefevrii	
Aphanocapsa sp.	Scenedesmus quandricula	
Lyngbia sp.	Selenastrum gracile	
Merismopedia sp.	Spirogyra sp.	
Oscillatoria sp.	Tetradron sp.	
	Cyanophyta	
	Chodatella sp.	
	Oscillatoria sp.	
	Pleurotenium sp.	
	Quandri gulalucustris	

Oocystis sp., *Staurastrunarctiscon* sp. and *Volvox* sp. were identified. In the *Cosmarium* species, *Cosmarium reniformae* and *Cosmarium contractum* were identified. In the Chlorophyceae, *Scenedesmes* sp. were the most predominant form. In the cyanophyceae,

Table 8.5: Zooplanktons in Riverine, Estuarine and Marine Environment during the Period January to August 2006 at Thrissur District

Riverine Environment	Estuarine Environment	Marine Environment
Rotifera	**Rotifera**	**Rotifera**
Branchionus falcatus	*Notholca* sp.	*Notholca* sp.
B. caudatus	*Branchionus diversicornie*	*Trichocerca* sp.
Monostyla sp.	Monostyle	**Claclocera**
Cladocera	*Keratella* sp.	*Bosmina*
Daphnia sp.	*Platyias quadricornis*	*Moina*
Simocephalus sp.	*Lecane* sp.	*Penila* sp.
	Cladocera	
Protozoa	*Bosmina* sp.	
Tintinidium sp.	*Daphnia* sp.	
Meroplanktons	*Moina* sp.	
Nymph of May fly		
Water spider	**Protozoa**	
	Euglena sp.	
	Copepoda	
	Calanoid copepod	
	Ectocyclops	
	Meroplanktons	
	Brittle worm	
	Water spider	

four species were identified, which include *Oscillatoria* sp., *Lynbia* sp., *Merismopedia* sp. And *Aphanocaspa* sp.

In the riverine environment zooplanktons consisted of rotifera, protozoa, cladocera and meroplanktons. In rotifera, *Branchionus fulcatus, Branchionus caudatus* and *Monostyla* sp. were identified. Rotifers were the dominant zooplankton in theriverine environment.In Cladocera, *Daphnia* sp. and *Simocephalus* sp. were observed. In protozoa, *Tintiniduim* sp. was identified. Nymph of May fly and Water spider were also found. Similar observations were reported by Venketeswarlu and Sampath (1990) in their studies in the river Moosi.

Estuarine Environment

In the estuarine environment phytoplanktons consisted of bacillariophyceae, chlorophycea and cyanophyceae. In baillariophyceae, *Navicula* sp., *Fragellria* sp., *Melosira* sp., *Cyclotella* sp., *Synedra* sp., *Pinnularia* sp, *Nitzshia* sp. and *Gomphonema* sp. were observed. *Navicula* sp. and *Melesira* sp. were the most predominant form.In *Chlorophyceae, Eustrum* sp., *Tetradron* sp., *Staurastrum* sp., *Closterium* sp., *Staurodesmus* sp., *Scenedesmees* sp., *Pediastrum* sp., *Selenastrumgracile* sp., *Spirogyra* sp., *Maugeotia* sp.

and *Cosmariumdecoratum* sp. were identified. *Pediastrum* speciesand *Scenedesmus* were more dominant forms.In *Cyanophyceae* only one species was identified. *Oscillatoria* sp. was the common form. *Melosira* sp.and *Nitzshia* sp. were included in the estuarine pollution algae.

In estuarine environment zooplanktons were included in rotifera, cladocera, protozoa, and copepoda. In rotifer, *Nothalca* sp., *Branehionus* sp., *Monostyla* sp., *Keratella* sp., *Platyias* sp. and *Lecane* sp. were identified. *Beanchionus* species is the most predominant in Rotifers. In Cladocera *Daphnia* sp., *Bosmina* sp., *Moina* sp. and in protozoa *Euglena* sp. were identified. In Copepoda, Calonoidcopepodiswas the predominant form. Ectocyclops and Meroplanktons, brittle worm and water spider were identified.

Marine Environment

In marine environment phytoplanktons composed of Bacillariophyceae, Chlorophyceae and Dinophyceae. Bacillariophyceaewere the dominant one.In Bacillariophyceae, *Thalla riosira* sp., *Pleuro sigma* sp., *Nitzshia* sp., *Biddulpia* sp., *Chaetocerus* sp. and *Asterionella* sp. were identified. In Chlorophyceae only one species was identified–*Actinastrum,* – which was very rare. In *Dinophyceae,Ceratium* sp. was identified. Simillar observations were made by Asha *et. al.* (2002) in Cochin marine environment.

In the marine environment zooplanktons composed of Rotifera, and Cladorcera. In Rotifers *Trichocerca* sp. and *Notholca* sp. were noticed and in Cladocera, *Bosmina* sp., *Moina* sp. and *Penila* sp. were identified.

Conclusion

Almost all hydrographic parameters are exhibited in a characteristics pattern of fluctuations in monsoon and premonsoon months. Comparatively high temperatures, dissolved oxygen, Nitrate – Nitrogen and free carbon dioxide were observed at monsoon months in all the three environments. Inthe riverine environment salinity was below 1 per cent but the estuarine environment showed great variation in salinity during pre monsoon and monsoon months. In the marine environment salinity was almost constant and it was above 30 per cent. Salinity, alkalinity, calcium, chloride content, total dissolved solids, total suspended solids and total solids were very high in estuarine and marine environment compared to those in the riverine environment. But temperature, pH, free Carbon dioxide, ammonia- nitrogen, nitrate-nitrogen and phosphate-phosphorus showed slight changes among freshwaters, backwaters and marine environment.

Ecosystem degradation is a common scene. So monitoring the ecosystem is very essential. With the increase of population, urbanization and industrialization, it has become difficult to provide adequate amount of water to the mankind. At the same time, the prominence of water quality has become more pressing than the quantity. The increasing rate of pollution has degraded our water resources. Water pollution problems are very well related to human health. In all aspects the physicochemical parameter study is essential to analyze whether the system is polluted or not. It also gives an idea regarding whether the system is fit for aquaculture practices.

Acknowledgements

The authors are grateful to the Principal, S.N. College, Nattika, Thrissur for providing the necessary facilities during this work.

References

Anil Kumar, N. C. and Abdul Aziz, P. K. 1995. Status of pollution in the Akathumuri Estuary with reference to retting of Coconut husk. *Proceeding of 7th Kerala Science Congress*, Palakkad. 177pp.

Asha, B. S., Satheeshkumar N. C. and Ouseph, P. P. 2002. Plankton characteristics in the marine environment ofCochin. *Proceeding of Kerala Science Congress*, Kochi. 435-438pp.

Battish, S. K. 1992. Freshwater Zooplankton of India. Oxford and IBH Publishing Cor. Privt. Ltd. pp. 233.

Madhukumar A. and Anirudhan T. S. 1995. Plankton distribution in sediments of Edava- Nadayara and Paravur backwater system along the Southwest Coast of India. *Indian J. of Mar. Science.* 24: 186-191pp.

Venketeswarlu, V. and Sampath 1969. A ecological study of the algae of the river Moosi, Hyderabad. With special reference to water pollution and factors influencing the distribution of algae. *Hydrobiologia.* 33: 352-368pp.

RajvaiyaNeelima and MarkandeyDilip1998. *Advance in Environmental Science and technology of water characteristic and properties.* APH Publishing Corporation, New Delhi.

Trivedy, R. K. and Goel, P. K. 1986. *Chemical and biological methods for water pollution studies.* Environmental Publications. 138-146pp.

Neal, C., Helen, P. J., Andrew, J. W. and Paul, G. W. 2002. Water quality functioning of low land permeable catchments: Interferences from an intensive study of river Kennet and upper river Thames. *Science of the Total Environment.* 282-283: 471-490pp.

Menon, N. N., Balachand, A. N. and Menon, N. R. 2000. Hydrobiology of the Cochin backwater system- A review. *Hydrobiologia.* 430: 149-183pp.

Chapter 9

Comparative Study on Water Quality Parameters of Freshwater Habitats in Selected Areas of Thiruvananthapuram City, Kerala

☆ *V. Rajani and G. Aswathy Krishnan*

ABSTRACT

Water is one of the five elements among air, water, fire and earth. Availability of good quality water is very essential for the existence and sustenance of plant and animal life. The present study was carried out to study the main water quality parameters of selected streams of Thiruvananthapuram district namely Akkulam Lake, Parvathy Puthanar, Kannammoola Thodu and Murinjapalam Thodu (Amaezhanjan Thodu). Among the studied water quality parameters all were not in the prescribed limits. In terms of pH all sites showed the values within the limits. Samples from Parvathy Puthanar and Kannammula Thodu showed zero values for dissolved oxygen which indicated high pollution and high activity of aerobic bacteria. Akkulam Lake samples showed biochemical oxygen demand values within the permissible limit. Akkulam Lake was found to be better in quality of water compared with the other three selected areas.

Keywords: *Pollution, Water quality, Dissolved oxygen, Conductivity, Hardness.*

Introduction

Water is one of the five great elements among air, water, fire and earth. Water is also intermediate between all pervasive air and localized earth. The whole universe is made up of water. All begins are made up of water. About 97 per cent of the total available water on earth is contained in oceans and is, hence, saline or salty in nature. Out of the balance, 3 per cent, which is available as freshwater, about 2 per cent is contained as ice on poles, and 0.75 per cent is ground water. Out of the remaining, about 0.25 per cent, only about 0.01 per cent is found available in lakes and rivers at any given time, as the rest occurs as glaciers and snow. The total water contained in the atmosphere is still less, and is of the order of 0.001 per cent of the total available water. The total quantity of water in that system at any given time is also dependent on many factors including storage capacity in lakes, wetlands and artificial reservoirs, the permeability of the soil beneath these storage bodies, the runoff characteristics of the land in the watershed, the timing of the precipitation and local evaporation rates. Temperature of water depends on water depth besides solar radiation, climate and topography. It also reflects the dynamics of living organisms (Kant and Raina, 1990). All of these factors also affect the proportions of water lost. Human activities can have a large and sometimes devastating impact on these factors. The natural aquatic resources are causing heavy and varied pollution in aquatic environment leading to pollute water and leads to depletion of aquatic biota (Simpi *et al.*, 2011). Poor water quality may favour the phytoplankton blooms (including harmful algae) and occasional hypoxic/anoxic events (fish kills) that affect our estuaries and other water bodies.

Kerala gets an average of 307 cm rainfall, the bulk of which (70 per cent) is received during the South-West monsoon which sets in June and extends up to September. The state also gets rain from the North-East monsoons during October to December. The state experiences severe summer from January to May when the rainfall is minimum. The two monsoons have a direct bearing on the ground water potential of the state which also follows the same seasonal trends. Kerala is a land abundant in water resources which includes rivers, lakes, back waters, big and small ponds etc. Kerala has 44 rivers of which 41 are west flowing and 3 east flowing. Many of these rivers serve as inland water ways in many parts of the state. Water from these rivers is used for irrigational purpose, drinking, hydroelectric power production etc. They also serve ground for inland fishing. Most of the streams and other water bodies are under threat of pollution due to various anthropogenic activities.

The aim of the present study was to analyze the water quality of selected streams in the vicinity of Thiruvananthapuram city which included Akkulam Lake, Parvathy Puthanar, Kannammoola Thodu and Murinjapalam Thodu (Amaezhanjan Thodu).

Materials and Methods

Four sites were selected for the present study which included Akkulam Lake (A), Parvathy Puthanar (P), Kannammoola Thodu (K) and Murinjapalam Thodu (Amaezhanjan Thodu) (M). All the four streams receive various anthropogenic wastes from the city. Water samples were collected from the specified locations for the analysis during the months of March, April and May. Water samples were collected from different sites in bottles and were transported to the laboratory for the chemical analysis. Water quality parameters such as conductivity, dissolved Oxygen, biochemical oxygen demand, chemical oxygen demand, total hardness (as $CaCO_3$), calcium hardness (as $CaCO_3$), magnesium hardness (as $CaCO_3$), salinity, total dissolved solids, total alkalinity (as $CaCO_3$), phosphate-p and nitrite-N were analyzed using the standard analytical methods prescribed by Trivedy and Goel (1986) and APHA (1998).

Results and Discussion

The results of the present study are given in Figures 9.1–9.11. The pH values were found to lie within the range of 6.7-7.4. The water samples from all sites were slightly acidic. According to APHA (1998), the desirable limit for pH in drinking water ranges from 6.5-8.5. The pH of water samples from all sites were found to lie within the range of desirable limit. The highest pH value was observed in the month of May in the sample of Kannammoola Thodu. Low value for pH for surface and bottom water was reported during monsoon (Kaushik and Saksena, 1999). The conductivity values were found to lie within the range of 285-1525 μmohs/cm. According to APHA (1998), the desirable limit for conductivity is 150-1500 μmohs/cm. The values obtained from 4 sites showed that the water from Akkulam Lake and Parvathy Puthanar had high conductivity and Kannammola Thodu and Murinjapalam Thodu had low conductivity.

The DO was maximum during the month of March water for the water sample collected from Akkulam Lake. In Parvathy Puthanar and Kannammola the DO values were zero due to high pollution load. The permissible limit is 4-6 mg/l and no samples

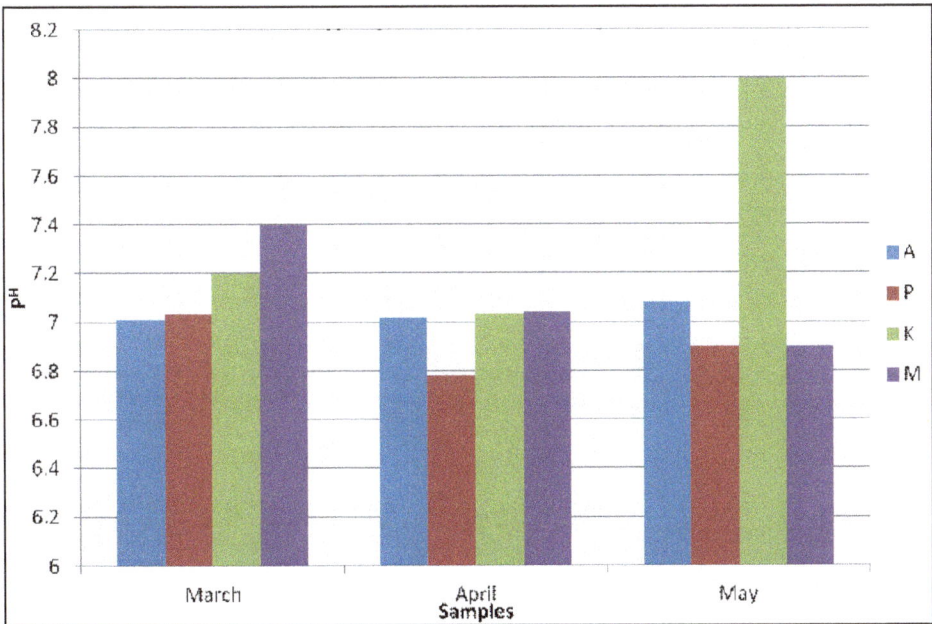

Figure 9.1:Variations in pH of Water Samples.

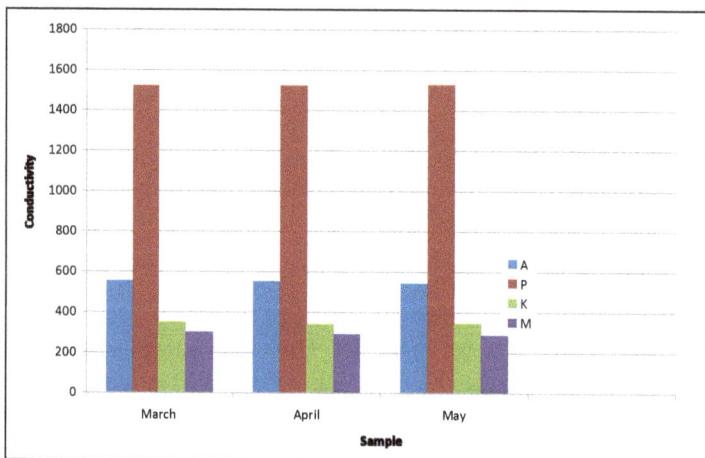

Figure 9.2: Variations in Conductivity (µmohs/cm) of Water Samples.

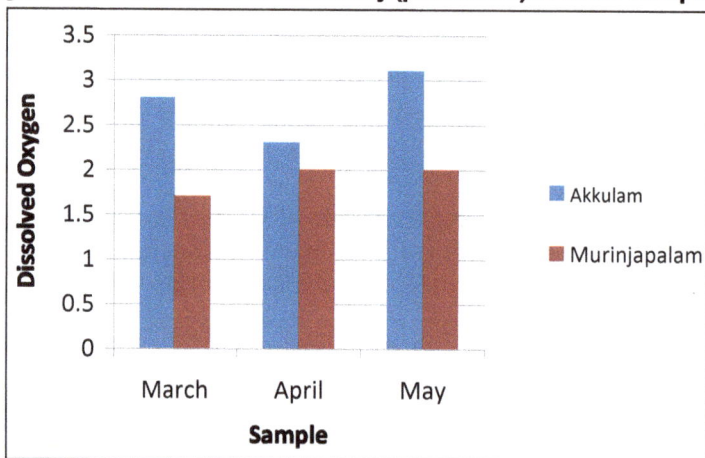

Figure 9.3: Variations in Dissolved Oxygen (mg/L) of Water Samples.

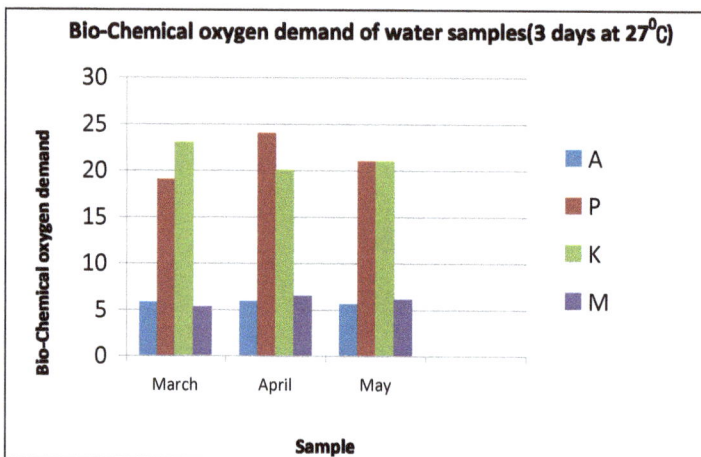

Figure 9.4: Variations in BOD (mg/L) of Water Samples.

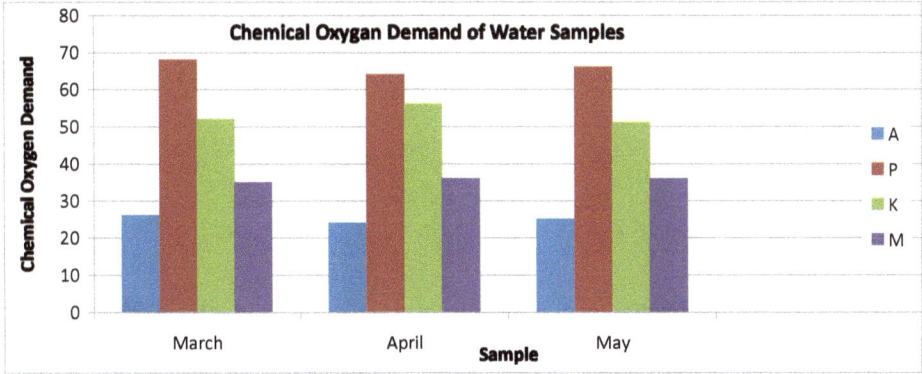

Figure 9.5: Variations in COD (mg/L) of Water Samples.

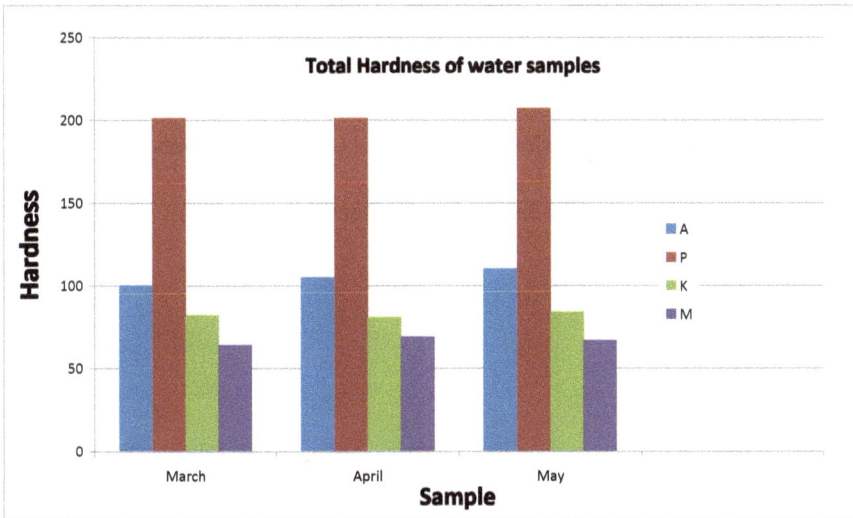

Figure 9.6: Variation in Hardness (mg/L) of Water Samples.

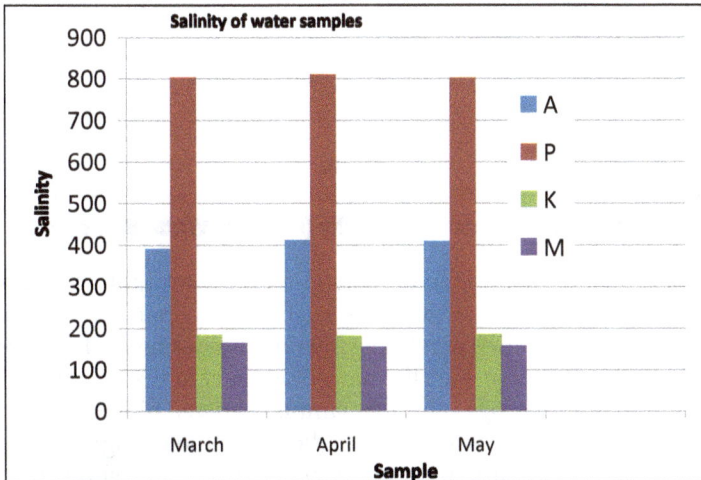

Figure 9.7: Variations in Salinity (mg/L) of Water Samples.

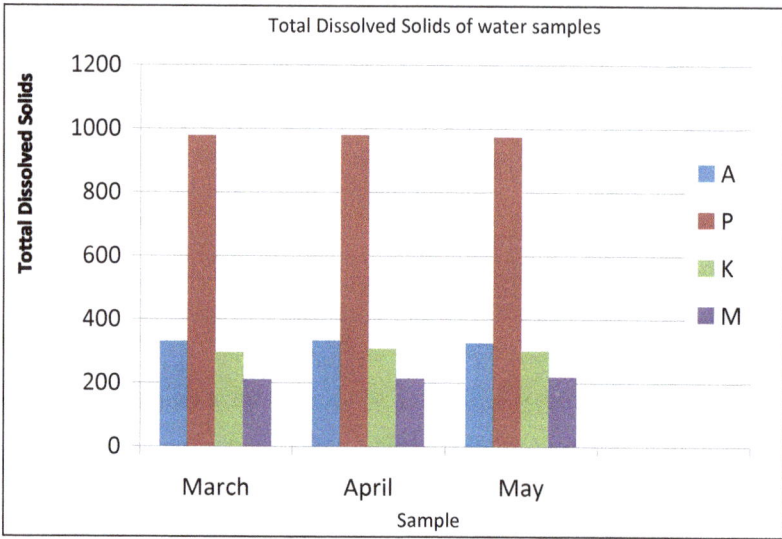

Figure 9.8: Variation in TDS (mg/L) of Water Samples.

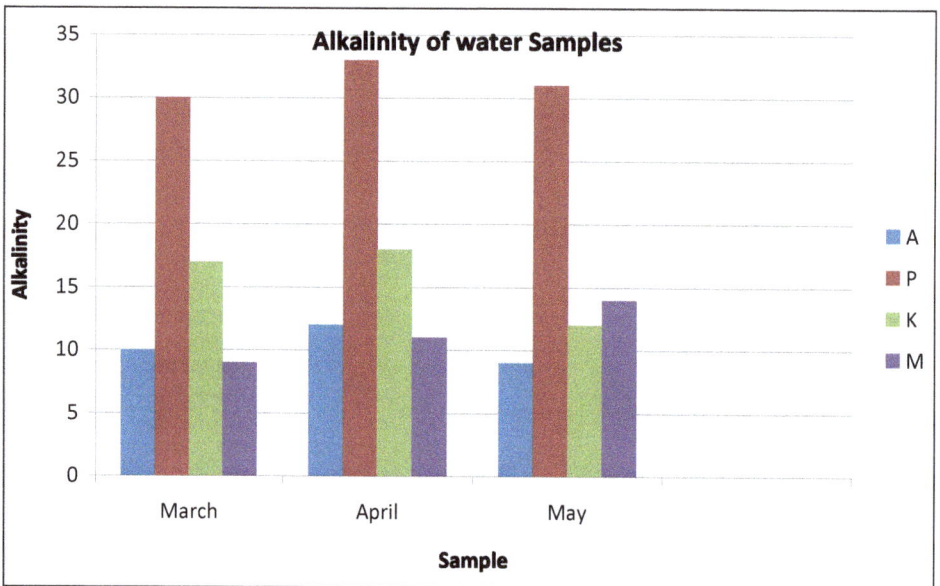

Figure 9.9:Variations in Alkalinity (mg/L) of Water Samples.

had DO values with in this range.Dissolved oxygen and pH affects directly or indirectly other limnological parameters such as transparency, viscosity, total dissolved solids and conductivity (Whitney, 1942). Comparing the BOD values of the 4 sites, the Parvathy Puthanar had the high BOD during the month of April. The permissible limit is only 2-3 mg/L, and all samples showed higher BOD values than the permissible limits. BOD is an excellent indicator of the strength of domestic and industrial pollutants in aquatic regimes (APHA, 1998). This showed that all the

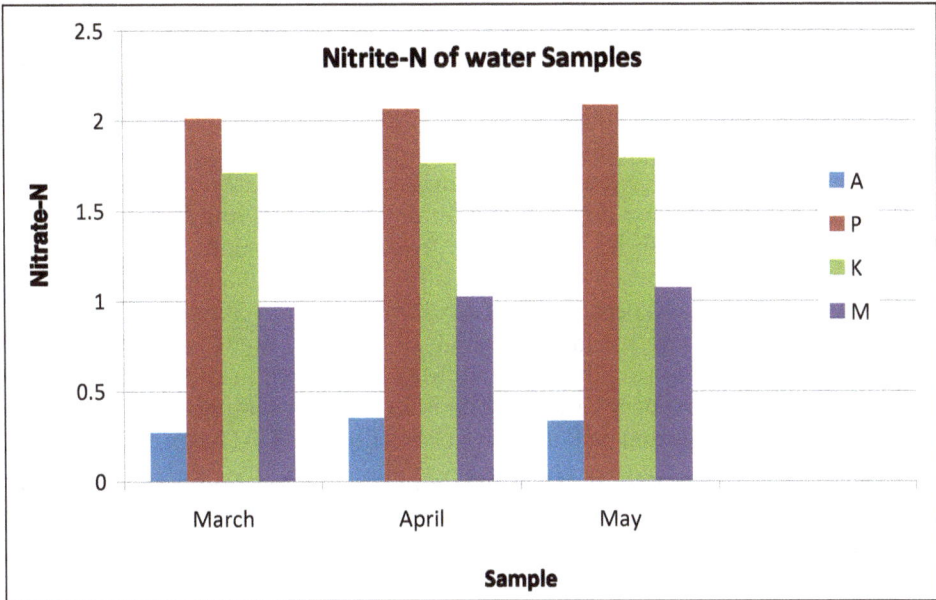

Figure 9.10: Variations in Nitrate (mg/L) of Water Samples.

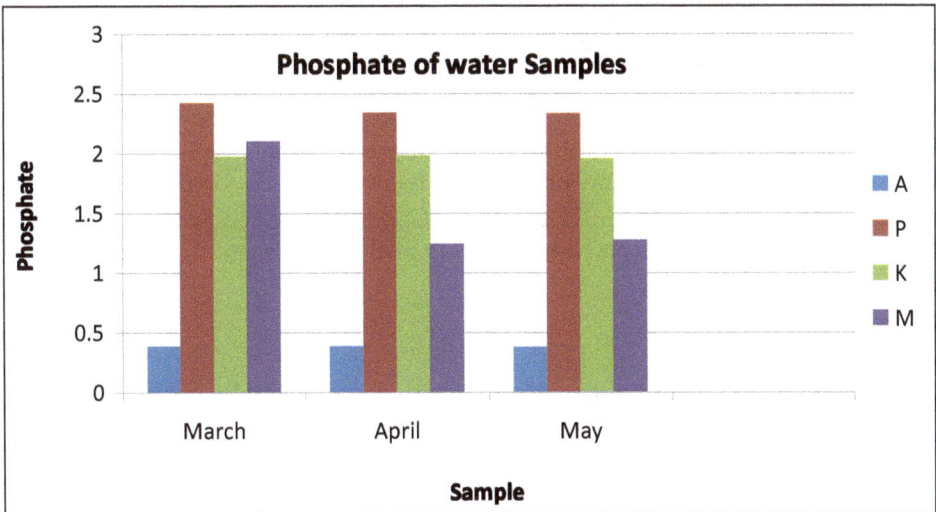

Figure 9.11: Variations in Phosphate (mg/L) of Water Samples.

areas were polluted. COD values were greater in Parvathy Puthanar than the other three sites. Akkulam Lake had low COD value compared to other sites. It indicated that water quality was comparatively poor in Parvathy Puthanar.

Hardness of the water was maximum at Akkulam and Parvathy Puthanar during the months April and May. Minimum hardness value was obtained for the water sample of Murinjapalam Thodu during the month of March (64 mg/L). According to the APHA (1998), the desirable limit for total hardness is 300 mg/L. Compared to the desirable limit, the values of the samples were found to lie within the limit. Hardness of water is mainly due to the presence of Ca and Mg ions and is an important indicator of the toxic effect of poisonous elements (Tiwari, 2001). Water samples from Parvathy Puthanar showed higher salinity (809mg/L) than other sites during the month April. During the same month, Murinjapalam Thodu had minimum salinity value (154mg/L). The TDS values were maximum in Parvathy Puthanar (979mg/L), it was above the standard values (500 mg/L). Sample from Murinjiapalam Thodu had the minimum TDS value (211mg/l) than the other three sites. The alkalinity of the samples is only due to bicarbonates because phenolphthalein alkalinity is zero (APHA, 1998). Alkalinity was maximum during the month of April (12 mg/L) at Kannammola Thodu and minimum during the month of May and March in Akkulam Lake and Murinjapalam Thodu respectively. The Alkalinity of water samples were within the permissible limits suggested by Pollution Control Board SPCB (200 mg/L).

The phosphate concentration was maximum during the month of April (2.338 ppm) at Parvathy Puthanar and minimum during the month of May at Akkulam Lake (0.377 ppm). Phosphate and nitrate in natural drinking water may be derived through allochthonous input through rainwater, leaching of soil and weathering of rocks (Jhingran, 1988). The maximum nitrate content was shown by water sample from Parvathy Puthanar (2.08ppm). It was greater than the standard values. Akkulam Lake possessed permissible level of nitrite content. The water from Akkulam Lake showed less nitrite content. High concentration of nitrate in drinking water is toxic (Umavathi *et al.,* 2007).

The correlation study was also done to identify the inter-relation between water quality parameters (Figures 9.12–9.16). The results showed that the pH and phosphate were positively correlated and pH and hardness negatively correlated. DO and

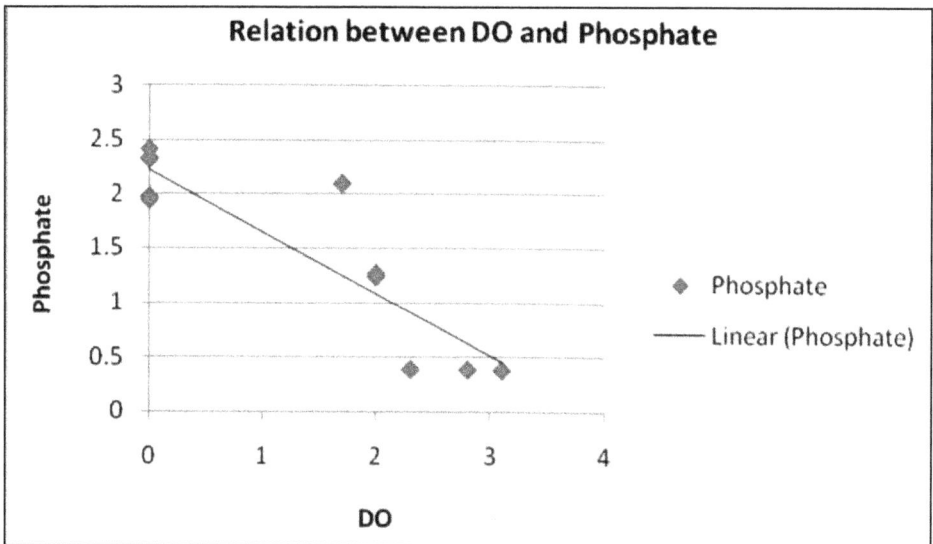

Figure 9.12: Relation between DO and Phosphate.

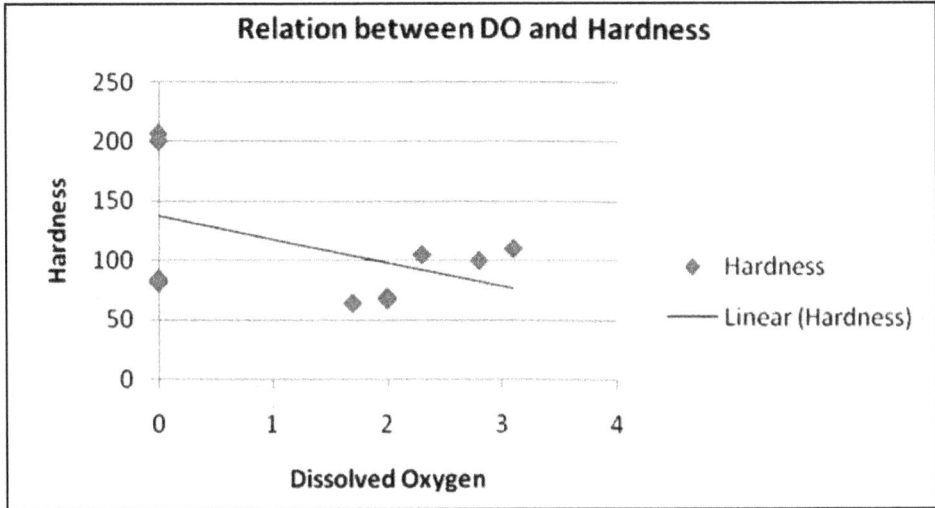

Figure 9.13: Relation between DO and Hadness.

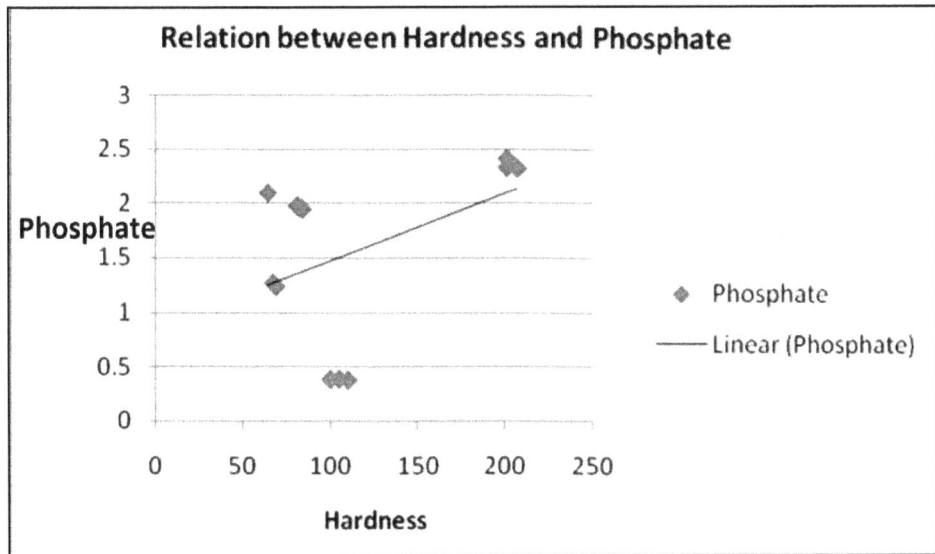

Figure 9.14: Relation between Hadness and Phosphate.

phosphate showed strong negative correlation but at the same time hardness and phosphate were strongly correlated. Dissolved oxygen and hardness showed slight negative correlation.

Conclusion

Among the studied water quality parameters, all were not in the prescribed limits. In terms of pH, all sites showed the values within the limits. Samples from Parvathy puthanar and Kannammula Thodu showed zero values for dissolved oxygen

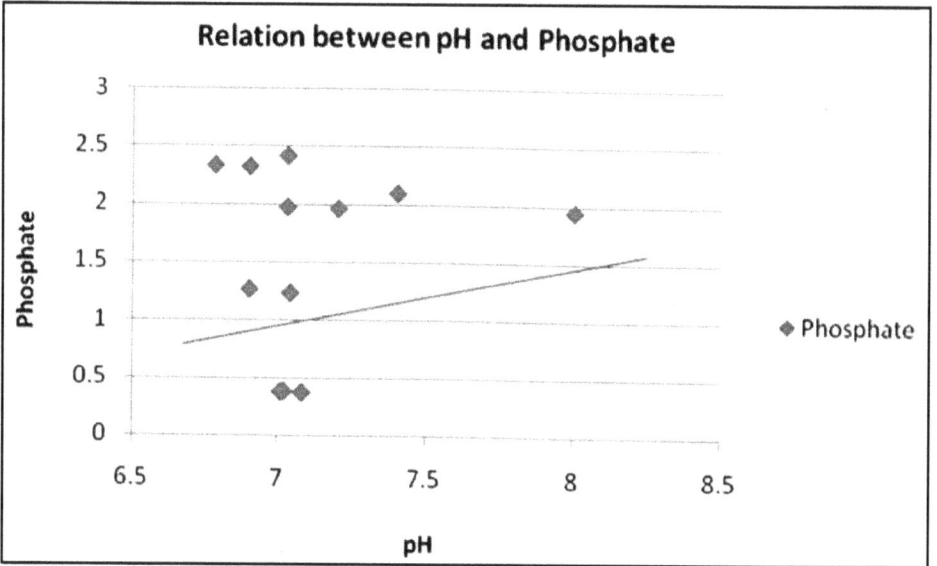

Figure 9.15: Relation between pH and Phosphate.

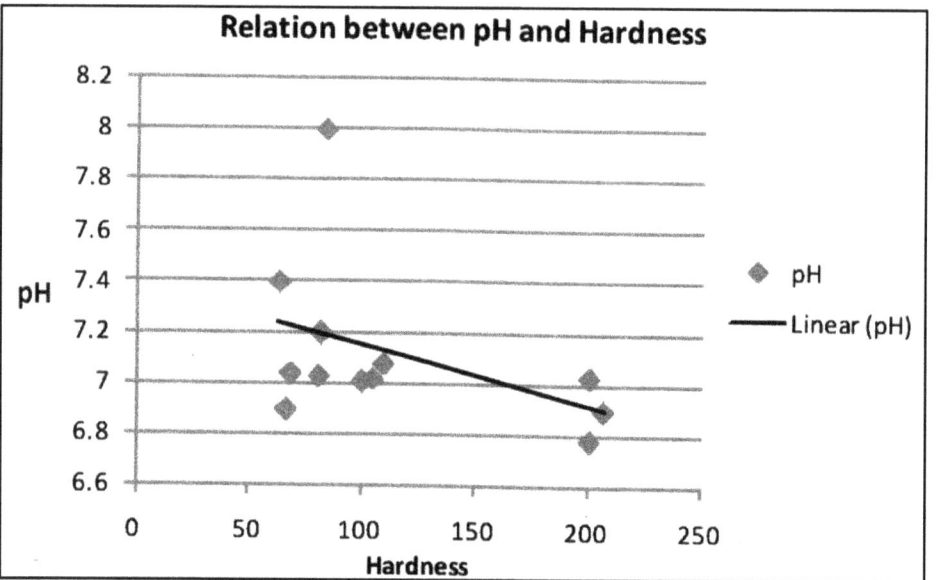

Figure 9.16: Relation between pH and Hardness.

which indicated high pollution and high activity of aerobic bacteria. Akkulam Lake samples showed BOD values within the permissible limit, but water from Parvathy Puthanar and Kannammula Thodu showed higher values up to 24mg/l which indicated the extent of water pollution. By looking into the parameters checked in the present study, it was clear that Akkulam Lake was better in quality of water when compared with the other three selected areas. By controlling the dumping of waste into these water bodies, the extent of pollution can be decreased up to certain level. The various treatment processes like aeration, coagulation, flocculation, sedimentation, filtration, and disinfection should be adopted to treat the waste before being dumped into the nearby water bodies.

References

APHA, 1998. Standard Methods for Examination of Water and Wastewater, 20th Edition, American Public Health Association, Washington D. C.

Jhingran, V. G. *et al.,* 1988. Methodology of reservoir fisheries investigations in India, Bulletin No. 12, CICFRI, India.

Kant, S. and Raina, A. K. 1990. Limnological studies of two ponds in Jammu-11. Physico- chemical parameters. *Environ. Biol.* 11: 137-44

Kaushik, S. and Saksena, D. N. 1999. Physico -chemical limnology of certain water bodies of central India. In, K Vismayam edited fresh eater ecosystem in India. Daya Publishing House, Delhi. 336.

Simpi Basavaraja, S. M. *et al.,* 2011. Analysis of Water Quality Using Physico-Chemical Parameters Hosahalli Tank in Shimoga District, Karnataka, India. *Global Journal Science Frontier Research.* 11(3).

Tiwari, D. R. 2001. Hydrogeochemistry of underground water in and around Chatarpur city. *Indian Envirn Health.* 43(4): 176.

Trivedy, R. K. and Goel P. K. 1986. Chemical and biological methods for water pollution studies, Environmental Publication, Karad, Maharashtra.

Umavathi, S. *et al.,* 2007. Studies on the nutrient content of Sulur pond in Coimbator, Tamil Nadu, *Ecology and Environmental Conservation*, 13(5): 501-504.

Whitney, R. J. 1942. Diurnal fluctuations of oxygen and pH in two small ponds and a stream. *J. Expt. Biol.* 19: 92-99.

Chapter 10

The Study of Benthic Fauna in Canoli Canal in Relation to Environmental Parameters, Thrissur

☆ *P. Nimisha and S. Sheeba*

ABSTRACT

The aim of this study is to document the distribution and abundance of benthic animals and physicochemical parameters of water in three selected stations of Canoli canal. The study provides the preliminary details of the benthic fauna. Benthic production is of importance in assessing the biological productivity and demersal fishery of that region. So the baseline data generated from this investigation will be of much use in all future assessment.

Introduction

The benthic organisms play an important role in the food chain at the primary, secondary and tertiary levels. The demersal fishery depends mainly on benthic productivity. As benthic animals lead a relatively sedentary mode of life, any change in the environment is reflected in the benthic organisms of that region (Harkantra *et al.*, 1980 and Sheeba, 2000). According to the feeding habit benthos are mainly divided into two namely phytobenthos and zoobenthos. Phytobenthos is the collective name for all primary producers (*i.e.* various algae and aquatic plants), whereas zoobenthos comprises all consumers (*i.e.* benthic animals and protozoa). Both of them are found on hard substrates, such as rocks, woods and shells and are very different from those of the soft sediments such as sand and mud. The makeup of a benthic community is controlled mainly by light, temperature, salinity, nature of bottom and by weather. Environmental stability seems to favour evolution of highly diverse communities, in which many kinds of plants and animals coexist. Abundance of benthic organism is controlled by productivity of surface waters and by water depth. Numbers of benthic organisms decrease with increasing water depth and distance from land. The present study focuses on the qualitative and quantitative analysis of benthic fauna and the analysis of physicochemical parameters of water of Canoli canal at three selected stations.

Materials and Methods

Canoli canal has an approximate length of 80 Km extending between Kottappuram near Kodungallur and Chettuva near Guruvayoor. It has an average width of 150M and depth of 10M. Four major rivers- Kechery, Karuvannur, Chalakkudy and Periyar river join this canal. Kechery and Karuvannur rivers join at Chettuva region; Chalakkudy and Periyar river join at Kottappuram region. This canal opens to the Arabian sea through Azhikode and Chettuva estuaries. A vast area of Kole lands lie around this canal. The surface soil of the major part of the canal is formed of soft clay mixed with decaying organic matter. The present investigation was carried out in Canoli canal at three Stations namely Thriprayar, S.N. Puram and Mathilakam. Station I- Thriprayar is located at Thriprayar in Chavakkad Thaluk, Thrissur District, Kerala at latitude of 10°30′N and longitude 76°5′E. The water becomes turbid during monsoon months. The sediment of the area is sandy silt and black in colour. Coconut plantation and sparse human habitation is present on the bank of this station. Fishermen catch fish from this area using traditional crafts and gears such as odum, cast net, pole net and line net. The fishes generally found are *Etroplus suratensis, Mugil cephalus, Belone cancila, Gerrus filamentosus* and clam— *Villoritta cyprinoids*. Sand mining is prevalent in this area. Station II- S.N. Puram is at 12 Km away from Station I, located in Kodungallur Thaluk, Thrissur District at latitude of 10°25′N and longitude 76° 10′E. The sediment of this area is muddy. Coconut plantations are dominant on the bank. Majority of local population is engaged in fishing using traditional crafts odum, vallum and gears like cast net, pole net and line net. Fish like *Villorita cyprinoid* is abundant in this region. Indiscriminate sand mining activities are common issues. Station III- Mathilakam is at 18 Km away from Station I, located in Kodungallur Thaluk, Thrissur District of 10°20′N and longitude 76°15′E. The sediment of the area is silty. Riparian people engaged in fishing using cast net, pole net and line net. The coconut palms dominate the vegetation on the banks. Sand mining is prevalent in this region also. *Villorita cyprinoids*, the edible clam is abundant in this region. For the present study water and sediment samples are collected at three different stations of Canoli canal from April 2009 to July 2009. Physico-chemical parameters of water were analyzed as per the methods suggested by Trivedy and Goel (1986). Benthic organisms were studied as per the observance of standard references from Battish, (1992).

Results and Discussion

Data on the monthly variations of physicochemical parameters of three stations are illustrated in Table 10.1. The temperature of atmosphere and water shows some difference. The atmospheric temperature ranged between 26°C (July) to 37°C (April) and water temperature varied between 25°C (July) to 36°C (April). The present study shows a direct relationship between temperature and benthic production. Similar observations were recorded by Suresh *et al.* (1978).

pH is an important factor that determines the quality of water. pH ranged between 5 (May) to 7.5 (April). Rainfall, river discharge, exchange from the sea and flow from coconut retting zones are important factors that influence pH variation in the backwater system (Nair *et al.*, 1984c).

Dissolved oxygen is a major constituent in the water quality parameters. Optimum concentration of dissolved oxygen is essential for maintaining aesthetic qualities as well as for supporting aquatic life. The dissolved oxygen concentration shows monthly variations. The oxygen concentration was fluctuated between 5 mg/l (April) to 9 mg/l (July). The distribution and abundance of zoobenthos is primarily determined by the oxygen concentration and type of bottom sediments (Devi *et al.*, 1994). In the present study there was a negative relation between faunal abundance and oxygen distribution.

Monthly variation of carbon dioxide concentration in three stations ranges between 4.9 mg/l (May) and 6 mg/l (April). The higher level of carbon dioxide may be due to the fast decomposition of organic matter.

Monthly variations of salinity ranged between 0.25 ppt (July) and 11.45 ppt (April) at three stations under study. The highest salinity was noticed during pre monsoon months and the lowest salinity was during monsoon months. It may be due to heavy rainfall and the subsequent dilution in the medium. Many benthic organisms have a high degree of euryhalinity and so they are able to withstand such wide fluctuation in salinity. A direct positive relationship between faunal abundance and salinity was noticed in the present study which can be paralleled with reports of Aseeda (1997).

Monthly variation of nitrite nitrogen ranges from 4.50μg/l (April) to 13.50μg/l (June). Most of the nitrogen comes in the form of organic and ammoniacal nitrogen. Maximum nitrite concentration was observed in June may be due to the effect of southwest monsoon. The rain water washes off fertilizers from the terrestrial environment into the aquatic medium which enhances nitrite concentration. Relatively low nutrient levels observed in summer months could probably be attributed to the utilization of nutrients by phytoplankton (Sarugunam, 1994). Phosphorous is present in very low concentration in all types of natural waters.

In the present study phosphate concentration ranges from 0.12μg/l (July) to 0.63μg/l (April). Similar observations were also made by Nair *et al.* (1984) in Ashtamudi estuary. Monthly variation of ammonia ranges from nil value (June and July) to 1.75 μg/l (April). The lower value of ammonia may be due to the consumption by plankton and it is considered the most acceptable form of nitrogen nutrients (Vaccaro, 1963).

Table 10.1: Monthly Variations of the Physico-chemical Parameters of Canoli Canal at Three Stations from April to July 2009

Parameters	April			May			June			July		
	Station I	Station II	Station III	Station I	Station II	Station III	Station I	Station II	Station III	Station I	Station II	Station III
AtmosphericTemperature(°C)	37	35	37	36	35	35	28	30	32	26	27	27
Water Temperature(°C)	35	36	34	35	34	33	27	26	28	26	26	25
pH	6.5	7.5	7	5.5	6.5	5	6.5	6	5.5	6.5	6.5	6
Dissolved O_2 (mg/l)	5	5.5	6	6.5	5.5	6.2	7.5	8	7	8.5	8.5	8.8
CO_2 (mg/l)	6	5.5	5	5.2	5.1	4.9	5.2	5.4	5.3	5.6	5.6	5.5
Salinity (ppt)	11.45	8.5	7.75	5.5	6.2	5	0.65	0.65	0.34	0.25	0.25	0.36
Nitrite (µg/l)	4.5	5.8	7.2	5.4	7.5	8.4	11.1	12	13.5	11.2	11.2	10.7
Phosphate (µg/l)	0.6	0.63	0.57	0.58	0.48	0.55	0.15	0.26	0.34	0.2	0.2	0.23
Ammonia (µg/l)	1.75	1.35	1.2	1.2	1.25	0.59	0	0	0	0	0	0
Total suspended solid (mg/l)	22	23.5	21	23.5	24.2	23.5	44.8	45	38.5	42.5	42.5	49.2
Turbidity	0.01	0.02	0.01	0.01	0.02	0.02	0.02	0.03	0.07	0.05	0.05	0.06

Table 10.2: Monthly Variations of Benthic Organisms (N/m²) of Canoli Canal at Rhree Stations from April to July 2009

Name of Group or Species	Station I				Station II				Station III			
	April	May	June	July	April	May	June	July	April	May	June	July
Polychaetes												
Neries sp.	75	65	55	35	80	75	70	–	75	45	–	30
Arenicola	70	60	20	10	85	70	40	20	50	35	15	–
Capitellidae	75	55	–	20	70	45	25	10	–	–	–	–
Prinospio polybrnchiata	–	–	–	–	–	–	–	–	35	30	10	–
Amphipods												
Melita zeylanica	70	85	45	30	60	70	55	25	65	50	–	30
Unidentified sp.	50	35	25	–	40	35	–	10	40	15	20	–
Molluscs												
Villorita cyprinoides	80	70	60	50	85	80	75	60	90	75	80	40
Nucula sp.	60	50	40	20	55	45	25	10	40	45	60	–
Crustaceans												
Crab	40	30	20	15	50	35	30	25	50	35	20	15
Prawn	55	65	35	30	35	25	15	10	70	55	30	20
Juvenile fishes	20	15	10	–	20	15	10	5	–	–	–	–
Total no. of benthic organisms	665	560	310	225	620	525	355	175	515	385	235	135

Table 10.3: Monthly Variations of Biomass of Benthic Organisms (gram/m²) at Three Stations from April to August 2009

Name of Group	Station I				Station II				Station III			
	April	May	June	July	April	May	June	July	April	May	June	July
Polychaetes	200	185	150	120	200	180	110	70	140	120	60	70
Amphipods	75	67	56	38	70	78	40	30	70	60	15	25
Molluscs	4240	3875	3330	1554	3800	4000	2200	1600	3025	2910	3400	1300
Crustaceans	2520	2340	2120	2052	2600	1800	1550	2300	3500	2400	1700	2100
Juvenile fishes	400	357	276	0	400	240	400	0	–	–	–	–
Total Biomass	7435	6824	5932	3770	7070	6298	4300	4000	6735	5490	5175	3495

Table 10.4: Monthly Variations in the Percentage Composition of Benthic Organisms at Three Stations from April to August 2009

Name of Group	Station I				Station II				Station III			
	April	May	June	July	April	May	June	July	April	May	June	July
Polychaetes	43.61	37.5	24.19	35.55	44.35	41.9	40.84	17.14	31.07	28.57	10.63	22.22
Amphipods	18.05	21.43	22.58	13.33	16.12	20	15.49	20	20.39	16.88	8.51	20.22
Molluscs	21.05	21.43	32.26	31.11	22.58	23.81	28.17	40	25.24	31.16	59.57	29.62
Crustaceans	14.29	16.96	17.74	20	13.71	11.43	12.67	20	23.3	23.37	21.27	25.93
Juvenile fishes	3	2.67	3.22	0	3.23	2.85	2.81	2.7	0	0	0	0

Monthly variations of total suspended solids differed from 20 mg/l (March) to 49 mg/l (July). The highest value of TSS in July may be due to heavy river discharge of wastes and also indiscriminate sand mining activities. In the present study there was a negative relationship between faunal abundance and TSS. Similar observation was reported by Nooja (2004). Turbidity means cloudiness or muddiness in the water system. In the present study it ranges from 0.01 (April) to 0.07 (June).

The highest concentration of turbidity in June may be due to the effect of monsoon showers.

Monthly variations of benthic organisms, biomass and percentage of composition of benthic organisms of three selected stations of Canoli canal are presented in Tables 10.2–10.4.

Five major groups of benthic fauna were obtained from the study areas such as Polychaetes, Amphipods, Molluscs, Crustaceans and Juvenile fishes. Polychaetes such as *Arenicola, Nereis* sp., *Prionospio polybranchiata* and *Capitellidae* were observed. Amphipods include *Melita zeylanica* and Molluscs such as *Villorita cyprinoids, Nucula* sp. Crustaceans such as Crabs, Prawns and Juvenile fishes of *Gobidae* family were obtained. One or two unidentified species were observed throughout the period of the study. In the present investigation, total density of benthic organisms (April, May, June and July) observed at Station I, II and III was 440 No/m^2, 418 No/m^2 and 317 No/m^2 and Total biomass density of benthic organisms (April, May, June and July) was 5990g/m^2, 5417 g/m^2 and 5223 g/m^2 respectively. From the result, total density of benthic organisms and their biomass density were high at Station I compared to those in the other two stations and this may be attributed to nature of sediment and the influence of salinity in this station. From these findings it can be concluded that the three selected study sites were not highly polluted because they have high benthic production.

Conclusion

Investigation of benthic fauna gives a clear indication that the environmental parameters like salinity, temperature, total suspended solids and dissolved oxygen have a major influence upon them. Of these salinity and temperature have a direct positive relationship with faunal abundance and dissolved oxygen has a negative relationship. The distribution of fauna exhibited considerable variation at different stations. The highly complex, yet delicate aquatic system, whose features are influenced and controlled by unique properties of water is under great stress caused by anthropogenic activities. Stringent measures should be taken for the revival of the system and continuous monitoring and documentation of the organisms are highly essential.

Acknowledgements

The authors are grateful to the Principal, S.N. College, Nattika, Thrissur for providing the necessary facilities during this work.

References

Aseeda, A. A. 1997. A study of interstitial fauna of Nattika Beach, Thrissur. M.Sc. Dissertation submitted to University of Calicut.

Battish, S. K. 1992. Freshwater zooplankton of India. Oxford and IBH Publishing Cor. Privt. Ltd. pp. 233.

Devi, K. S., Sankaranarayan, V. N. and Venugopal, P. 1994. Benthos of Beypore and Korapuzha estuaries of North Kerala. *Proceedings of sixth Kerala Science Congress*, pp. 64-67.

Harikantra, S. N., Ayyappan Nair, Ansari, Z. A. and Parulekar, A. H. 1980. Benthos of the shelf along the west coast of India. *Indian. J. Mar. Sci.* 9: 106-110.

Nair, N. B., Krishnakumar, K., Rajasekharan Nair, J., Abdul Aziz, P. K., Dharmaraj, K. and Arunachalam, M. 1984c. Ecology of Indian estuaries; Part VI Physico-chemical conditions in Kadinamkulam backwater. Southwest coast of India. *Indian J. Mar. Sci.* 7: 15-17.

Sarugunam, A. 1994. Studies of microfouling in the coastal waters of Kalpakkam, Eastcoast of India with reference to diatoms. Ph. D thesis, Sci. India. 19: 257-297.

Sheeba, P. 2000. Distribution of benthic fauna in the Cochin backwaters in relation to environmental parameters. Ph. D Thesis. CUSAT.

Trivedy, R. K. and Goel, P. K. 1986. *Chemical and biological methods for water pollution studies*. Environmental Publications. 138-146pp.

Vaccaro, P. F. 1963. Available nitrogen and phosphorous and biochemical cycle in Atlantic of New England. *J. Mar. Sci.* 282-300.

Chapter 11

The Impact of Rainfall on Fecal Coliform Bacteria in Shallow Lakes of Silicon City, India

☆ *H. Krishna Ram*

ABSTRACT

Fecal coliform bacteria are the most common pollutant in lakes. In Silicon city, 80 per cent lakes, water had some level of contamination. The objective of this research was to assess the effect of surface runoff amounts and rainfall amount parameters on fecal coliform bacterial densities in Byramangala lake, Hebbal lake and Yelahanka lake in silicon city. Samples from all the lakes were collected monthly and analyzed for the presence of fecal coliforms. Fecal coliforms isolated from these samples were identified to the species level. The analysis of the bacterial levels was performed following standard test protocols levels was performed following standard test protocols as described in standard methods for the examination of water and wastewater. Information regarding the rainfall amounts and surface runoff amounts for the selected years was retrieved from the state pollution control board. It was found that a significant increase in the fecal coliform numbers may be associated with average rainfall amounts. Possible sources of elevated coliform counts could include sewage discharges from municipal treatment plants and septic tanks and runoff from pastures and range lands. It can be concluded that non point source pollution that is carried by surface runoff has a significant effect on bacterial levels in water resources.

Introduction

Fecal coliform bacteria in surface water are a concern because of the risks to human health when infectious bacteria are present in the environment. It occurs most commonly in fecal waste of homeotherms, including man, and has been found to be the best biological indicator of fecal contamination in waters (Edberg *et al.,* 2000). Possible sources of elevated coliform count include sewage discharges from municipal treatment plants and septic tanks, stormwater overflows, and runoff from pastures and range lands. Faust and Goff (1978) found highest levels of fecal coliform bacteria in arms of their Chesapeathe Bay study area receiving runoff from pasture areas. Intermediate coliform contamination was associated with forested regions; and least coliform contamination course from "urbanized" areas.

Coliforms, fecal coliforms and fecal streptococci are established ecological indicators of fecal contamination in water, and are used for determination of water suitability for human use (Dkawasili and Akujobi, 1996). In Urban and densely populated rural areas, the microbiological of quality of freshwaters in frequently threatened by contamination with untreated domestic wastewaters (Grisel and Jagals, 2002). Recently, it has been realized that in addition to various anthropogenic activities, a changing global climate is also exerting a significant impact on storm flow and water availability (Milly *et al.,* 2005). This study was designed to analyze the possible impact of monthly rainfall amounts on fecal coliform bacteria levels found in Byramangala lake, Hebbal lake and Yelahanka lake located in north and south region of silicon city.

Materials and Methods

Method

A monthly of the fecal coliform level data collected from Byramangala lake, Hebbal lake and Yelahanka lake for period of the years of 2003, 2004 and 2005 was done. Measurements regarding rainfall level for the three R.G. of the lakes city. All test conducted were referenced from the 20th Edition of the Standard Methods for the Examination of water and wastewater.

Collection of Samples

All water samples were collected with a volume of not less than 100 ml. The containers used were in accordance to the 20th edition of Standard Methods for the examination of water and wastewater. Samples were collected in non-reactive glass or plastic bottles that had been cleaned and rinsed carefully, given a final rinse with distilled water, and sterilized. Containers were lowered to a depth of not greater than 2 ft below the surface to fill. The samples were placed immediately on ice in order to have a temperature of less than 10°C during a maximum transport time of 6hours. Five duplicate samples from each site were collected once monthly for a period of three years. The area of sample collection used for Byramangala, Hebbal and Yelahanka lakes had a minimal flow rate.

Fecal Coliform Count

100 ml of water sample were vortexes and filtered onto a membrane filter using

a sterile filtration unit. The approved technique used was from Clesceli *et al.*, 1999. After filtration, forceps were used to place the membrane filter on an MFC agar plate. The plate was then incubated in an incubator at a temperature of 45°C for 24 hours. The plates were then checked for bacteria colony growth.

Results

The mean annual rainfall level in Byramangala RG highest was found (974.6mm) in 2005 and lowest was found (577.8mm) in the year 2003. In Hebbal RG highest was found (1094.3 mm) in 2005 and lowest was found (562.1 mm) in the year 2003, and Yelahanka RG highest was found (1103 mm) in 2005 and lowest was found (592.8 mm) in the year 2003 (Figure 11.1). As shown in Figure 11.2 the greatest monthly rainfall of Byramangala RG during the years 2003, 2004 and 2005 occurred during the months of October, September and October respectively, and Hebbal RG during the years occurred during the months of August, September and October (Figure 11.3) and Yelahanka RG during the years highest occurred during the months of October, September and October respectively (Figure 11.4).

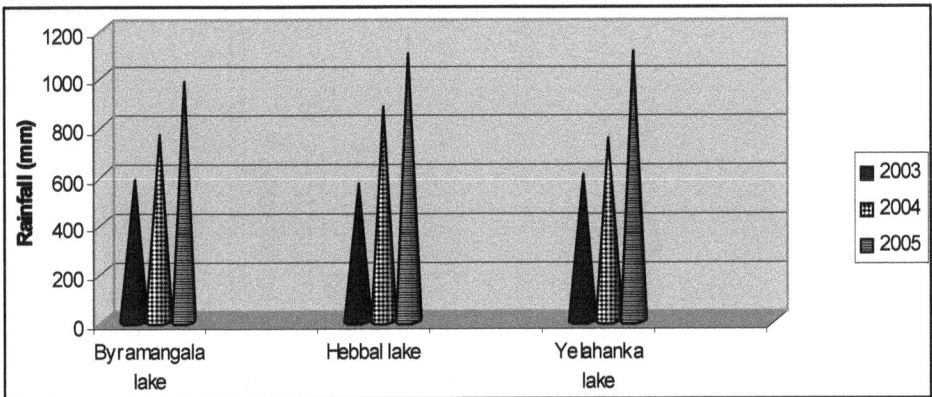

Figure 11.1: Mean Value of Rainfall Amounts for the Years 2003–2005.

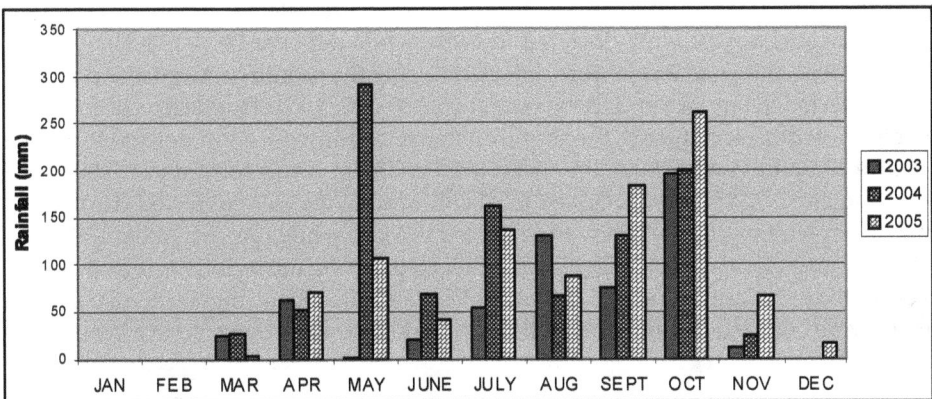

Figure 11.2: Rainfall Amounts for the Years 2003–2005 at Byramangala RG.

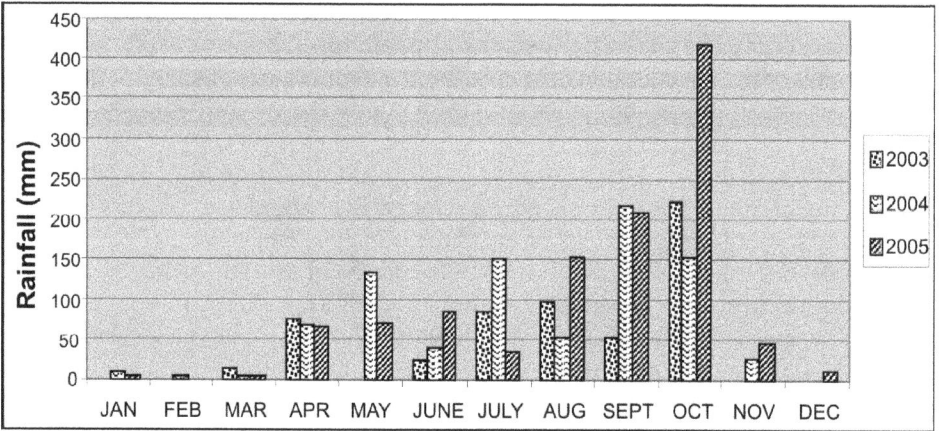

Figure 11.3: Rainfall Amounts for the Years 2003–2005 at Hebbal RG.

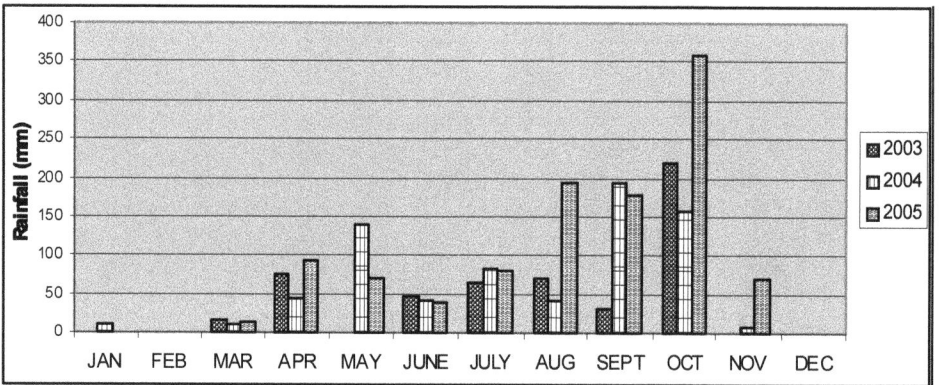

Figure 11.4: Rainfall Amounts for the Years 2003–2005 at Yelahanka RG.

The mean fecal coliform levels for the year 2003 at Byramangala lake was found to be 162-420/100ml, highest fecal coliforms levels were detected during the months of August and lowest were detected in May, while the year 2004 160-310/100ml, highest fecal coliforms levels were detected during the month of August and lowest were detected in January and in the year 2005, 130-300/100 ml highest fecal coliform levels were detected during the month August and lowest were detected in May (Figure 11.5). In Hebbal lake, the mean fecal coliform levels for the year 2003 was found to be 127-245/100ml highest fecal coliform levels were detected during the month September and lowest were detected is May, while the year 2004, 138-230/100ml, highest fecal coliform levels were detected during the month June and lowest were detected in January, and in the year 2005, 139-230/100 ml, highest fecal coliform levels were detected during the month July and lowest were detected in May (Figure 11.6).

In Yelahanka lake, the mean fecal coliform levels for the year 2003 was found to be 123-262/100 ml, highest fecal coliform levels were detected during the months

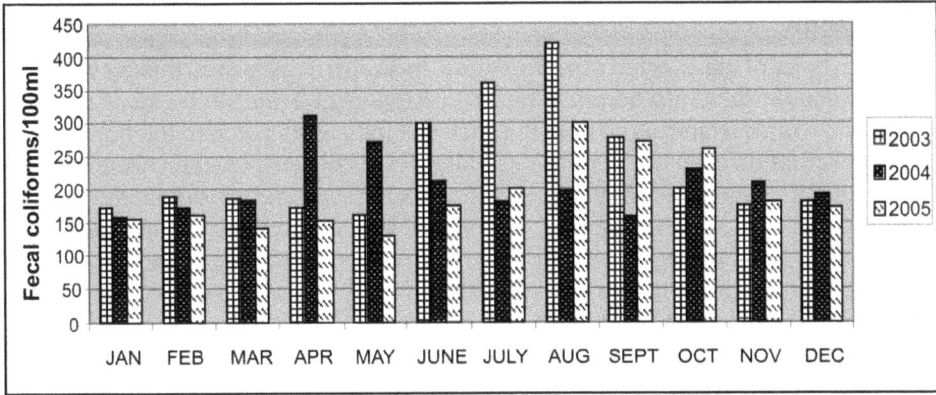

Figure 11.5: Fecal Coliforms Level in Byramangala Lake during 2003–2005.

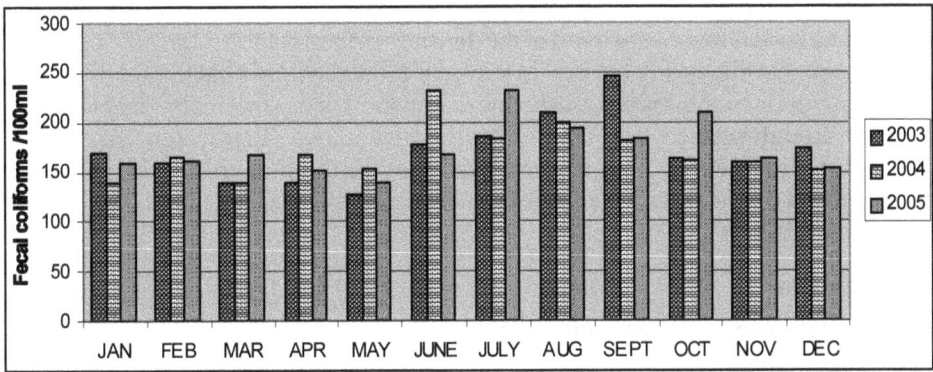

Figure 11.6: Fecal Coliforms Level in Hebbal Lake during 2003–2005.

September and lowest detected in May, while the year 2004, 139-273/100 ml, highest fecal coliform levels were detected during the months September and lowest were detected in June, and in the year 2005, 98-238/100 ml, highest fecal coliform levels were detected during the months October and lowest were detected during the month of December (Figure 11.7).

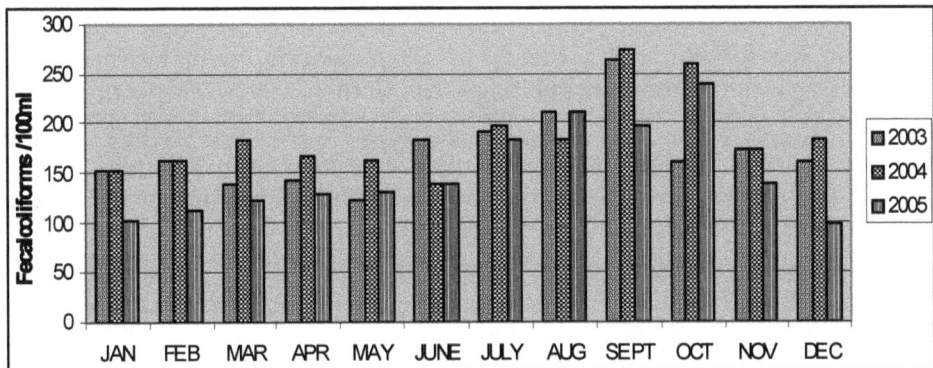

Figure 11.7: Fecal Coliforms Level in Yelahanka Lake during 2003–2005.

Discussion

The choice of the selected microbiological indicates to include in this preliminary analysis should take into account the specificity of the water to be analyzed and requires the understanding of how the bacterial indicators relate to the presence of pathogens that directly impact public health. Fecal coliform bacteria are routinely used to monitor aquatic systems for sewage contamination and considerable attention has been directed at educating the survival of fecal coliforms in aquatic systems. These water having uses such as primary contact recreation, secondary contact recreation, propagation of fish and wildlife, agriculture and as being out standings natural resource water (LDEQ, 2002).

Although the mean monthly rainfall levels for three years with in the study were very similar the fecal coliform levels were not. The mean bacterial level for the year 2005 was four times higher than the bacterial level for the year 2003, 2004. Unknown factors such as increased recreational fishing during this time could possibly account for elevated level. The general trend of the densities of fecal coliforms greatly increased in monsoon months and decreased in summer months, higher bacterial population during monsoon months is obviously due to transport of organic matter from various sources through surface run off from the catchment area other factors that have a bearing on the bacterial population density are (a) the human activities causing pollution and (b) the run off water from catchment areas flowing into the lakes with abundant nutrients.

Conclusion

It was proven by these observations that the fecal coliform levels at all lakes are not significantly linked to the rainfall level. It can therefore be concluded from the findings of this research that although non-point source pollution has a significant effect on bacterial levels to runoff water and in water resources, this effect would be the result factors other than just the mean monthly rain fall level and not this alone.

References

Clesceli, L. S., Greenberg, A. E. and Eaton, A. P. (Eds) 1999. Standard methods for the examination of water and wastewater, 20th Edn., American Public Health Association. American Public Health Association, Washington, D. C.

Edberg SC, Rice EW, Karlin R. J, Allen M. J. 2000. *Escheeichia Coli:* the best biological drinking water indicator for public health protection. *J. Appl. Microbiol.* 88: 1065-1165.

Faust, M. A., and N. M. Goff, 1978. Sources of bacterial pollution in an Estuary, P. 819-839 (In) coastal zone 78. Symposium on Technical Environmental, Socio economic and Regulatory aspects of coastal zone management, Vol. II, American Society of Civil Engineers, New York.

Geldreich, E. E. 1970 Applying bacteriological parameters to recreational water quality. *J. Amer. Water. Works Ass.* 62: 113-120.

Grisel, M. and Jagals, P. 2002. Fecal indictor organisms in the Renoster Spruit System of the Modder-Riet River catchment and implications for human users of the water, Water SA 28, 227-234.

Joyce, G., and H. H. Weiser. 1967. Survival of Entero viruses and bacteriophage in farm Pond waters. *J. Amer. Water. Works Ass.* 59: 491-501.

Louitiana Department of Environment Quality. 2002. Title 33 Environmental quality part IV Water Quality. http: / /www. deg. state. la. us/planning/regs/title 33/ 33u 09. pdf.

Milly, PCD, Dunn K. A., Vecchia AV. 2005. Global pattern of trends in Steam flow and water availability in a changing climate, *Nature* 438: 347-350.

Okpowasili G C, Akujobi, T. C. 1996. Bacteriological indicators of tropical water quality. *Int. J. Environ Toxicol. Water* 11: 77-81.

Tchounwou, P. B. and Warron, M. 2001. Physicochmical and bacterial assessment of water quality at the Ross Barnett Reservoir in Central Mississippi, *Reviews on Environmental Health,* 16: 203-212.

Van Donsel, D. J., and E. E. Geldreich 1971. Relationship of Salmonellae to fecal coliforms in bottom sediments, *Water Res.* 5: 1979-1087.

Chapter 12

Hydrochemistry of some Lentic Water Bodies of Karwar-Uttara Kannada District, Karnataka

☆ *B. Vasanthkumar, S.V. Roopa and B.K. Gangadhar*

ABSTRACT

Present work was pertaining to studies conducted at lentic perinial water bodies at Karwar. The period of study undertaken it one year from October 2012 to September 2013. This study is intended to monitor the water quality of lentic water bodies and also helpful for using this water for making fisheries activity and agricultural purpose. The present investigation encompassed collection of data pertaining to various aspects such as meteorological condition, physico-chemical parameter of these water bodies.

Keywords: *Hydrochemistry, Lentic water body, Meteorology, Karwar.*

Introduction

Present work was pertaining to studies conducted at lentic perinial water bodies at Mudageri and Hotegali pond, Karwar. The period of study undertaken it one year from Oct-2012 to Sept-2013. Uttara Kannada district (formerly North Kanara) is located between 13°55' to 15°32' N lat and 74°05' to 75° 05' E long. Its geographic area is 10,291 km². The district has boundaries with Goa and Belgaum towards the north Dharwar, Haveri and Shimoga towards the east and Udupi towards the south. The

Arabian Sea borders it on the west creating a long continuous, though narrow coastline, of 120 km. Uttara Kannada has a tropical climate with a well-defined rainy season between June and October, when the South-west Monsoon winds bring dowon an average 2500 mm rainfall annually. The remaining part of the year has hardly any rains. Whereas the coastal and crestline taluks receive high rainfall, the north-eastern taluks, Haliyal and Mundgod have very low rainfall.

Karwar is the administrative headquarters of an eponymous taluk and of Uttara Kannada district in the Indian state of Karnataka. It was the chief town of the North Kanara district in British India. Karwar is a sea side town situated on the banks of the Kali river which is on the west coast of the Indian peninsula. The town lies about 15 kilometres (9.3 mi) south of the Karnataka–Goa border.

The hydrobiological investigations were limited primarily in the Mudageri pond and Hotegali with a view to understand the ecobiology of water bodies in details water body is known to be utilized for fisheries activity and at the same time dumping sink of for the agricultural waste. However, this pond is mesotrophic. And present study of various physico-chemical parameters of these ponds were made to establish to know the safe for fisheries.

Materials and Methods

In the present study, water samples were collected on monthly basis from Mudageri pond and Hotegali pond. Surface water samples were collected using clean BOD bottles for the study of various physicochemical parameter. All collections and observations were made between 08.30 to 10.30 am throughout the period of study. Physico-chemical parameters like Air and water temperature, pH, dissolved oxygen, total dissolved solids, salinity, conductivity, turbidity, colorimetric were recorded at the sampling site using systronics water analyzer (Model 371). Phosphate, Nitrate, Nitrite, silicate were analyzed in the laboratory titrimetric method as per standard methods for examination of water (APHA AWWA. 1985, APHA 2000).

Results and Discussion

Atmospheric Temperature

The air temperature showed variation during the study period or among the seasons. The atmospheric temperature varied between 28°C to 35°C. A lowest temperature was recorded during November-December (28°C) and highest during May month (35°C) in the Mudageri pond (Table 12.1) and in Hotgali pond, lowest temperature was recorded during December-January (29.0° C) and highest (33° C) during May month (Table 12.2).

Water Temperature

The water temperature did not show any drastic variation either during the study period or among the seasons. In Mudageri pond lowest temperature was recorded during January (26.0° C) and highest (33° C) during May month (Table 12.1). A lowest temperature was recorded during August-September (25°C) and highest during May month (31°C) in the Hotegali pond (Table 12.2).

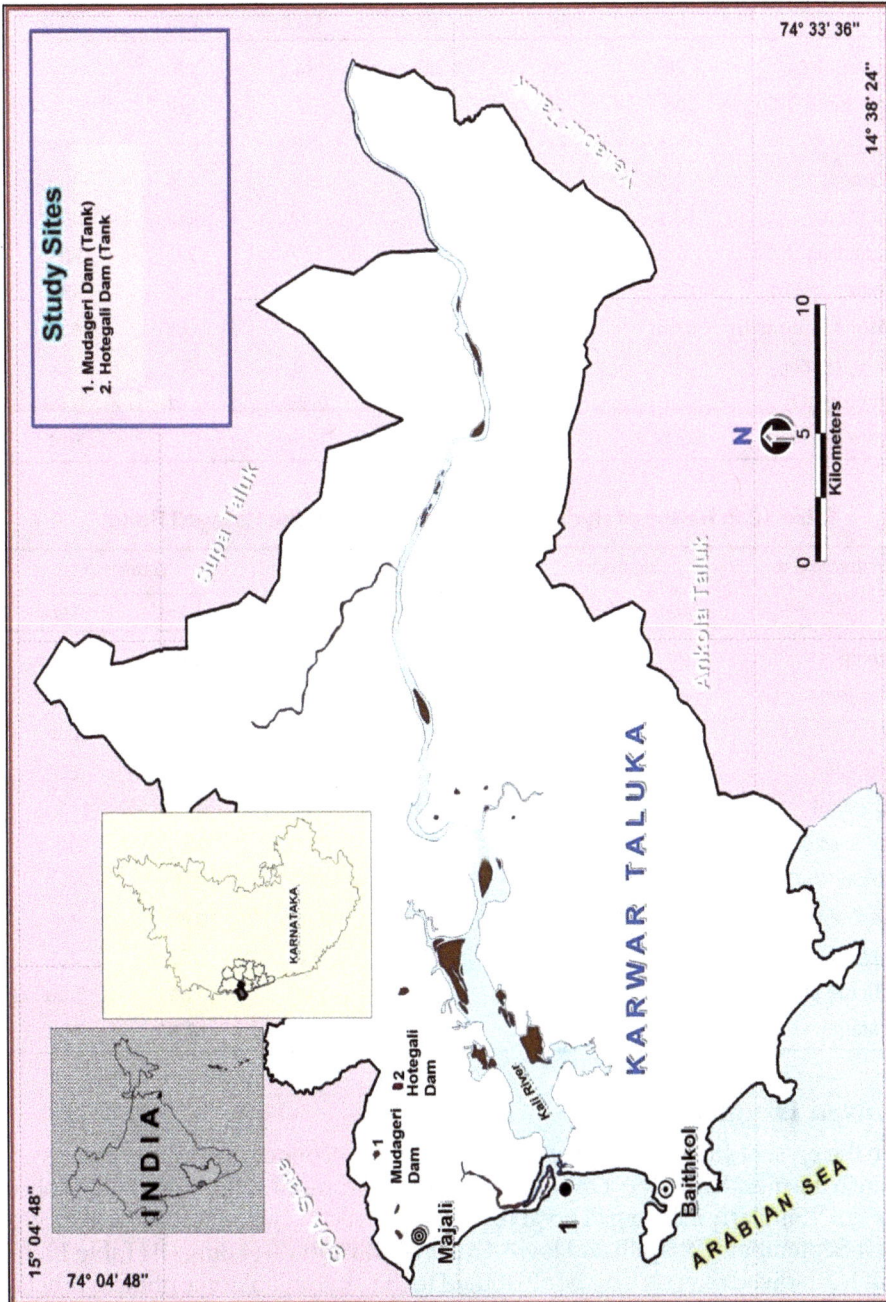

Figure 12.1: Map of Kali River Showing the Position of Study Sites.

Table 12.1: Range of Hydrological Parameters in the Mudageri Pond

Parameters	Station 1		Station 2		Station 3	
	Min	Max	Min	Max	Min	Max
Air temp. (°C)	28	35	29	34	28	34
Water temp. (°C)	26	33	25	32	25	31
pH	6.3	7.9	6.2	7.9	6.5	8.4
DO (mg/l)	5	6.4	5	6.9	4.8	6.8
TDS (ppm)	72.4	110	64.9	124	66.2	124.1
Conductivity (uS)	65	105.2	68	157	74	99.3
Turbidity (NTU)	3.2	21.05	3.5	45.7	4.2	45.2
Phosphate (µg at/l)	0.56	1.9	0.56	1.8	0.65	1.99
Nitrate (µg at/l)	1	2.8	1	2.2	1	3.2
Nitrite (µg at/l)	0.4	0.98	0.4	0.98	0.41	0.99
Silicate (µg at/l)	20.14	79.8	20.14	79.8	21	691

Table 12.2: Range of Hydrological Parameters in the Hotegali Pond

Parameters	Station 1		Station 2		Station 3	
	Min	Max	Min	Max	Min	Max
Air temp. (°C)	29	33	27	33	28	34
Water temp. (°C)	27	30	26	32	26	32
pH	6.5	8.3	6.2	8.3	6.3	8.4
DO (mg/l)	5	6.5	4.6	6.9	4.8	6.7
TDS (ppm)	63.4	96.3	63.1	114	62.3	119.2
Conductivity (uS)	70.14	98.5	70.1	98.2	70.2	100.4
Turbidity (NTU)	4.1	607	4.2	49.1	5.1	56.8
Phosphate (µg at/l)	0.59	1.95	0.7	2.01	0.69	1.9
Nitrate (µg at/l)	1.02	3.2	1.5	3.5	1.13	2.9
Nitrite (µg at/l)	0.23	0.98	0.38	0.99	0.39	0.98
Silicate (µg at/l)	20.4	80.1	23	80.4	19.8	71.4

Dissolved Oxygen

In the present study it was observed that Mudageri pond shows more dissolved oxygen in the month of July 6.4 mg/l and minimum 5.1 mg/l in the month of October (Table 12.1). Similarly in Hotegali pond highest dissolved oxygen was recorded during August-September (6.8mg/l) and lowest during May month (4.8mg/l) (Table 12.2). The high dissolved oxygen may be attributed to the phytoplanktonic photosynthetic activity and in addition temperature and high salinity similar observations were made by various workers on dissolved oxygen productivity and attributed to the water temperature, phytoplankton and degree of pollution (Ganapathi, 1960, Gosh *et al.*, 1974, George *et al.*, 1986).

Hydrogen ion Concentration (pH)

Range of the hydrogen ion concentration in Mudageri and Hotegali pond are presented in Tables 12.1 and 12.2. pH was low during July month (6.9) and high in the month of Dec (7.8) in the Mudageri pond. In Hotegali pond low during month of July (6.5) and high during May (8.4) In the present study higher concentration of pH was observed during summer and South West monsoon season could be attributed to enhanced rate of evaporation coupled with human interference are partly to enhanced photosynthetic activity. Maximum pH during summer season also observed by David *et al.* (1974), Ayyappan and Gupta (1980), Vijaykumar (1991). They also related the high pH with photosynthetic activity and more conductive for net production.

Total Dissolved Solids

Total dissolved solids ranged from 64.9 to 124.1 mg/L, the minimum was recorded in August (72 mg/L) and maximum in May (112 mg/L) at Mudageri pond (Table 12.1). In Hotegali pond TDS values ranged from 62.3 to 119.2 mg/L (Table 12.2). The minimum value may be due to the stagnant condition of the water body. The values are within permissible limits of 1500 mg/L (BIS, 1993). High values of TDS and sulphates in drinking water are generally not harmful to human beings but high concentration of these may affect persons, who suffering from kidney and heart diseases (Gupta *et al.*, 2004).

Turbidity

Turbidity measures the suspended and inorganic matter in the water. The values fluctuated between 3.2 to 45.5 NTU (Table 12.1). In the Mudageri pond minimum value was recorded in January (3.2 NTU) and maximum in June (45.5 NTU). Minimum turbidity recorded 4.2 NTU in October and maximum in 21.5 NTU (Table 122). The variations of turbidity depend on the inflow of rain water carrying suspended particles (Nafeesa Begum *et al.*, 2006). In natural water bodies, turbidity may import a brown color to water (George, 1997).

Nutrients

Phosphate Phosphorus

In the present investigation it was observed that the minimum phosphate was in the month of June (0.56 µg at/l) and maximum in the month of March (1.8 µg at/l) (Table 12.1) while in Hotegali pond minimum phosphate was in the month of June (0.65 µg at/l) and maximum in the month of July (1.8 µg at/l) (Table 12.2) high phosphate in the pond favours for productivity of phytoplankton. According to Welch (1952) Hutchinson (1957) and Horne (1978) studies main source of phosphate is sewage. Ganapathi (1960) in tropical water, phosphates are always present in sufficient quantity. In present study the frequent occurrence of cyanophycean blooms in the pond attributes to high content of phosphates.

Nitrate

In present work the nitrate content of Mudageri pond varies from a minimum of 1.0 µg at/l (August) and maximum of 2.11 µg at/l l (June) (Table 12.1). Similarly nitrate content of Hotegali pond varies from a minimum of 1.0 µg at/l (August) and maximum of 3.2 µg at/l (May) (Table 12.2). The high content of nitrate in this tank can be attributed to sewage inflow. And also other polluting agents observed similar observation was made by of Venktateshwarlu (1969) have shown that organic pollution increase the nitrogen content of natural water bodies. Munwar (1970) indicated an inverse relation between nitrates and phosphate. Role of nitrate and phosphorus ratio information of blue green algal blooms has been studies by Jayangoudar (1964), Zafar (1967). According to them high nitrate and low phosphate favors the formation of cyanophyccae bloom.

Nitrite

In present work the nitrite content of Mudageri pond varies from a minimum of 0.40 µg at/l (August) and maximum of 0.98 µg at/l (March) (Table 12.1). Similarly nitrite content of Hotegali pond varies from a minimum of 0.41 µg at/l (August) and maximum of 0.95 µg at/l (January and December) (Table 12.2).

Silicate

In the present work it was observed in the Mudageri pond shows maximum silicate in the month of June (79.8 µg at/l) and minimum in the month of August (20.14 µg at/l) (Table 12.1). silicate was observed in the Hotegali pond shows maximum in the month of August (79.2 µg at/l) and minimum in the month of october (21.0 µg at/l) (Table 12.2). Munwar (1970) based on the work on freshwater ponds found direct correlation of silicates with temperature. However, in the present investigation it was not correlated. Atkin (1926) is opinion that increase in silicates is due to increases in pH.

Conclusion

The study revealed that there were variations in certain physico-chemical properties of Mudageri and Hotegali pond in Karwar, Uttara Kannada district of Karnataka state due to the surface run-off and other excessive human activities. All the physico-chemical characteristics were found within permissible limits as suggested by Zafer (1964) and Khan and Siddiqui (1971). Therefore, the present investigation based on scientific methodology clearly shows that the said study tank water can be easily used for drinking, agricultural purpose and aquaculture purpose.

Acknowledgements

The authors are grateful to University Grants Commission, New Delhi for granting the Major Research Project and thankful to the Principal, Government Arts and Science College, Karwar for providing laboratory facilities.

References

APHA, 2000. Standard methods for water and wastewater. American Public Health Association.

Atkins, W. R. G. 1926: Seasonal changes in the silica content of natural waters in relation to phytoplankton. *J. Mar. Biol. Assoc. U. K.* 14: 89-99.

Ayyappan, S. and Gupta, T. R. C., 1985: Limnology of Ramasamudra tank - Primary production. *Bull. Bot. Soc., Sagar,* 32: 82-88.

Cole, G. A., 1975. Text book of Limnology, Pub. C. V. Mosby, Co., pp. 283.

Das, S. M. and Srivastava, V. K., 1956. Quantitative studies on the freshwater plankton part I, plankton of the fish tank in Lucknow, India. *Proc. Nat. Acad. Sci,* 26: 82-89.

David, A., Rao, N. G. S. and Ray, P., 1974. Tank fishery resources of Karnataka. Bull. Cent. Inland Fish. Res. Inst., Barrackpore, 20: 87.

Ganapathi, S. V. 1960. Ecology of tropical waters. Proc. Symp. Algology. ICAR., New Delhi, PP. 204-218.

George, J. P., Venugobal, G. and Venkateshwaran 1986. Anthropogenic eutrophication in a perennial tank: effect on the growth of *Cyprinus carpio communis. Ind. J. Environ. Hlth.* 218 (4): 303-313.

Gosh, A., Rao, L. H and Banerjee, S. C. 1974. Studies of the hydrological condition of a sewage fed pond with a note on their role in fish culture. *J. Inland fish soc. India.* V: 51-61.

Herne, A. J. 1978. Nitrogen fixation in eutrophic lakes. In water pollution microbiology Vol. 2 (Ed. R. Mitchell) John Wiley and Sons. New York. PP. 1-30.

Hutchinson, G. E. 1957. A Treatise on limnology: Geography, Physics and Chemistry. Vol. I, John Wiley and Sons. Inc. (U. S. A), p. 1015.

Jayangowdar, I., 1964. The bioecologicla study of nuggikere lake in Dharwar, Mysore state South India *Hydrobiologia* 23(3-4): 515-532.

Karuppasamy, P. K. and P. Perumal. 2000. Biodiversi ty of zooplankton at Pichavaram mangroves, South India. *Adv. Biosci.,* 19: 23-32.

Kumar, H. D., Bisaria, G. P., Bhondari, C. M., Rana, B. C. and Sharma, V. 1994. Ecological studies on algae isolated from the effluent of an oil refinery, a fertilizer factory and a brewary. *Ind. J. Environ. Hlth.* 16: 243-255.

Munawar, M. 1970. Limnological studies on freshwater ponds of Hyderabad, India-II. The Biocoenose, distribution of unicellur and colonical phytoplankton in polluted and unpolluted environments. *Hydrobiology,* 36: 105-128.

Nafessa Begum., Purushothama, R., Naranaya, J. and Ravindra Kumar, K. P. 2006. Water quality Studies of TV Station Reservoir at Davangere City, Karnataka (India), *Journal of Environmental Science and Engineering,* 48: 281!284

Prasad, B. N. and Manjula, S. 1980: Ecological study of blue-green algae in river Gomati, *Indian J. Envt. Hlth.,* 22 (2): 151-168.

Ruttner, F. 1953. Fundamentals of Limnology. Uni. of Toronto Press; Toronto.

Schindler, D. W. 1971. Light temperature and oxygen regimes of selected lakes in the experimental lakes area, North West Ontario. 4. Fisch. Res. Bd. Can., 28(2): 157-169.

Senthilkumar, S., P. Santhanam and P. Perumal. 2002. Diversity of phytoplankton in Vellar estuary, southeast coast of India. The 5th Indian fisheries forum proceedings (Eds.: S. Ayyappan, J. K. Jena and M. Mohan Joseph). Published by AFSIB, Mangalore and AeA, Bhubanewar, India. pp. 245-248.

Swarnalatha, N. and Narsing Rao, A. 1998. Ecological studies of Banjara lake with reference to water pollution. *J. Environ. Biol.* 19 (2): 179-186.

Vamos, R. 1964: The release of hydrogen sulphide from mud. *J. Soil. Sci.* 15 (1): 103-109.

Venkateshwaralu, V., 1969: An ecological study of the algae of the river Mossi. Hyderabad, (India) with special reference to water pollution. I. Physico-chemical complex. *Hydrobiologia.* 33: 117-143.

Vijaykumar, K. and Paul, R 1990. Physico-chemical studies on the Bhosga Reservoir in Gulbarga. Karnataka. *J. Ecobiol.* 2(4): 332-335.

Welch, P. S. 1952. Limnology. McGraw Hill Book Co. INC, New York.

Zafar, A. R., 1967. On the ecology of algae in certain fish ponds of Hyderabad, India III-The periodicity. *Hydrobiologia.* 30(1): 96-112.

Chapter 13

Physico-chemical and Bacteriological Parameters of Hemavathi River at Holenarasipura, Hassan District, Karnataka

☆ *B.M. Sreedhara Nayaka, S. Ramakrishna and Jayaprakash*

ABSTRACT

The study of physico-chemical and bacteriological aspects is important in evaluating the tropic status of water bodies. The river Hemavathi starts in the Western Ghats and is approximately 245 km long. About 09 physico-chemical and 02 bacteriological parameters were studied from April 2012 to March 2013. Significant variations and interesting correlations among the parameters were observed throughout the period of the study.

Keywords: *Physico-chemical, Bacteriological, Health impact, Parameters.*

Introduction

Water pollution is widely spreading throughout the word due to increased human activity. Water used for drinking should be of potable nature which could be consumed in desired amount without any adverse effect on health. The importance of water in the control of diseases had long been reported (Hofkes, 1981). It is an essential part of protoplasm and creates a state for metabolic activities to occur smoothly; therefore, no life can exist without water (Dubey and Maheshwari, 2006). Water plays an essential role in the ecosystem.

The quality of natural water is generally governed by various physico-chemical and microbiological parameters.It is very necessary to understand the physico-chemical and bacteriological qualities of water. Presence of coliforms (T-coli and F-coli), total dissolved solids (TDS), total alkalinity (TA), pH,hardness, dissolved oxygen (DO), biological oxygen demand (BOD) and nitrate (NO_3) are some of the significant parameters to study the water quality. The present study was aimed to know the seasonal variations in physico-chemical and bacteriological parameters of the river Hemavathi in Holenarasipura belt. Hassan District.

Materials and Methods

The river Hemavathi takes its birth at Ballalarayanadurga in the Chikmagalur District Western Ghats of the State of Karnataka in southern India at an elevation of about 1.219 meters. It is approximately 245km long and has a drainage area of about 5,410km². A large reservoir has been built on the river at *Gorur* in the Hassan District, about 60 kms from Hassan

The investigations of physico-chemical (pH, DO, BOD, TDS, Cl⁻, TA, NO_3, NH_4-N) and bacteriological parameters (T-coli, F-coli) were carried out from April 2012 to March 2013. The water samples were collected on monthly basis between 7 AM and 9 AM and brought to laboratory for further analysis of physico-chemical and bacteriological parameters, following the standard method (APHA, 2005).

Results and Discussion

The physico-chemical and Bacteriological parameters of Hemavathi river water at Holenarsipura, Hassan summarized in Table 13.1.

Temperature plays an important role in water system. Change in temperature was observed in water due to biotic and abiotic reactions and water temperature changes were according to change in atmospheric change as reported by Kundangar *et al.* (1996). In the present investigation, the minimum water temperature recorded during July was 21.1°C and maximum recorded in the month of May was 29.1°C. Vijajkumar *et al.* (2006) while working on the water quality of Bennithora River in Karnataka and many other workers have also observed similar trend in different water bodies (Geroge *et al.,* 1986).

The slight seasonal fluctuation of pH can be attributed to the combined effect of temperature, liberation of ions and buffering capacity of water (Agarwall, 1999).In the present study, the pH concentration has been recorded from April-2012 to March-2013. During this period the pH values of water ranged from 7.8 to 8.5. In the study, it is interesting to note that the pH values were more or less towards alkaline throughout the period of study. Maximum values during summer of March, April and mays may be due to the precipitation of carbonates of calcium and magnesium from bicarbonates causing higher alkalinity (Agarwal and Rozgar, 2010).

Table 13.1: Monthly Variations in the Physico-chemical and Bacteriological Parameters in Hemavathi River Water

Month/ Year	Water Temp. °C	pH	DO mg/L	BOD mg/L	TDS mg/L	Cl⁻ mg/L	TA mg/L	NO_3 mg/L	NH_4N mg/L	T.coli MPN/ 100ml	F.coli MPN/ 100ml
Apr-12	28.2	8.42	5.76	3	192	36	116	0.19	ND	1600	350
May-12	29.1	8.50	5.60	3	200	40	120	0.16	ND	1600	350
Jun-12	24.5	8.40	6.00	2	184	32	112	0.12	ND	1600	540
Jul-12	21.1	8.20	6.20	1	172	28	104	0.16	ND	>1600	920
Aug-12	21.4	8.25	6.40	1	160	20	100	0.19	ND	>1600	920
Sep-12	22.0	8.20	6.50	1	162	24	100	0.20	ND	>1600	920
Oct-12	22.3	8.15	6.80	1	164	24	96	0.16	ND	1600	540
Nov-12	23.0	8.00	7.00	1	170	28	92	0.18	ND	920	240
Dec-12	22.5	7.80	7.10	1	170	28	88	0.14	ND	920	240
Jan-13	23.4	7.90	7.00	2	172	32	88	0.16	ND	1600	350
Feb-13	24.1	8.00	7.00	2	172	32	92	0.18	ND	1600	350
Mar-13	25.2	8.40	6.15	3	190	36	116	0.19	ND	1600	350

DO: Dissolved oxygen; BOD: Biological oxygen demand, TDS: Total Dissolved Solids; Cl⁻: Chlorides; TA: Total Alkalinity; NO_3: Nitrate; NH_4N: Ammonical Nitrogen; ND: Not detected.

Dissolved oxygen (DO) is a very important parameter of water quality and an index of physical and biological process occurring in water. In the present study, the value for DO ranged from 5.6-7.1 mg/L from April-10 to March-11 respectively. The low values of DO noticed during higher temperatures was a similar trend as observed by Bhattarai *et al.* (2008). When temperature increases gas solubility of water decreases and microbial activity increases; both these changes can reduce DO in water.

Bio-chemical Oxygen Demand (BOD) is the amount of oxygen required by the living organisms engaged in the utilization and ultimate destruction or stabilization of organic water (Hawkes, 1993). In the present study, the BOD ranges from 1-3 mg/L. The values of BOD clearly showed higher concentration during summer and comparatively low during other seasons respectively (Bhattarai *et al.,* 2008).

Present study on total dissolved solids (TDS) found to be increased during summer. During summer amount of solutes were high due to decrease in the water level in the river (Bhattaraj *et al.,* 2008). The Hemavathi water contains less TDS (160-200 mg/L). Theses values are far below the permissible level of drinking water standards of WHO (1000 mg/L). This result also supports the studies of Agarwal and Rozgard (2010).

Chloride (Cl⁻) is one of the important indicators of pollution. The value of chloride concentration in water was highest during summer season. These values are usually in the lower range of values in different rivers of India (Sabata and Nayar, 1995). In our study, chloride value was below permissible limit which may be attributed to the absence of major pollutants.

It could be observed that the total alkalinity (TA) values in Hemavathi river water indicated fluctuations with minor differences. The higher values of total alkalinity in the river water reached up to 120 mg/L during the month of May 2012.

Nitrate (NO_3) is generally known to show no seasonal variations our study showed similar result as observed by *Pearsall* (1930). Ammonical nitrogen (NH_4^-N) estimated in the present investigation remains not detected throughout the study period.

Coliforms Total coli (TC) and Fecal Coli (FC) are normal inhabitants of digestive tracts of animals, including human and are found in their wastes, besides soil material (SabbaRao, 2004). They are also considered as indicator organisms of water pollution caused by faecal contamination which is a serious problem due to the potential for contacting diseases from pathogens.

The range of total coli form (TC) and Fecal coli form (FC) defected from the river water sample > 1600-1600 and 350-920 MPN/100ml respectively. The present study revealed average TC and FC counts highest in mid-monsoon and lowest in winter. Similar results were observed by Agarwal and Rozgar (2010).

References

Agarwal, A. K., and Rozgar, G. R. 2010. Physico-chemical and Microbiological study of Tehri Dam Reservoir, Garhwal Himalaya, India, *Journal of American Science,* 6.

Agrawal, S. C. 1999. Limnology. A. P. H. Publishing Corporation, New Delhi.

APHA. 2005. Standard methods for the examination of water and wastewater, 21st ed., Washington, DC: American Public Health Association.

Bhattarai, K. R., Shrestha, B. B., and Lekhak, H. D. 2008. Water Quality of Sundarijal Reservoir and its Feeding Streams in Kathmandu. *Scientific World*, 6(6): 99-106.

Dubey, R. C. and Maheshwari, D. K. 2006. Text Book of Microbiology. 1st ed., S. Chand and company Ltd, New Delhi.

George, J. P., Venugopal, G., and VenkateshWaran, I. P. 1986. Anthropogenic eutrophication in a perennial tank: effect on the growth of cypsiscarpio communities. *Indian J. Environ. Health*. 28(4): 303-313.

Hawkes, H. A. 1993. The ecology of wastewatertreatment. Pergamon press, Oxford.

Hofkes E. H, Huisman. L, Sundaresan. B. B, AzevedoNetto De J. M, Lanoix J. N. 1981. Small Community Water, John willet and Sons. 1-299.

Kundangar, M. R. D., Sarwar, S. G., and Hussain, J. 1996. Zoo plankton population and nutrient dynamics of wetlands of wular lake Kashmir, India". In Environment and Biodiversity: In context of South Asia, Jha, P. K., Ghimire, G. P. S., Karmacharya, S. B., Baral, S. R., and Lacoul. 1st ed., Ecological Society (Ecos), Kathmandu, Nepal. 128-134.

Pearsal, W. H. 1930. Phytoplankton in the English Lakes I: The water of dissolved substances of biological importance. *J. Ecol*. 18: 306-320.

Sabata, B. C. and Nayar, M. P. 1995. River Pollution in india: A case study of Ganga River", 33.

SubbaRao, N. S. 2004. Soil Microbiology: Oxford and IBH publishing Co. Pvt. Ltd., New Delhi.

Chapter 14

Qualitative and Quantitative study of Zooplankton in Ganikere Tank of Anandapura Village, Karnataka

☆ *H.A. Sayeswara, H.M. Ashashree*
and R. Purushothama

ABSTRACT

In the present study efforts have been made to ascertain the seasonal abundance and population dynamics of zooplankton community in the Ganikere tank. The study was conducted during January to December 2013. Zooplankton diversity is one of the most important ecological parameters in water quality assessment. Zooplanktons are good indicators of the changes in water quality because they are strongly affected by environmental conditions and respond quickly to changes in water quality. Zooplankton is the intermediate between phytoplankton and fish. Hence qualitative and quantitative studies of zooplankton are of great importance. A total of twenty one species of zooplankton belonging to four groups were recorded. Rotifera constituted the main dominant group in this tank contributing 33.33 per cent of the total zooplankton population followed by cladocera (28.57 per cent), copepoda (23.8 per cent) and protozoa (14.28 per cent). The highest density of total zooplankton was recorded in the month of April, May, November and December being 405, 385, 380 and 372 org./lit. respectively, while the lowest density was recorded during June and July being 105 and 88 org./lit. The presence of three species of Brachionus indicates that the tank is approaching towards eutrophic action and is organically polluted. The information contributed by this investigation will be highly significant and useful in order to create a general awareness in the people to prevent further water pollution and improve aquaculture.

Keywords: *Zooplankton diversity, Ganikere tank, Shivamogga, India.*

Introduction

Zooplanktons are the major trophic link in food chain and being heterotrophic organisms it plays a key role in cycling of organic materials in aquatic ecosystem (Patra *et al.,* 2011). In addition, their diversity has assumed added importance during recent years due to the ability of certain species to indicate the deterioration in the quality of water caused by pollution or eutrophication. Zooplanktons are aquatic microscopic organisms which do not have the power of locomotion (Pandey *et al.,* 2013). Rotifers, cladocerans, copepods and ostracods constitute the major groups of zooplankton. They are also indicating the tropic status of a water body and some of them are also acting as bio-indicators of organic and inorganic pollution of the aquatic environment. Ecologically zooplanktons are one of the most important biotic components influencing all the functional aspects of an aquatic ecosystem such as food chains, food webs, energy flow and cycling of organic matter (Park and Sin, 2007). They usually act as primary consumers and constitute an important link between phytoplanktons and the consumers like fishes in the aquatic food chain. Without these primary consumers, herbivores and other levels of food chain would collapse. Their qualitative and quantitative estimate provides good indices of water quality. Zooplankton is a good indicator of changes in water quality because it is strongly affected by environmental conditions and responds quickly to changes in environmental quality (Sheeba and Ramanujan, 2005). Among zooplankton, rotifers are apparently the most sensitive indicators of the water quality. The density and diversity of zooplankton in freshwater ecosystem is controlled by several factors. Temperature, dissolved oxygen and organic matter are important factors which controls the growth of zooplanktons (Hannazato and Yasuno, 1955). A number of researchers have studied the density and diversity of zooplankton lentic water bodies (Purushothama *et al.,* 2011; Thirupathaiah *et al.,* 2011; Swati *et al.,* 2012; Sayeswara *et al.,* 2013; Anil *et al.,* 2014). Hence study of zooplankton is of great importance. In view of the importance of study of zooplankton, the present study was undertaken to assess the biodiversity and density of zooplankton in Ganikere tank of Anandapura village, Sagara, Shivamogga.

Materials and Methods

Study Area

Ganikere tank (Anandapura village) is a perennial freshwater body situated at about 16 km away from the Sagara town, located between 14° 4' N Latitude and 75° 12' E Longitude. This is medium sized tank, with total water spread of 14.38 hectare, where rain is the main source of water. The river basin of the tank is Krishna. The water has undergone moderate changes in the physico-chemical properties due to overflowing of water from adjacent paddy fields and other excessive human activities. The water is used for agricultural purpose.

Collection of Samples

Zooplanktons were collected monthly from Ganikere tank for a period of one year from January to December 2013. Sampling was made between 9.00 am to 11.30 am. 20 L of water was filtered by passing water through plankton new made up of bolting silk cloth having mesh size of 25 µm. Filtrate was stored in 20 ml plastic bottles and 5 per cent formalin was added for sample preservation. These samples were then brought to laboratory for further studies.

Identification and Estimation of Zooplankton

The zooplanktons were observed under research microscope and photographed using digital camera. For the quantitative study of zooplankton, a Sedgwick Rafter Counting Cell was used and identification of zooplankton was done with the help of monographs and workshop manual (Battish, 1992 and Needham and Needham, 1941). The results expressed as number of organisms/litre.

Analysis of Zooplankton Community

To under biotic community it is very important to work out some indices of species structure. Environmental indices have been used to monitor the quality of the environment. Six indices, Shannon diversity index, Simpson's diversity index, Margalef's index, Equitability index, Fisher's alpha index and Berger-Parker Dominance index were employed and estimated in the present investigation.

Results and Discussion

Zooplankton recorded in present study was represented by four taxonomic groups – rotifera, cladocera, copepoda and protozoa. In the present study the zooplankton populations of Gowrikere tank have been presented in Table 14.1. Statistical analysis of groups of Zooplankton is given in Table 14.2. Twenty one species of Zooplankton were recorded belonging to rotifera (7 species), cladocera (6 species), copepoda (5) and protozoa (3 species). Rotifera was the dominant group composing 33.33 per cent of total zooplanktons, cladocera constitutes 28.57 per cent, copepoda constitutes 23.80 per cent and protozoa composing 14.28 per cent of total zooplankton population. The highest density of total zooplankton was recorded in the month of April, May, November and December being 405, 385, 380 and 372 org./ lit. respectively, while the lowest density was recorded during June and July being 105 and 88 org./lit. Zooplanktons of different groups and comparison of different groups is depicted in Figure 14.22 and 14.23. In the present investigation, rotifers invariably constitute a dominant component of the studied tank.

Rotifers are small zooplankton that occurs in freshwater, brackish and marine environments. Rotifers feed on microalgae and are consumed by a wide variety of fishes. They are used extensively in aquaculture and aquariums because their very high reproductive rates. Rotifera was represented by 5 genera and 7 species. *Branchionus* were represented by 3 species, *Fillinia, Keratella, Lepadella* and *Trichocera* by a single species each. Presence of *Brachionus* sp. Indicates that the pond is approaching towards

Table 14.1: Seasonal Variation of Individual Zooplankton Species in Ganikere Tank from April 2011 to March 2012

Taxonomic Groups and Species	Months – 2011													
	Jan.	Feb.	Mar.	Apr.	May	June	July	Aug.	Sep.	Oct.	Nov.	Dec.	Mean	S.E
ROTIFERA														
Branchionus caudatus	12	23	19	27	31	2	5	1	1	22	26	37	17.16	± 12.60
Branchionus longirostris	8	11	12	9	10	–	1	1	2	3	7	16	7.27	± 4.98
Branchionus calciflorus	10	9	16	8	18	2	1	3	5	8	13	19	9.33	± 6.12
Fillinia longiseta	5	6	6	7	8	2	1	3	4	5	3	9	4.91	± 2.42
Keratella tropica	20	19	39	56	33	3	2	1	4	24	31	18	20.83	± 17.02
Lepadella ovalis	13	17	10	26	10	2	3	1	2	13	21	14	11	± 8.01
Trichocera similis	12	9	17	26	21	2	1	2	3	9	14	11	10.58	± 7.99
Total	80	94	119	159	131	13	14	12	21	84	115	124		
CLADOCERA														
Alona pulchella	12	16	6	1	2	9	22	13	2	11	13	9	9.66	± 6.24
Diaphanosoma sarsi	9	13	2	3	1	1	2	6	2	9	11	7	5.5	± 4.23
Daphnia carinata	7	13	4	1	1	9	3	4	2	7	13	5	5.75	± 4.18
Macrothrix goeldi	9	8	6	5	2	6	4	12	14	7	11	10	7.83	± 3.51
Moina bachiata	15	13	10	9	7	2	3	6	11	15	13	19	10.25	± 5.11
Moina micrura	24	32	45	47	39	5	4	19	3	19	25	31	24.41	± 15.25
Total	76	95	73	66	52	32	38	60	34	68	86	81		

Contd...

Table 14.1–Contd...

Taxonomic Groups and Species	Jan.	Feb.	Mar.	Apr.	May	June	July	Aug.	Sep.	Oct.	Nov.	Dec.	Mean	S.E
COPEPODA														
Cyclops vicinus	11	13	9	14	20	5	4	2	17	12	15	9	10.91	±5.40
Heliodiaptomus vidus	6	8	6	7	9	2	1	–	–	3	6	5	5.3	±2.73
Mesocyclops leuckarti	9	13	11	14	19	3	1	7	15	11	19	21	11.91	±6.25
Neodiaptomus stregilipes	18	15	11	19	12	5	4	2	9	13	13	14	11.25	±5.36
Tropocyclops prasinus	19	12	13	17	9	1	1	3	7	11	17	13	10.25	±6.19
Total	63	61	50	71	69	16	11	14	48	50	70	62		
PROTOZOA														
Euglena gracilis	30	42	51	49	62	20	11	16	26	29	45	39	35	±15.52
Paramecium caudatum	26	39	56	41	48	18	9	19	25	34	41	40	33	±13.74
Vorticella compunulata	11	9	14	19	23	6	5	9	29	32	23	32	17.66	±10.01
Total	67	90	121	109	133	44	25	44	80	95	109	111		
Grand total	**286**	**340**	**363**	**405**	**385**	**105**	**88**	**130**	**183**	**297**	**380**	**378**		

eutrophication and is organically polluted. This is in agreement with Ahmad (2011), Mola (2011) and Uzma Ahmad (2012) they mentioned that rotifers, especially, *Branchionus* sp. are the major component of zooplankton in eutrophic water bodies. The population density of rotifers reached its peak in April 2013 with 159 org./lit., while in August 2013 it was least with 12 org./lit. The seasonal variation of individual Rotifers is depicted in Figure 14.1 to Figure 14.7.

Table 14.2:Statistical Analysis of different Groups of Zooplankton in Ganikere Tank

Taxonomic Groups	Taxa_ S	Indivi- duals	Shannon_ H	Simpson_ I-D	Mar- galef	Equitability_ J	Fisher- Alpha	Berger- Parker
Rotifera	7	81	1.848	0.8282	1.365	0.9495	1.838	0.22
Cladocera	6	60	1.671	0.7448	1.221	0.9328	1.665	0.21
Copepoda	5	47	1.607	0.7643	1.039	0.9987	1.415	0.23
Protozoa	3	85	1.059	0.6366	0.4502	0.9641	0.606	0.20

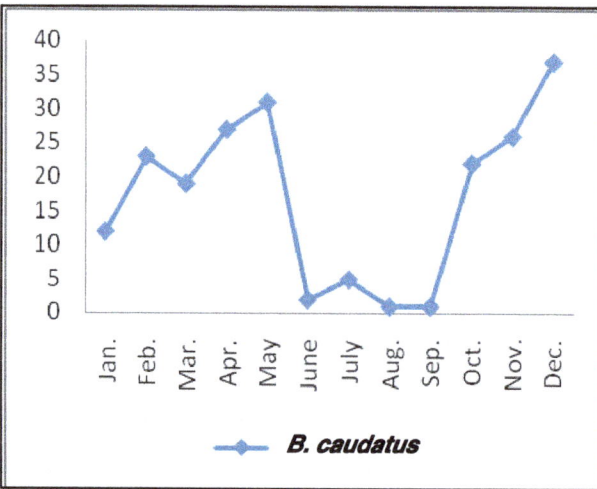

Figure 14.1: Monthly Variation of *B. caudatus*.

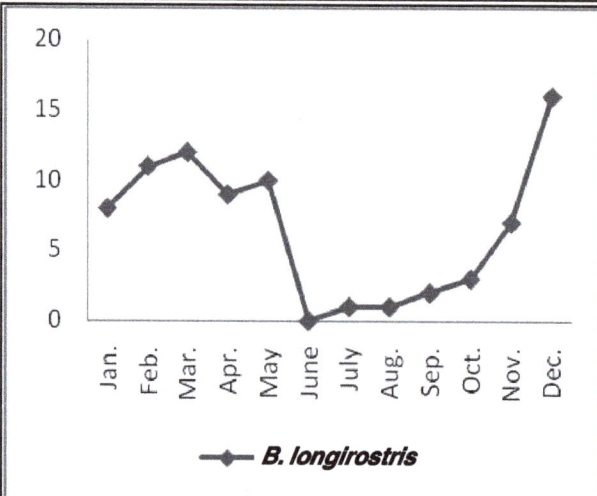

Figure 14.2: Monthly Variation of *B. longirostris*

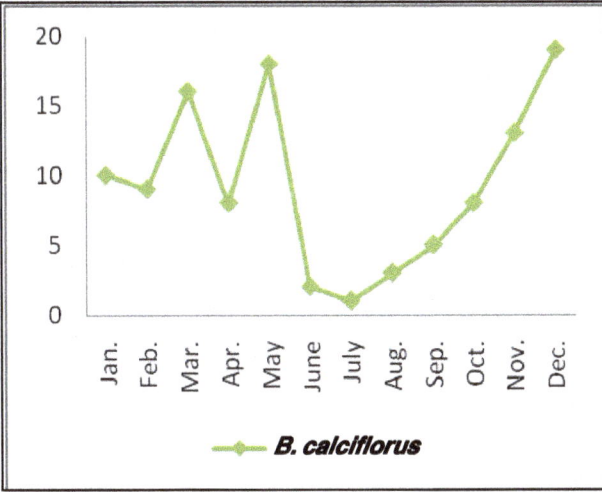

Figure 14.3: Monthly Variation of *B. calciflorus*.

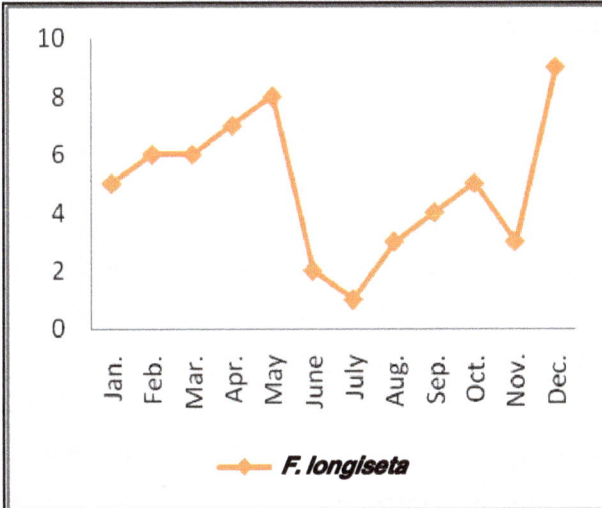

Figure 14.4: Monthly Variation of *F. longiseta*.

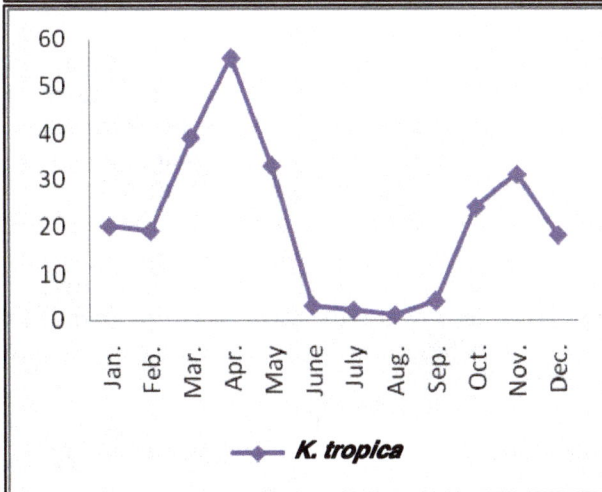

Figure 14.5: Monthly Variation of *K. tropica*.

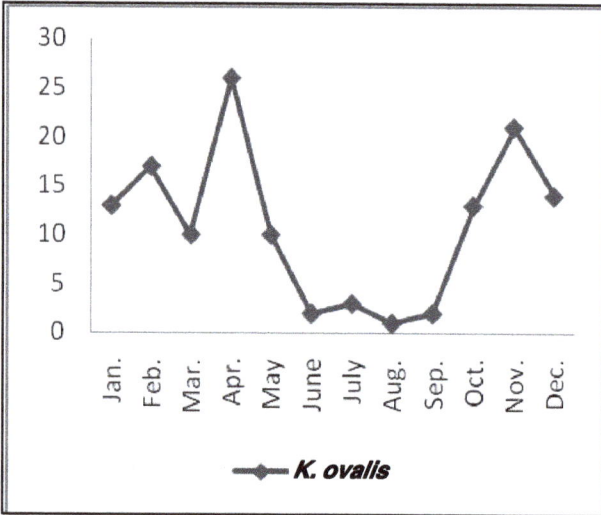

Figure 14.6: Monthly Variation of *K. ovalis*.

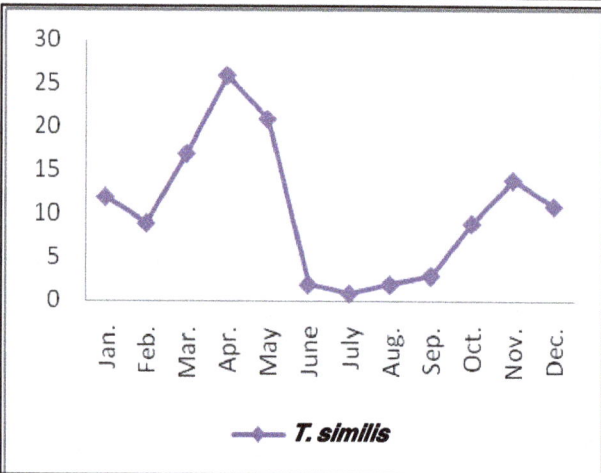

Figure 14.7: Monthly Variation of *T. similis*.

Cladocera is an order of small crustaceans commonly called water fleas. Cladocera formed the second most abundant group of zooplankton. Cladocera was represented by 5 genera and 6 species. *Moina* were represented by 2 species, *Alona*, *Diaphanosoma*, *Daphnia* and *Macrothrix* by a single species each. Abundance of cladoceran ranges from minimum of 32 org./lit. in June 2013 to a maximum of 95 org./lit. in February 2013. The seasonal variation of individual Cladoceran is depicted in Figures 14.8–14.13. Cladocerans are known to be abundant in water with good littoral vegetation, while ponds and lakes without vegetation have fewer cladoceran species (Idris and Fernando, 1981). Decay of vegetation during summer may serve as food, thus maximum during that season. Low densities during other season may be due to predation by copepoda (Hassen, 2003).

Copepoda formed the third most abundant group of zooplankton. Copepods are a group of tiny crustaceans, so they are cousins of crayfish and water fleas. You can

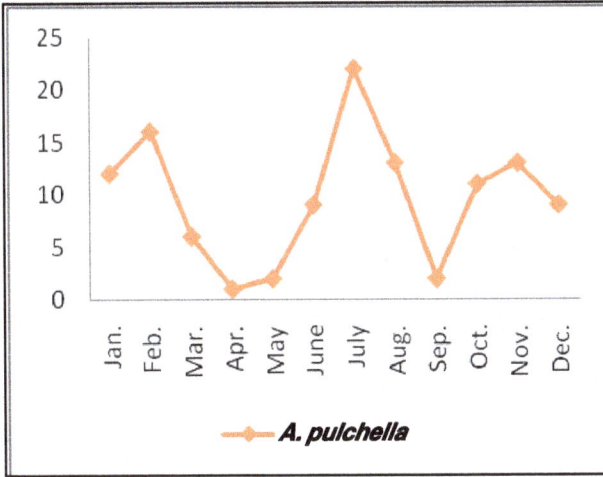

Figure 14.8: Monthly Variation of *A. pulchella*.

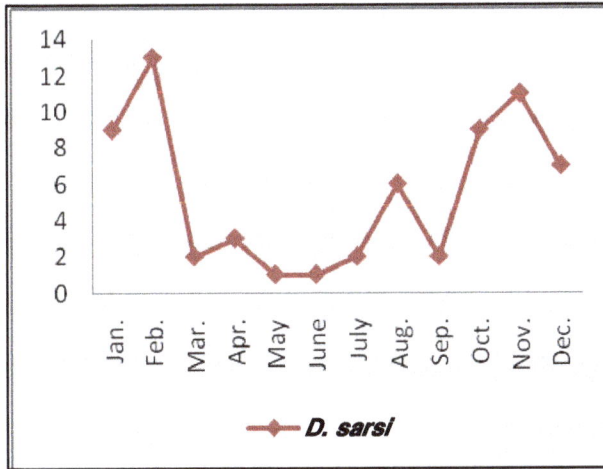

Figure 14.9: Monthly Variation of *D. sarsi*.

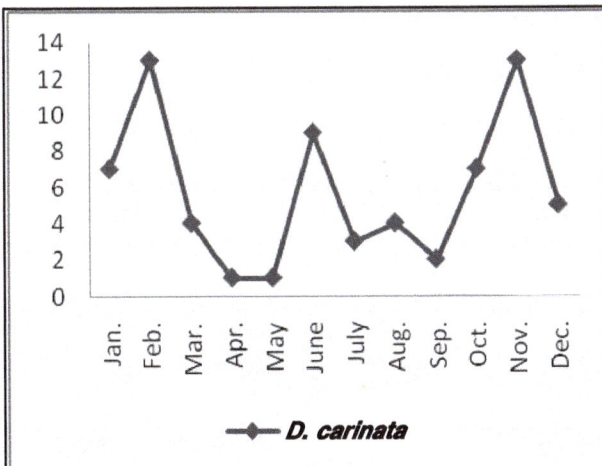

Figure 14.10: Monthly Variation of *D. carinata*.

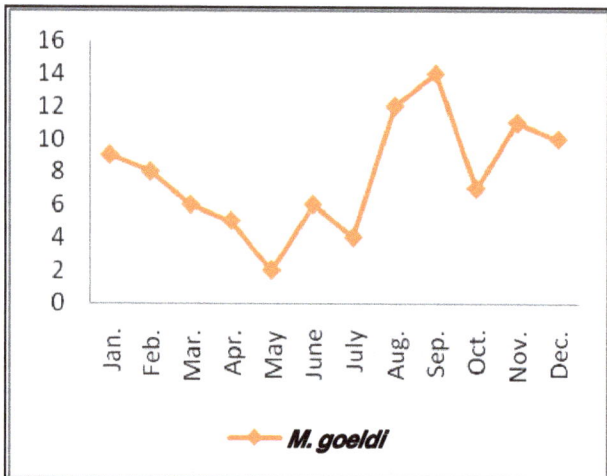

Figure 14.11: Monthly Variation of *M. goeldi*.

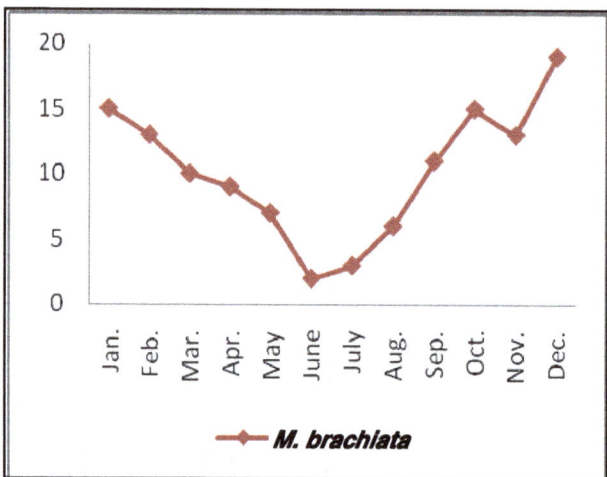

Figure 14.12: Monthly Variation of *M. brachiata*.

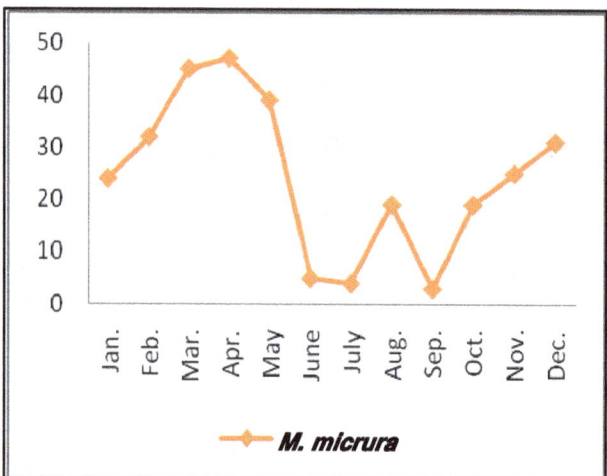

Figure 14.13: Monthly Variation of *M. micrura*.

see them with your eyes. Copepoda was represented by 5 genera and 5 species. *Cyclops, Heliodiaptomus, Mesocyclops, Neodiaptomus* and *Tropocyclops* were represented by a single species each. Monthly density of copepods recorded a minimum 11 org./lit. during July 2013 and maximum of 71 org./lit. during April 2013. The seasonal variation of individual copepods is depicted in Figures 14.14–14.18. Copepods are regarded as pollution sensitive zooplankton as they disappear from polluted water (Verma *et al.,* 1984).

Protozoa are a diverse group of unicellular eukaryotic organisms, many of which are motile. Comparatively protozoan population was less which represent 3 genera and 3 species. *Euglena, Paramecium* and *Vorticella* were represented by a single species each. When monthly variations of protozoans are considered, Gowrikere tank recorded a minimum of 25 org./lit. during July 2005 and maximum of 133 org./lit. during May22 2013. The seasonal variation of individual Protozoans is depicted in Figures 14.19–14.21.

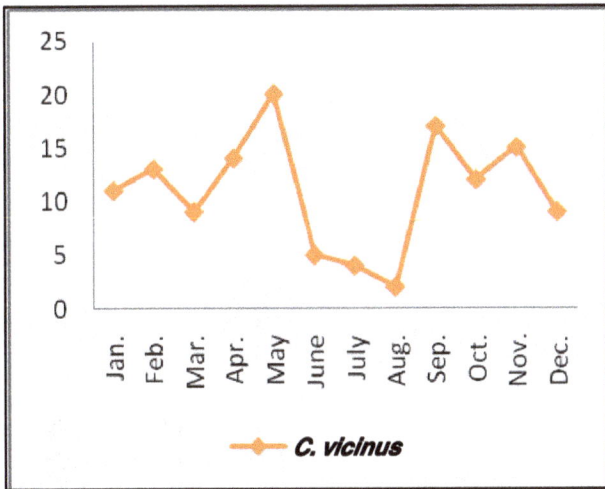

Figure 14.14: Monthly Variation of *C. vicinus.*

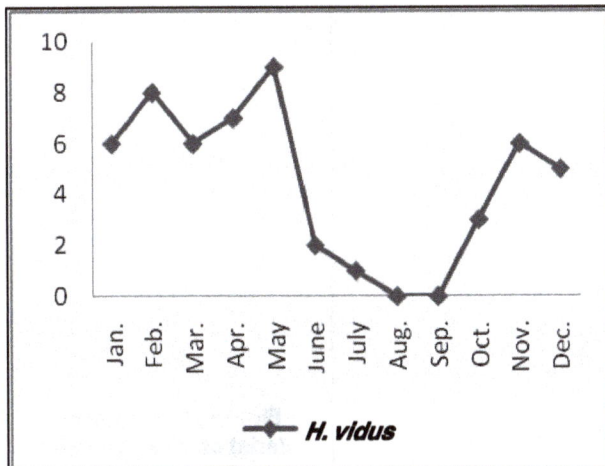

Figure 14.15: Monthly Variation of *H. vidus.*

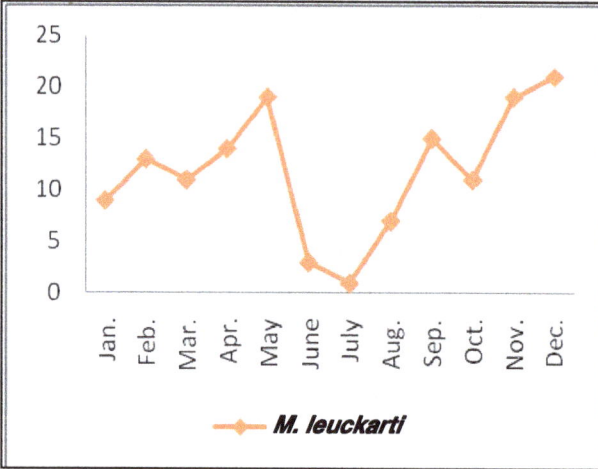

Figure 14.16: Monthly Variation of *M. leuckarti.*

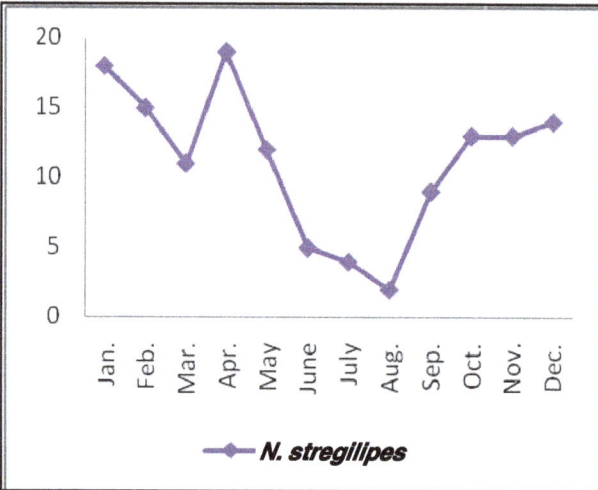

Figure 14.17: Monthly Variation of *N. stregilipes.*

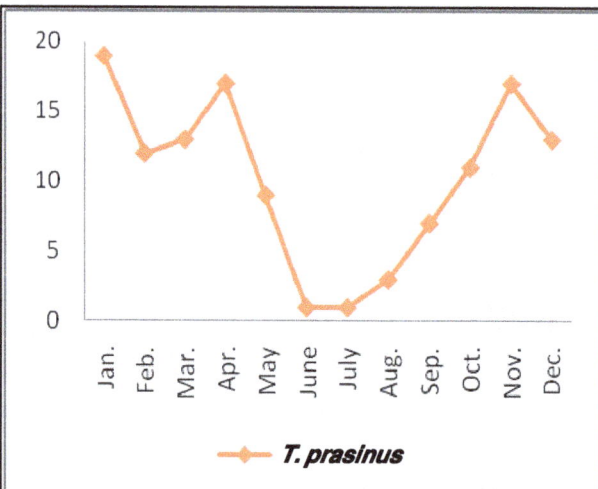

Figure 14.18: Monthly Variation of *T. prasinus.*

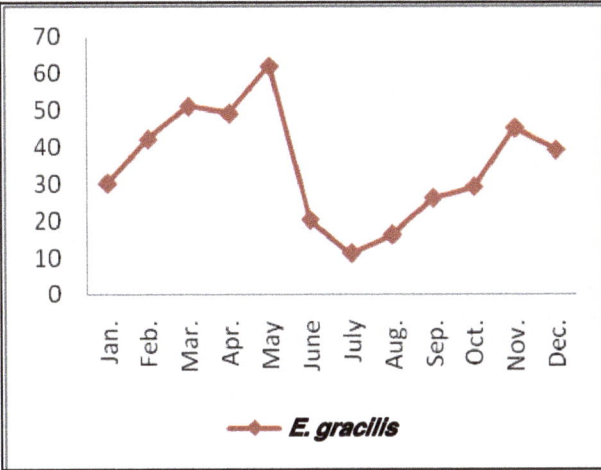

Figure 14.19: Monthly Variation of *E. gracilis*.

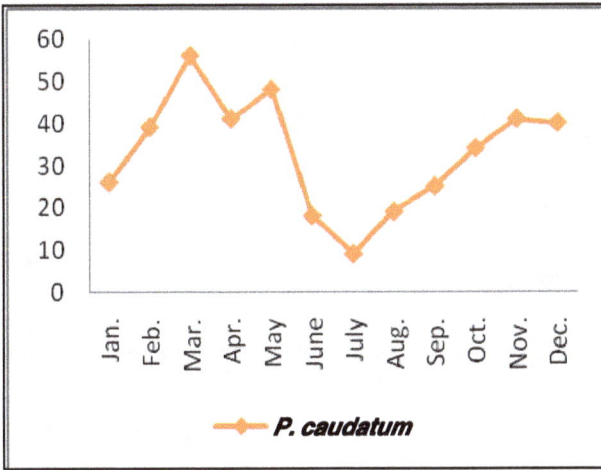

Figure 14.20: Monthly Variation of *P. caudatum*.

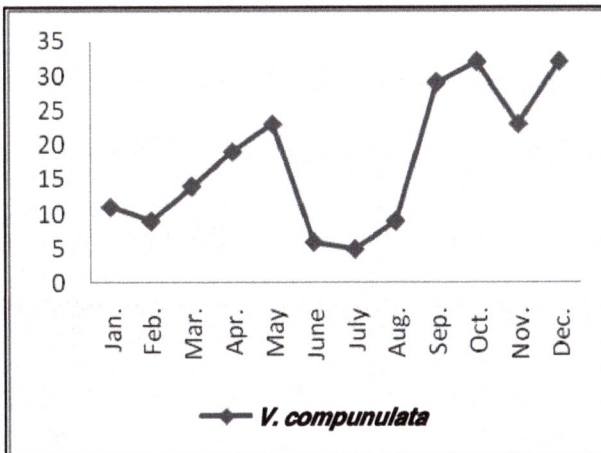

Figure 14.21: Monthly Variation of *V. compunulata*.

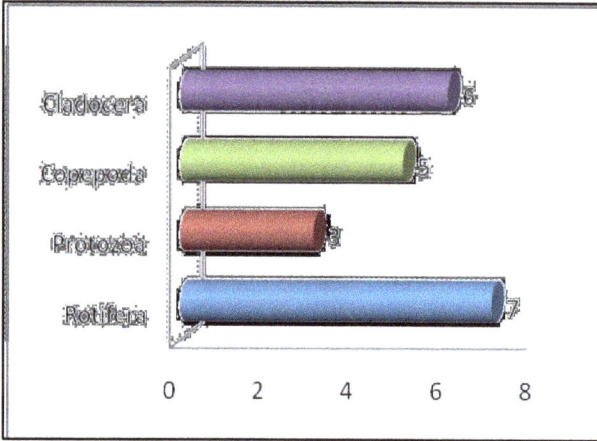

Figure 14.22: Zooplankton of different Groups.

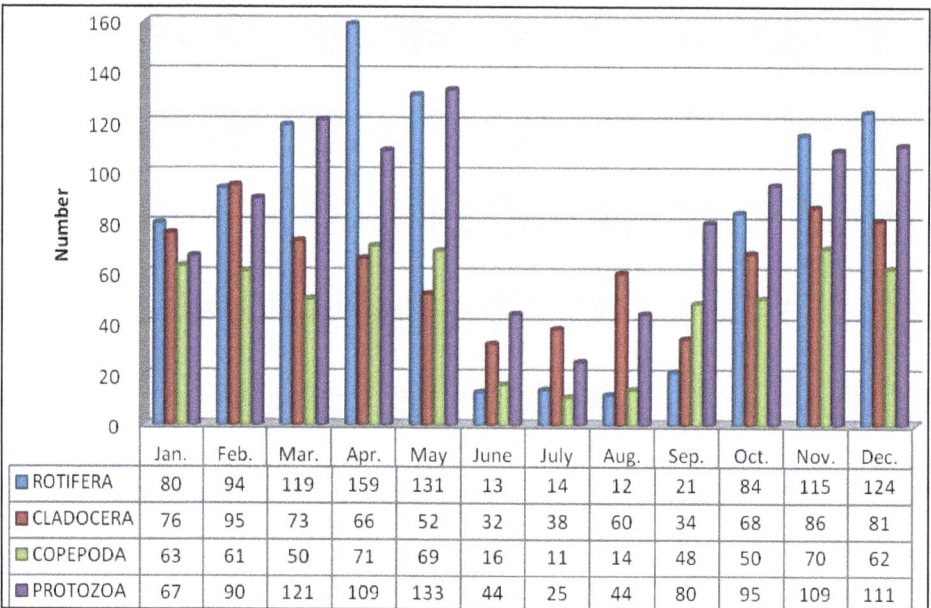

	Jan.	Feb.	Mar.	Apr.	May	June	July	Aug.	Sep.	Oct.	Nov.	Dec.
ROTIFERA	80	94	119	159	131	13	14	12	21	84	115	124
CLADOCERA	76	95	73	66	52	32	38	60	34	68	86	81
COPEPODA	63	61	50	71	69	16	11	14	48	50	70	62
PROTOZOA	67	90	121	109	133	44	25	44	80	95	109	111

Figure 14.23: Monthly Variation of different Groups.

For the present data we carried out basic diversity indices and we found rotifera is the dominant taxa followed by cladocera, copepoda and protozoa. Shannon diversity index (H) has been a popular index in the ecological literature. It is commonly used to characterize species diversity in a community. Shannon diversity index accounts for both abundance and evenness of the species present. The values of H were in between 1.059 to 1.848, being maximum in Rotifers (1.848) and minimum in protozoa (1.059). High values of H would be representative of more diverse communities. The decrease in the value of H is considered as an evidence of pollution. In the present study, Shannon diversity index (H) value was more than 1. Hence, it indicates the more diversity of zooplankton in the tank with moderate pollution.

Simpson's diversity index (D) is measure of diversity. In ecology, it is often used to quantify the biodiversity of a habitat. It takes in to account the number of species present, as well as the abundance of each species. The higher the value of D, the greater the diversity. The maximum value is the number of species in the sample. The values of D fluctuated between 0.6366 (protozoa) to 0.8282 (rotifera). Simpson's diversity index gives more weight to the more abundant species in a sample. The addition of rare species to a sample causes only small changes in the value of D.

Margalef's index was used as a simple measure of species richness. Margalef's species richness was maximum in rotifers (1.365) and minimum in protozoa (0.4502). The higher the index the greater the diversity. The Equitability index concept is a useful one and permits considerable refinement in diversity studies. The Equitability index was maximum in Copepoda (0.9987) and minimum in cladocera (0.9328).

Fisher's alpha is a parametric index of diversity that assumes that the abundance of species follows the log series distribution. Fisher's alpha was maximum in rotifers (1.838) and minimum in protozoa (0.606). Berger-Parker Dominance (d) is simple measure of the numerical importance of the most abundant species. Berger-parker Dominance was maximum in copepoda (0.23) and minimum in protozoa (0.20). An increase in the value of Berger-Parker Dominance accompanies an increase in diversity and a reduction in dominance. Basic diversity indices indicate that rotifers and cladocera species may be mutualistic during the study period.

Conclusion

Depending upon the study it can be concluded that the diversity and density of zooplanktons from Ganikere tank exhibited by four major groups. Rotifera constituted the main dominant group in this tank contributing 33.33 per cent of the total zooplankton population followed by cladocera (28.57 per cent), copepoda (23.8 per cent) and protozoa (14.28 per cent). Although zooplankton exists under a wide range of environmental conditions, yet many species are limited by physical and chemical factors of the aquatic environment. In the present study, Shannon diversity index (H) value was more than 1. Hence, it indicates the more diversity of zooplankton in the tank with moderate pollution. The presence of three species of *Brachionus* indicates that the Ganikere tank is approaching towards eutrophic action and is organically polluted.

Different species of zooplankton showed their abundance according to the favorable condition, so they disappear in unfavorable conditions and reappeared on return of favorable condition. Statistical analysis reveals the status of the tank and it needs regular monitoring to know the influence of biotic, abiotic and unnoticed factors. The information contributed by this investigation will be highly significant and useful in order to create a general awareness in the people to prevent further water pollution and improve aquaculture.

Acknowledgements

The authors express their gratitude to Dr. K.L.Naik, Chairman, Department of Zoology and Prof. G.Shankunthala, Principal, Sahyadri Science College, Shivamogga for providing facilities and encouragement.

References

Ahmad, U., Praveen, S., Khan, A. A., Kabir, H. A., H. R. A. Mola, H. R. A., Ganai, A. H. 2011. Zooplankton population in relation to Physico-chemical factors of a sewage fed pond of Aligarh (UP), India. *Biology and Medicine* 3: 336-341.

Anil, K. T., Girish Chopra and Seema Kumari. 2014. Zooplankton diversity in shallow lake of Sultanpur National park. Gurgaon (Haryana). *International Journal of Applied Biology and Pharmaceutical Technology* 5(1): 35-40.

Ashis Patra, Kalyan, B. S., Chanchal Kumar, M. 2011. Ecology and diversity of zooplankton in relation to Physico-chemical characteristics of water of Santragachi Jheel, West Bengal, India. *J. Wetlands Ecology* 5: 20-39.

Battish, S. K. 1992. Freshwater zooplankton of India. Oxford and IBH Publ. Co., New Delhi.

Hannazato, Yasuno. (1985). Population dynamics and production cladoceran zooplanktons in the highly eutrophic lake Kasumigaura. *Hydrobiologia* 124: 13-22.

Haseen, D. O. 2003. Phyytoplankton constibution to seston mass and elemental ratio in lakes: Implication for Zooplankton nutrition. *Limnology and Oceanography* 48: 1289-1296.

Idris, B. A. H., Fernando, C. H. 1981. Cladocera of Malaysia and Singapore with new records, redescription and remarks on some species, *Hydrobiologia*, 77: 233-256.

Mola, H. R. 2011. Seasonal and spatial distribution of Brachionus, a bioindicator of eutrophication in lake EI-Manzalah, Egypt. *Biology and Medicine* 3: 60-69.

Needham, J. M., Needham, P. R. 1941. A guide to the study of freshwater biology, Comstock, Theca, New York, USA.

Pandey, B. N., Siddhartha, R., Tanti, K. D., and Thakur, A. K. 2013. Seasonal variation is Zooplanktonic community in swamp of Purnia (Bihar), India. *Aquatic Biology Research* 1(1): 1-9.

Park, K. S., Sin, H. W. 2007. Studies on Phytoplankton and Zooplankton composition and its relation to fish productivity in west coast fish pond production. *Journal of Environmental Biol*ogy 28: 415-422.

Purushothama, R., H. A. Sayeswara, H. A., Goudar, M. H. 2011. Dynamics of zooplankton diversity in relation to water Heggere tank, Kanale, Sagara, Karnataka, India. *Environment Conservation Journal* 12(1): 29-34.

Sayeswara, H. A., Vasantha Naik, T., Nafeesa, B. 2013. Zooplankton biodiversity study of Gowri tank, Anandapura, Sagara, Shivamogga, Karnataka, India. *Research Journal of Pharmaceutical Biological and Chemical Sciences*, 4(3): 580-584.

Sheeba, S., Ramanujan, N. 2005. Qualitative and quantitative study of Zooplankton in Ithikkara river, Kerala. *Poll. Res.* 24(1): 119-122.

Swati, J., Sunitha, B., Dilip, J., Atul, H. 2012. Seasonal variations of Zooplankton community in Sina Kolegoan Dam Osmanabad district, Maharashtra, India. *Journal of Experimental Sciences* 3(5): 19-22.

Thirupathaiah, H., Samatha, C. H., Sammaiah, C. 2011. Diversity of Zooplankton in freshwater Lake of Kamalapur, Karimnagar district (A. P.) India. *The Ecoscan* 5(1 and 2): 85-87.

Uzma Ahmad, Saltanat Praveen, Hesham, R. A. M., Habeeba, A. K., Altaf, H. G. 2012. Zooplankton population in relation to Physico-chemical parameters of Lal Diggi pond in Aligarh, India. *Journal of Environmental Biology* 33: 1015-101.

Verma, S. R., Sharma, P., Tyagi, A., Rani, S., Gupta, A. K., Dalela, R. C. 1984. Pollution and saprobic status of Eastern Kalinandi. *Limnologica.* 15: 69-133.

Chapter 15

Diversity of Zooplanktons of Nalganga Reservoir in Buldana District, Maharashtra

☆ *G.A. Malthane*

ABSTRACT

The plankton plays a very important role for maintaining the productivity of the water body. During present investigation 20 species of zooplankton belonging to protozoa, rotifera, cladocera and copepoda were identified from Nalganga Reservoir of Buldana district in Maharashtra.These groups are represented in order of dominance as rotifera>cladocera>copepoda>protozoa.

Keywords: *Zooplankton diversity, Nalganga reservoir, Buldana district.*

Introduction

The zooplankton organisms occupy a central position in the food webs of aquatic ecosystem. They play a significant role in aquatic system as consumers. They are capable of affecting the entire aquatic biota. The studies of zooplanktons have been described by Welch (1948), Ruther (1963), Govind (1963), Franklin (1969), Singh and Desare (1980), Sharma and De-wan (1989). The aim of the present study was to investigate the species composition, distribution of diversified zooplankton assemblages in Nalganga Reservoir, Nalgangapur. The fluctuation of the zooplanktonic assemblages in Nalganga Reservoir was studied in the Year 2011-2012.

Materials and Methods

Nalganga Reservoir is located at Malkapur-Buldana Road, about 17-Km from Malkapur city. It is a man-made reservoir situated in the Buldana District of Maharastra. The plankton samples were collected as per earlier workers, Lind (1974) and Welch (1953) by filtering 50 litres of water through a plankton net having pore size 64 m. concentrated plankton samples were fixed in 5 per cent formalin solution. They were identified with the help of Keys, provided by Pennak (1978); Needham and Needham (1962), Tonapi (1998) and APHA (1991).

Results and Discussion

In the present study total zooplankton comprises of five groups protozoa, rotifera, cladocera, ostracoda and copepoda. Out of these groups rotifera is dominant group and is represented in the order of dominance as rotifera >cladocera >copepoda > ostracoda > protozoa.

The rotifers are microscopic soft-bodied freshwater zooplanktons. They indicate trophic status of water body. In the present study the major peak in rotifer populations is recorded during March and April, and minor peak in October. Many workers Choubey (1991) Ganpati and Pathak (1969), Sharma (1993) Moitra and Bhowmick (1968) reported the rotifer groups in their studies on the different water bodies in India. George (1966); Michael (1968) observed maximum number of rotifers during summer months. The present study also showed this condition in Nalganga Reservoir. The high rotifer densities in summer seasons may be due to reduced water volume and their by increased concentration of nutrients.

The cladocerans are of commonly occurrence in almost all the freshwater bodies. They represent an important link in the aquatic food chain. This group also showed major peak in May and June, and minor peak in September and October. It is second dominating group of zooplankton in the present study. Govind (1978), Ganpati and Pathak (1979), Sharma (1993) reported cladoceran population as second dominant from various freshwater bodies.

The copepods are major links in the aquatic ecosystem. The copepod population ranked third in order of dominance during present study. This group showed major peak in April and May and the minor peak in January and December. Sharma (1980) reported the bimodal pattern in copepod population as reported in the present study.

The ostracoda also form a major link in the aquatic ecosystem. The ostracoda population ranked fourth in order of dominance during present study. This group showed major peak in summer and minor peak in winter.

Protozoans are also important members in food chain in an aquatic ecosystem. In the present study the maximum protozoan population was observed during March and April, and minimum in June. This group ranked fifth in order of dominance.

Thus the present study deals with the abundance and dominance of zooplanktonic groups, which revealed rotifers as dominant group of zooplanktons in Nalganga Reservoir, Nalgangapur.

Table 14.1: Occurrence of Zooplankton in the Nalganga Reservoir, Nalgangapur

Name of Group and Species	J	F	M	A	M	J	J	A	S	O	N	D
PROTOZOA												
1. *Arecello* sp.	–	+	+	+	+	–	–	–	–	–	+	+
2. *Ceratium* sp.	+	+	+	+	+	+	–	–	–	–	+	+
3. *Euglypha* sp.	+	+	+	+	+	+	–	–	–	–	+	+
4. *Opercularia*	–	–	+	+	+	–	–	–	–	–	+	+
5. *Diffugia* sp.	–	–	+	+	+	–	–	–	–	–	–	+
ROTIFERA												
1. *Branchious auadridentatus*	+	+	+	+	+	+	–	+	+	+	+	+
2. *Branchious divesicornis*	+	+	+	+	+	+	–	–	–	–	+	+
3. *Branchious caudatus*	+	+	+	+	+	+	–	–	–	–	+	+
4. *Branchious forficula*	+	+	+	+	+	+	–	–	–	–	+	+
5. *Asplanchna* sp.	+	+	+	+	+	+	–	–	+	+	+	+
6. *Testinella* sp.	+	+	+	+	+	+	–	–	–	–	–	–
7. *Horella* sp.	+	+	+	+	+	+	+	–	–	–	+	+
8. *Filina* sp.	+	+	+	+	+	+	+	–	–	–	+	+
9. *Hexarthra* sp.	+	+	+	+	+	+	+	–	–	–	+	+
10. *Conochilus* sp.	+	+	+	+	+	+	–	+	+	+	+	+
11. *Monostyla* sp.	+	+	–	–	–	–	–	–	–	–	–	+
12. *Keratella tropica*	+	+	+	+	+	+	–	–	–	–	+	+
13.*Notholca* sp.	+	+	+	+	+	+	–	–	–	–	–	+
CLADOCERA												
1. *Leydigia* sp.	+	+	+	+	+	+	–	–	–	–	+	+
2. *Chydrous sphaerieus*	+	+	+	+	+	+	–	–	+	+	+	+
3. *Bosmina* sp.	+	+	+	+	+	+	–	–	–	–	+	+
4. *Macrothria laticoruis*	+	+	+	+	+	+	–	–	+	+	+	+
5. *Monia Brachiata*	+	+	+	+	+	+	+	–	+	+	+	+
6. *Diaphenosoma* sp.	+	+	+	+	+	+	+	–	–	+	+	+
7. *Diaphnia* sp.	+	+	+	+	+	+	–	–	–	+	+	+
OSTRACODA												
1. *Cypris* sp.	+	+	+	+	+	–	–	–	–	+	+	+
2. *Cyprinotus* sp.	+	+	+	+	+	–	–	–	–	+	+	+
COPEPOD												
1. *Mesocyclops hyalins*	+	+	+	+	+	+	–	–	–	–	+	+
2. *Mesocyclops* sp.	+	+	+	+	+	+	–	–	–	+	+	+
3. *Phyllodiaptomus* sp.	+	+	+	+	+	+	–	–	–	+	+	+

+: Present; –: Absent.

References

Ayyappan, S. and Gupta, T. R. 1980. Limnology of Ramasamundra tank; J. *Inland Fish Society India,* 12 (2): 1-12.

Choubey, V. 1991. Studies on physico-chemical and biological parameters of Gandhi Sagar Reservoir Ph. D Thesis Vikram University Ujjain pp. 244.

Das, S. M. 1989. Handbook of Limnology and water pollution. South Asian Publisher Pvt. Ltd, 174Pp.

Ganapati, S. V. and C. H. Pathak, C. H. 1969. Primary productivity in the Sayaji Sarovar (a man made lake) at Baroda. Proc. Sem. Eco. And Fish. Freshwater Reservoir ICAR at CIFRI, Barrackpore 27-29.

George, M. G. 1966. Comparative plankton ecology of five tanks in Delhi, *Hydrobiologia* 27(2): 81-108.

Govind, B. V. 1978. Planktonological studies in the Tungabhadra Reservoir and its comparison with other stor-age reservoirs in India: Proc. Semi. Eco. And Fish freshwater reservoir, ICAR at CIFRI, Barrackpore 66-72.

Hutchinson, G. E. 1967. A treatise on Limnology Vol. II Introduction on lake biology and the limnoplankton. John Wiley and Sons Inc. New York 1115.

Lind, O. T. 1979. Handbook of common methods in Limnology C. V. Moshy Co. 2nd Edition, St Louis, Missouri, pp. 199.

Pathan, S. *et al.* 2002. Some physicochemical parameters and primal productivity of river Ramgang: (Uttaranchal). *Him. J. Env. Zool,* 16(2): 151-158.

Sunder, S. 1998. Ecology and Pollution of Indian Rivers. (Ed) R. K. Trivedy, Ashih Publ. House, New Delhi pp. 131.

Chapter 16

Zooplankton Composition in Reservoirs of Buldhana District, Maharashtra

☆ B.V. Patil

ABSTRACT

Zooplankton composition of 19 reservoirs in Buldhana district in Maharashtra state has been studied in post and pre monsoon season of 2011-2012. A total of 71 species were recorded, 38 of them belonging to rotifera, 22 cladocera, 7 copepoda and 4 ostracoda. Of these, five species i.e one cladoceran and four rotiferan were reported for the first time from this area. A maximum 29 species were recorded from Dahid reservior and only 5 species were recorded from Gyanganga reservoir. Frequently recorded rotiferans were *Keratella tropica. Habrotrocha bidens, Brachionus calyciflorus. B. diversicornis, Keratella cochlearis, Lecane luna, L. leontia, Fillinia longisela* and *F. opoliensis.* Common cladocerans were *Diaphanosoma excum, Ceriodaphnia cornuta, Moina micrura* and *M. macracopa.* The regular copepods were *Neodiaptomus strigilipes. Heliodiaptomus viduus, Paracyclops fimbiatus, Tropocyclops prasinus* and *Mesocyclops leuckarti.* Common ostracods were *Hemicypris fosuculla, Llyocypris gibba* and *Darwinula* sp. The species richness was high in monsoom and season compared to post monsoon season. In Dahid reservoir 15 species were found in post monsoon season whereas 29 species were recorded in premonsoon season. Other reservoirs also showed same trend.

Keywords: Cladocerans, Copepods, Buldhana, Maharashtra, Rotifers, Species composition.

Introduction

Buldhana district covers a geographical area of 9640 km^2. The average annual rainfall is 792.5 mm, and the temperature ranges from 18°C to 42°C. Studies on freshwater biodiversity in Buldhana district are scarce. There is only one survey report on the water quality of reservoirs in Buldhana district. (Uttangi, 2001).

Zooplankton occupies an important position in the tropic structure and play a major role in the energy transfer of an aquatic ecosystem. An inadequate knowledge of the zooplankton and their dynamics is a major handicap for better understanding of life processes of freshwater bodies. Pederson *et al.* (1976) have indicated the importance of such studies since eutrophication is bound with the components and production of zooplankton. Discharge of urban, industrial and agricultural wastes have increased the quantum of various chemicals that enters the waters, which considerably alter their physico-chemical chemical characteristics. Phosphorus and Nitrogen inputs from domestic wastes and fertilizers accelerate the process of eutrophication, which alter food chain sequences leading to production of commercially less valuable higher trophic organism (Rao *et al.,* 1994). One of the main difficulties in studying loss of biodiversity due to eutrophication is the absence of previous records of species composition. The present work was undertaken to study the zooplankton composition of some reservoirs of Buldhana district.

Materials and Methods

Zooplankton samples were collected from different reservoirs of Buldhana district (19.98'N and 76.51'E). The collection was made in October 2011 (post monsoon) and March 2012 (pre monsoon). As most of the reservoirs in Buldhana district dry in March only few sample were collected. The collection was made with horizontal net-tows in the littoral zone using plankton net of 68μm mesh and samples were fixed in 4 per cent formaldehyde. Organisms were identified to the possible taxonomic level (Genus/species), using an optical microscope and a specialized bibliography (Edmondson, 1959; Dhanapathi, 1974, 1976; Dumont and Velde, 1977; Dumont, 1983; Sharma, 1979, 1980, 1983, 1987; Sharma and Michael, 1980, 1987; Rajapaksha and Fernando, 1982; Fernando and Kanduru, 1984; Patil and Gouder, 1982a, 1982b, 1982c, 1989; Hudec, 1987; Sharma and Sharma, 1997). Only quantitative analysis of zooplankton was done.

Results and Discussion

A total of 71 species were identified (Table 16.1); 38 of them belonging to rotifera, 22 to cladocera, 7 to copepoda and 4 species to ostracoda. Of these, 5 species are new to this region (1 cladoceran and 4 rotifers). The highest number of species was obtained in the samples from Dahid reservoir (29 species) and the lowest number was found in the sample of Gyanganga reservoir. During postmonsoon season (October 2011) the zooplankton diversity of various regions (Table 16.2) was as follows: Dahid reservoir, 15 spp.; Nalganga reservoir, 10 spp.; Vishwganga reservoir, 21 spp.; Paldhag reservoir, 10 spp.; Takli reservoir, 18 spp.; Khandwa reservoir, 12 spp.; Borjawala reservoir, 13 spp.; Kardi reservoir, 6 spp.; Shekapur reservoir, 22 spp.; Zari reservoir, 6 spp.; Rajura reservoir, 12 spp.; Torna reservoir, 6 spp.; Gyanganga reservoir, 5 spp.; Vidrupa reservoir, 17 spp.; Hatnur reservoir, 17 spp.; Utawali reservoir, 9 spp.; Kolhigolar, 10 spp.; Botha reservoir, 8 spp.; and Wari reservoir, 20 spp. In premonsoon season (March 2011) species diversity as follows: Dahid reservoir, 29 spp.; Nalganga reservoir, 19 spp.; Paldhag reservoir, 14 spp.; Kardi reservoir, 12 spp.; Botha reservoir, 11 spp.

Table 16.1: Systematic Account of Zooplankton

Cladocera

Family: sididae

1. *Diaphanosoma excisum*
2. *Diaphanosoma sarsi*

Family: Daphnidae

1. *Daphnia carinata*
2. *Ceriodaphnia cornuta*

Family: Moinidae

1. *Moina brachiata*
2. *Moina macrocopa*
3. *Moina rectirostris*
4. *Moina micrura*
5. *Moina oryzae*

Family: Bosminidae

1. *Bosminopsis deitersi*

Family: Macrothricidae

1. *Macrothrix goeldi*
2. *Macrothrix laticornis*

Family: Chydoridae

1. *Leydiga acanthocercoides*
2. *Alona monocantha monocantha*
3. *Alona cambouei*
4. *Alona pulchella*
5. *Biapertura karua*
6. *Pleuroxus trigonellus*
7. *Pleuroxus denticulatus*
8. *Chydorus sphaericus*
9. *Chydorus barroisi barroisi*
10. *Chydorus reticulates*

Copepoda

Family: Diaptomidae

1. *Rhinediaptomus indicus*
2. *Heliodiaptomus viduus*
3. *Neodiaptomus stregilipes*

Family: Cyclopidae

1. *Tropocyclops prasinus*
2. *Paracyclops fimbriatus*
3. *Mesocyclops leuckarti*
4. *Mesocyclops hyalinus*

Rotifera

Family: Brachionidae

1. *Brachionus angularis*
2. *Brachionus bidentata*
3. *Brachionus caudatus*
4. *Brachionus falcatus*
5. *Brachionus calyciflorus*
6. *Brachionus forficula*
7. *Brachionus bennini*
8. *Brachionus quadridentatus*
9. *Brachionus rubens*
10. *Brachionus plicatilis*
11. *Brachionus urceolaris*
12. *Bra chionus diversicornis*
13. *Platyias quadricornis*
14. *Platyias putulus*
15. *Keratella tropica*
16. *Keratella cochlearis*
17. *Keratella quadrata*

Family: Lecanidae

1. *Lecane leontina*
2. *Lecane luna*
3. *Monostyla bulla*

Family: Euchlanidae

1. *Euchlanis dilatata*
2. *Euchlanis triquetra*

Family: Testudinella

1. *Testudinella patina*
2. *Pompholyx sulcata*

Contd...

Table 16.1–*Contd...*

Family: Notommatidae	Family: Philodinidae
1. *Cephalodella gibba*	1. *Callidina bidens*
Family: Asplanchnidae	2. *Rotifer tardus*
1. *Asplanchna priodanta*	3. *Rotatoria neptunia*
2. *Asplanchna brightwelli*	Family:Meliceratidae
Family: Mytilinidae	1. *Lacinularia socialis*
1. *Mytilina acanthophora*	Family: Trochospharidae
2. *Mytilina ventralis*	1.*Horella brehmi*
Family: Colurellidae	Ostracoda
1. *Lepadella rhomboids*	Family: Cyprididae
Family: Trichoceridae	1. *Hemicypris fossulata*
1.*Trichocera similes*	Family: Ilyocypridae
Family: Filinidae	1. *Ilyocypris gibba*
1.*Filinia opoliensis*	Family: Darwinulidae
2. *Filinia longiseta*	1. *Darwinula* sp.
	Family: Stenocyprinae
	2. *Stenocypri shislopi*

Obviously we found higher species diversity in these reservoirs in March. Rotifer was the richest group with 38 species, which accounts for 53 per cent of total zooplankton group. Taxonomic dominance of rotifers was reported in several water bodies (Nogueira, 2001; Cavalli *et al.,* 2001; Sampaio *et al.,* 2002; Neves *et al.,* 2003). This pattern is common in tropical and subtropical freshwaters, whether in lakes, ponds, reservoirs, rivers, or streams (Neves *et al.,* 2003; Rocha *et al.,* 1995; Starkweather, 1980; 1998 cited in Arora and Mehra, 2003a)

Most frequently collected rotifers were: *Keratella tropica (3 7 per cent), Habrotrocha bidens (32 per cent), Brachionus calyciflorus (26 per cent), and B. diversicornis, Keratella cochlearis, Lecane luna, L. leontina, Fillinia longiseta* and *F. opoliensis* (21 per cent each). The families brachionidae and lecanidae were represented by large number species, which is considered typical, and most frequent in tropical environment (Dumont, 1983). Of the seven genera of Brachionidae five genera *i.e., Brachionus, Keratella, Platyias, Anuraeopsis* and *Notholca* are found in India and they form a significant fraction of Rotifer (Sharma,1987). Of the 38 species of rotifers, 17 species belong to brachionidae and three species to Lecanidae. *Brachionus* was the prominent genus represented by 12 species *i.e., Brachionus angularis, B.caudatus, B. plicatilis, B. calyciflorus, B. diversicornis, B. quadridentatus, B. falcatus, B. bidentata, B. forficula, B. urceolaris, B. bennini* and *B. rubens.* Genus *Brachionus* is one of the most ancient genus of monogonont rotifers and is represented by 46 species in India (Sharma, 1983). The genus *Brachionus* is the index of eutrophic waters (Sladecek, 1983) and its abundance is considered as a biological indicator for eutrophication (Nogueira, 2001). The species *B. calyciflorus* is

Table 16.2: Occurrence of Zooplankton in Buldhana District during Wet and Dry Season

Species	Z1 W	Z1 D	Z2 W	Z2 D	Z3 W	Z3 D	Z4 W	Z4 D	Z5 W	Z5 D	Z6 W	Z6 D	Z7 W	Z7 D	Z8 W	Z8 D	Z9 W	Z9 D
Cladocera																		
Diaphanosoma excisum	–	+	+	–	–	–	–	–	–	–	–	–	–	–	+	+	–	+
Diaphanosoma sarsi	–	–	+	–	–	–	–	–	–	–	–	–	–	–	–	–	–	–
Ceriodaphnia cornuta	–	–	+	–	–	–	+	–	+	–	–	–	+	+	–	–	+	+
Daphnia carinata	–	–	–	–	–	–	–	–	–	–	–	–	–	–	–	–	–	–
Bosminopsis deitersi	–	–	–	–	–	–	–	–	–	–	–	–	–	+	–	–	–	–
Macrothrix goeldi	–	–	–	–	–	–	–	–	–	+	–	–	–	–	–	–	–	–
Macrothrix laticornis	–	–	–	–	–	–	–	–	–	+	–	–	–	–	–	–	–	–
Moina oryzae	–	–	–	–	–	–	–	–	–	–	–	–	–	–	–	–	–	–
Moina macrocopa	–	+	+	–	–	–	–	–	–	+	–	–	+	+	–	–	–	–
Moina brachiata	–	–	+	–	+	–	–	–	–	–	–	–	+	+	+	+	+	–
Moina rectirostris	–	–	–	–	–	–	–	–	–	–	–	–	–	–	–	–	–	–
Moina mirura	–	+	+	–	–	–	–	–	+	–	–	–	–	–	–	+	–	–
Leydigia acanthocercoides	–	–	–	–	–	–	+	–	–	+	–	–	–	–	–	–	–	–
Alonamonacantha monacantha	–	–	–	–	–	–	+	–	–	+	–	–	–	–	–	–	–	–
Alona cambouei	–	–	–	–	–	–	–	–	–	–	–	–	–	–	–	–	–	–
Alona pulchella	–	–	–	–	–	–	+	–	–	+	–	–	–	–	–	–	–	–
Biaptura karua	–	–	–	–	–	–	–	–	+	–	–	–	+	+	–	–	–	–
Pleuroxus trignonellus	–	–	–	–	–	–	–	–	–	+	–	–	–	–	–	–	–	–
Pleuroxus denticulatus	–	–	–	–	–	–	–	–	–	+	–	–	–	+	–	–	–	–
Chydorus sphericus	–	–	–	–	–	–	–	–	–	+	–	–	–	–	–	–	–	–

Contd...

170

Table 16.2–Contd...

Species	Z1 W	Z1 D	Z2 W	Z2 D	Z3 W	Z3 D	Z4 W	Z4 D	Z5 W	Z5 D	Z6 W	Z6 D	Z7 W	Z7 D	Z8 W	Z8 D	Z9 W	Z9 D
Chydorus reticulatus	–	–	–	–	–	–	+	–	–	+	–	–	–	–	–	–	–	–
Chydorus barroisi barroisi	–	–	–	–	–	–	–	–	–	+	–	–	–	–	–	–	–	–
Copepoda																		
Rhinediaptomus indicus	–	+	+	–	–	–	+	–	–	–	–	–	–	–	–	–	–	–
Heliodiaptomus viduus	–	+	+	–	–	–	+	–	–	+	+	–	–	–	+	–	+	+
Neodiaptomus strigilipes	+	+	+	–	+	–	+	–	+	+	+	–	+	+	+	+	+	+
Paracyclops fimbiatus	+	+	+	–	+	–	+	–	+	+	–	–	+	+	+	+	+	–
Tropocyclops prasinus	–	+	+	–	–	–	+	–	+	+	–	–	+	+	+	+	+	+
Mesocyclops leuckartii	–	+	+	–	–	–	+	–	+	+	–	–	+	+	+	+	–	–
Mesocyclops hyalinus	–	–	+	–	–	–	–	–	–	–	–	–	–	–	–	–	–	–
Rotifera																		
Brachionus angularis	–	–	–	–	–	–	–	–	–	–	–	–	+	–	–	–	–	–
Brachionus caudatus	–	–	+	–	–	–	–	–	–	–	–	–	+	–	–	–	+	+
Brachionus plicatilis	–	–	–	–	–	–	–	–	–	–	–	–	+	+	–	–	–	–
Brachionus calyciflorus	–	–	+	–	–	–	–	–	–	–	–	–	–	+	–	–	–	–
Brachionus diversicornis	–	–	–	–	–	–	–	–	–	–	–	–	–	–	–	–	–	–
Brachionus quadridentatus	–	–	–	–	–	–	–	–	–	–	–	–	–	+	–	–	–	–
Brachionus falcatus	–	–	+	–	–	–	–	–	+	–	–	–	–	–	–	–	–	–
Brachionus bidentata	–	–	+	–	–	–	–	–	–	–	–	–	–	–	–	–	–	–
Brachionus urceolaris	–	–	–	–	–	–	–	–	–	+	–	–	–	–	+	+	–	–
Brachionus forficula	–	–	–	–	–	–	–	–	–	+	–	–	+	–	–	–	–	–

Contd...

Table 16.2–Contd...

Species	Z1		Z2		Z3		Z4		Z5		Z6		Z7		Z8		Z9	
	W	D	W	D	W	D	W	D	W	D	W	D	W	D	W	D	W	D
Brachionus rubens	–	–	–	–	–	–	–	–	–	–	–	–	–	–	–	–	–	–
Brachionus bennini	–	–	–	–	–	–	–	–	–	–	–	–	–	–	–	–	–	–
Platias qudadricornis	–	–	–	–	–	–	+	–	–	–	–	–	–	+	–	–	–	–
Platias putulus	–	–	–	–	–	–	–	–	+	–	–	–	–	–	–	+	–	+
Kertella tropica	–	–	–	–	–	–	–	–	–	+	–	–	–	–	+	+	–	–
Kertella cochlearis	–	–	–	–	–	–	–	–	–	+	–	–	–	+	+	+	–	–
Kertella quadrata	–	–	–	–	–	–	+	–	–	–	–	–	–	–	–	–	–	–
Euchlanis dialata	–	–	–	–	–	–	–	–	+	–	–	–	–	–	–	–	–	–
Euchlanis triquetra	–	–	–	–	–	–	–	–	+	–	–	–	–	–	–	–	–	–
Mytilina ventralis	–	–	–	–	–	–	–	–	–	+	–	–	–	–	+	+	+	–
Mytilina acanthophora	–	–	–	–	–	–	–	–	–	–	–	–	–	–	–	+	–	–
Pompholyx sulcata	–	–	–	–	–	–	–	–	–	–	–	–	–	–	–	–	–	–
Monostyla bulla	–	–	–	–	–	–	+	–	+	+	+	–	–	+	–	–	–	+
Lecane luna	–	–	–	–	–	–	–	–	+	+	–	–	–	+	+	–	–	–
Lecane leontina	–	+	–	–	–	–	–	–	+	+	–	–	–	+	+	+	–	+
Fillinia longiseta	–	–	–	–	–	–	–	–	–	–	–	–	–	–	+	+	–	+
Fillinia opoliensis	–	–	–	–	–	–	–	–	–	+	–	–	–	–	–	–	–	–
Asplancha priodanta	–	–	–	–	–	–	–	–	–	+	–	–	–	+	–	–	–	–
Asplancha bright welli	–	–	–	–	–	–	–	–	–	+	–	–	–	+	–	–	–	–
Horella brehmi	–	–	–	–	–	–	–	–	–	–	–	–	–	–	–	–	–	–
Callidina bidens	+	–	+	–	+	–	–	–	–	–	+	–	–	–	–	–	–	–

Contd...

Table 16.2–*Contd...*

Species	Z1 W	Z1 D	Z2 W	Z2 D	Z3 W	Z3 D	Z4 W	Z4 D	Z5 W	Z5 D	Z6 W	Z6 D	Z7 W	Z7 D	Z8 W	Z8 D	Z9 W	Z9 D
Rotifer tardus	–	–	–	–	–	–	–	–	–	–	–	–	–	–	–	–	–	–
Rot atoria neptunia	–	–	–	–	–	–	–	–	–	–	–	–	–	–	–	–	–	–
Lacinularia socialis	–	–	–	–	–	–	–	–	–	–	–	–	–	–	–	–	–	–
Lepadella rhomobiodes	–	–	–	–	–	–	–	–	–	–	–	–	–	–	–	–	–	–
Triclocera sim iles	–	–	–	–	–	–	–	–	–	–	–	–	–	–	–	–	–	–
Cephalodella gibba	–	–	+	–	–	–	–	–	–	–	–	–	–	–	–	–	–	–
Testudinella patina	–	–	–	–	–	–	–	–	–	–	–	–	–	–	–	–	–	–
Ostacoda																		
Hemicypris fossulata	+	+	+	–	+	–	+	–	+	+	–	–	+	–	–	+	+	–
Llyocypris gibba	+	+	+	–	–	–	+	–	+	+	–	–	+	–	–	+	+	+
Stenocypris sp.	–	–	–	–	–	–	+	–	–	+	–	–	–	–	–	–	–	–
Darwinula sp.	+	–	+	–	+	–	–	–	+	+	–	–	–	–	–	–	–	–
Total species	**6**	**12**	**22**	**0**	**6**	**0**	**17**	**0**	**15**	**29**	**5**	**0**	**10**	**19**	**10**	**14**	**8**	**11**

W: Wet; D: Dry; +: Present; –: Absent; Z1: Kardi reservoir; Z2: Shekapur reservoir; Z3: Zari reservoir; Z4: Hatnur reservoir; Z5: Dahid reservoir; Z6: Gyanganga reservoir; Z7: Nalganga reservoir; Z8: Paldhag reservoir; Z9: Botha reservoir.

Table 16.3: Occurrence of Zooplankton in Reservoirs of Buldana District (Wet season only)

Species	Z1	Z2	Z3	Z4	Z5	Z6	Z7	Z8	Z9	Z10
Cladocera										
Diaphanosoma excisum	+	+	−	+	+	−	+	−	−	−
Diaphanosoma sarsi	−	+	−	+	−	−	−	−	−	−
Ceriodaphnia cornuta	+	+	−	+	−	−	+	−	−	−
Daphnia carinata	−	−	−	−	−	+	+	+	−	−
Bosminopsis deitersi	−	−	−	−	−	−	+	+	+	−
Macrothrix goeldi	+	−	−	−	−	−	−	+	+	+
Macrothrix laticornis	−	−	−	−	−	−	−	−	−	−
Moina oryzae	−	−	−	+	−	−	−	−	−	−
Moina macracopa	+	+	−	+	−	−	−	−	−	−
Moina brachiata	+	+	−	+	−	−	+	−	−	−
Moina rectirostris	−	−	−	+	−	−	−	−	−	−
Moina mirura	−	+	+	+	+	−	−	−	−	−
Leydigia acanthocercoides	−	−	−	−	−	−	−	+	−	−
Alonamonacantha monacantha	−	−	−	−	−	−	−	−	−	−
Alona cambouei	−	−	−	−	−	−	−	−	+	−
Alona pulchella	−	+	−	−	−	−	−	−	−	−
Biaptura karua	−	−	−	−	−	−	−	+	−	−
Pleuroxus trignonellus	−	−	−	−	−	−	−	−	−	−
Pleuroxus denticulatus	−	−	−	−	−	−	−	−	−	−
Chydorus sphericus	−	+	−	−	−	−	−	−	−	−
Chydorus reticulatus	−	−	−	−	−	−	−	−	−	−
Chydorus barroisi barroisi	−	−	−	−	−	−	−	−	−	−
Copepoda										
Rhinediaptomus indicus	+	−	−	+	+	−	+	−	−	−
Heliodiaptomus viduus	+	+	+	+	+	−	+	−	+	+
Neodiaptomus strigilipes	+	+	+	+	+	−	+	−	+	+
Paracyclops fimbiatus	+	−	−	+	+	+	+	+	+	−
Tropocyclops prasinus	−	+	+	+	+	+	+	−	+	+
Mesocyclops leuckartii	+	+	+	+	+	+	+	+	−	+
Mesocyclops hyalinus	−	−	−	−	−	−	−	−	−	−
Rotifera										
Brachionus angularis	−	−	+	−	−	−	−	−	−	−
Brachionus caudatus	−	+	−	−	−	−	−	−	−	−
Brachionus plicatilis	−	−	+	−	−	−	−	−	−	−
Brachionus calyciflorus	+	−	+	−	−	−	−	+	−	+
Brachionus diversicornis	−	−	−	−	−	−	−	+	−	−
Brachionus quadridentatus	−	+	−	−	−	−	−	−	−	−
Brachionus falcatus	−	−	−	+	−	−	−	−	−	−

Contd...

Table 16.3–*Contd...*

Species	Z1	Z2	Z3	Z4	Z5	Z6	Z7	Z8	Z9	Z10
Brachionus bidentata	–	–	–	–	–	–	–	+	–	–
Brachionus urceolaris	–	–	–	–	–	–	–	+	–	–
Brachionus forficula	–	+	–	–	–	–	–	–	–	–
Brachionus rubens	–	–	–	+	–	–	–	–	–	–
Brachionus bennini	–	–	–	+	–	–	–	–	–	–
Platias qudadricornis	–	–	–	–	–	–	–	–	–	+
Platias putulus	–	–	–	–	–	+	–	–	–	–
Kertella tropica	–	+	–	–	–	–	–	+	+	+
Kertella cochlearis	–	–	+	–	–	–	–	–	–	–
Kertella quadrata	–	–	–	–	–	–	–	–	–	–
Euchlanis dialata	–	–	–	–	–	–	–	–	–	–
Euchlanis triquetra	–	–	–	–	–	–	–	–	–	–
Mytilina ventralis	+	–	–	–	–	–	–	–	–	–
Mytilina acanthophora	–	+	–	–	–	–	–	–	–	–
Pompholyx sulcata	–	–	–	–	–	–	–	–	–	+
Monostyla bulla	–	–	–	–	–	–	–	–	–	–
Lecane luna	–	–	–	–	–	–	–	–	–	+
Lecane leontina	–	+	–	–	–	–	–	–	–	–
Fillinia longiseta	–	–	–	–	–	–	–	–	+	–
Fillinia opoliensis	–	–	–	–	–	–	–	–	–	+
Asplancha priodanta	+	–	–	–	–	–	–	–	–	–
Asplancha brightwelli	–	–	–	–	–	–	–	–	–	+
Horella brehmi	+	+	–	–	–	–	–	–	–	–
Callidina bidens	+	–	–	–	–	–	–	+	–	–
Rotifer tardus	–	–	–	–	–	–	–	+	–	–
Rot atoria neptunia	–	–	–	–	–	–	–	+	–	–
Lacinularia socialis	–	–	–	–	–	–	–	+	–	–
Lepadella rhomobiodes	–	–	–	–	–	–	–	+	–	–
Triclocera sim iles	–	–	–	+	–	–	–	–	–	–
Cephalodella gibba	–	–	–	–	–	–	–	+	–	–
Testudinella patina	–	+	–	–	–	–	–	–	–	–
Ostacoda										
Hemicypris fossulata	+	–	+	–	+	–	–	+	+	–
Llyocypris gibba	+	+	–	–	–	–	+	+	–	–
Stenocypris sp.	–	–	–	–	–	–	–	–	–	–
Darwinula sp.	+	–	+	–	–	–	–	–	–	–
Total species	**18**	**21**	**12**	**17**	**9**	**6**	**13**	**20**	**10**	**12**

W: Wet; D: Dry; +: Present; –: Absent; Z1: Takali reservoir; Z2: Khandari reservoir; Z3: Khandwa reservoir; Z4: Vidrupa reservoir; Z5: Utawali reservoir; Z6: Torna reservoir; Z7: Borjawali reservoir; Z8: Wari reservoir; Z9: Kolhigolar; Z10: Rajura reservoir.

considered to be a good indicator of eutrophication (Sampaio *et al.,* 2002). Presence of B. calyciflorus in Wari reservoir, Takli reservoir, Paldhag reservoir, Nalganga reservoir and Rajura reservoie suggest that these reservoirs have reached eutrophic stage. Increased nutrients in these reservoirs may be due to entry of sewage in Wari reservoir and Rajura reservoir, whereas in Takali reservoir, Paldhag reservoir and Nalganga reservoir it may be due to agricultural runoff.

The present study reveals that zooplankton species richness was high in pre-monsoon period when compared to post-monsoon season. Arora and Mehra (2003b) while analyzing seasonal dynamics of rotifers in relation to physico-chemical conditions of the river Yamuna, made the similar observations. In summer season, the absence of inflow of the water brings stability to the water body. The availability of food is more due to the organic matter production and decomposition. The above factors contribute for high species diversity in that season. Increased anthropogenic activities, siltation, sewge contamination, and high nutrients problems due to the indiscriminate use of fertilizers in the agricultural catchment area are the major cause for the eutrophication. The study indicates that many water bodies have already reached the eutrophication stage and the fates of other ponds are also not encouraging.

References

Arora, J. and Mehra, N. K. 2003a. Species diversity of planktonic and epiphytic rotifers in the backwaters of the Delhi segment of the Yamuna river, with remarks on new records from India. *Zoological Studies*, 42(2): 239-247.

Arora, J. and Mehra, N. K. 2003b. Seasonal dynamics of the rotifers in relation to physical and chemical conditions of the river Yamuna (Delhi), India. *Hydrobiol.* 491: 101-109.

Cavalli, L., Miquelis, A. and Chappaz, R. 2001. Combined effects of environmental factors and predatory prey interactions on zooplankton assemblages in five high alpine lakes. *Hydrobiol.* 455: 127-135.

Dumont, J. H. 1983. Biogeography of rotifers. *Hydrobiol.* 104: 19-30.

Dumont J. H. and Isabella Van De Velde. 1977. Report on a collection of cladocera and copepoda from Nepal. *Hydrobiol.* 53(1): 55-65.

Edmondson, W. T. 1959. Freshwater Biology, 2nd edn. John Wiley and Sons, New York, USA.

Fernando, C. H. and Kanduru, A. 1984. Some remarks on the latitudinal distribution of cladocera on the Indian sub continent. *Hydrobiol.* 113: 69-76.

Hegde, G. R. and Huddar, B. D. 1995. Limnological studies of two freshwater lentic ecosystems of Hubli-Dharwad Karnataka. pp. 35-43. In: Irfan A. Khan (eds), Frontiers in plant science, The book syndicate publication, Hyderabad.

Hudec, I. 1987. Moina oryzae n. sp. (Cladocera: Moinidae) from Tamil Nadu (South India). *Hydrobiol.* 145: 147-150.

Neves, I. F., O. Rocha, K. F. Roche and Pinto, A. A. 2003. Zooplankton community structure of two marginal lakes of the River Cuiaba (Mato Grasso, Brazil) with analysis of rotifera and cladocera diversity. *Brazil Journal of Biology*, 63(3): 329-343.

Nogueira, M. G. 2001. Zooplankton composition dominance and abundance as indicators of environmental compartmentalization in jurumirim reservoir (Paranapanema River), Sao Paulo, Brazil. *Hydrobiol.* 455: 1-18.

Patil C. S. and Gouder, B. Y. M. 1982a. Freshwater fauna of Dharwad (India) II: Rotifera. *Journal of Karnatak University Science*, 27: 93-114.

Patil C. S. and Gouder, . B. Y. M. 1982b. Freshwater fauna of Dharwad (India) II: Cladocera. *Journal of Karnatak University Science*, 27: 115-126.

Patil C. S. and Gouder, B. Y. M. 1982c. Freshwater copepods of Dharwad (Karnataka state, India). *Journal of Karnatak University Science*, 27: 130-141.

Patil C. S. and Gouder, B. Y. M. 1985. Ecological study of freshwater zooplankton of a subtropical pond (Karnataka state, India). *Int. Revue, ges. Hydrobiol.* 70(2): 259-267.

Patil, C. S. and Gouder, B. Y. M. 1989. Freshwater invertebrates of Dharwad, Prasaranga, Karnatak University, Dharwad.

Pederson, G. L., E. B. Wezch and Litt, A. H. 1976. Plankton, secondary production and biomass, their relation to trophic state of lake. *Hydrobiol.* 50: 129-144.

Pollard, A. I., M. J. Gonzalez, M. J. Vanni and Headworth, J. L. 1998. Effect of tubidity and biotic factors on the rotifer community in an Ohio reservoir. *Hydrobiol.* 387 / 3 88: 215-233.

Rajapaska, R. and Fernando, C. H. 1982. The cladocera of Srilanka (Ceylon) with remarks on some species. *Hydrobiol.* 94: 49-69.

Rao, V. N. R., R. Mohan, V. Hariprasad and R. Ramasubramanian. 1994. Sewage pollution in the high altitude Ooty lake, Udhagamandalam - cause and concern. *Poll. Res.* 13(2): 133-150.

Rocha, O., S. Sendacz and T. Matsumura-Tundisi. 1995. Composition, biomass and productivity of zooplankton in natural lakes and reservoirs in Brazil, Pp. 151-166. In: J. G. Tundisi, C. E. M. Bicudo and T. Matsumura-Tundisi (Eds.), Limnology in Brazil. ABC/SBL, Rio de Janeiro, 376pp.

Sampaio, E. V., Rocha, O., Matsumura-Tundisi, T. and Tundisi, J. G. 2002. Composition and abundance of zooplankton in the limnetic zone of seven reservoirs of the Paranapanema River, Brazil. *Brazil Journal Biology* 62(3): 525-545.

Santos-Wisniewski, M. J., O. Rocha, A. M. Guntzel and T. Matsumura-Tundisi. 2002. Cladocera chydoridae of high altitude water bodies (Serra Da Mantiqueira), in Brazil. *Brazil Journal Biology*, 62(4A): 681-6 87.

Sharma, B. K. 1979. Rotifers from West Bengal III. Further studies on the Euratatoria. *Hydrobiol.* 64(3): 239-250.

Sharma, B. K. 1983. The Indian species of the genus Brachionus (Euratatoria: monogononta). *Hydrobiol.* 104: 31-39.

Sharma, B. K. 1987. Indian Brachionidae (Euratatoria: Monogononta) and their distribution. *Hydrobiol.* 144: 269-275.

Sharma, B. K. and Michael, R. G. 1980. Synopsis of taxonomic studies on the Indian Rotatoria. *Hydrobiol.* 73: 229-236.

Sharma, B. K. and Michael, R. G. 1987. Review of taxonomic studies on freshwater cladocera from India with remarks on biogeography. *Hydrobiol.* 145: 29-3 3.

Sharma, B. K. and Sharma, S. 1997. Lecanid rotifers (Rotifera: Monogononta: Lecanidae) from Northeastern India. *Hydrobiol.* 356: 159-163.

Sharma B. K. and Sharma, S. 2001. Biodiversity of rotifers in some tropical flood plains lakes of the Brahmaputra river basin, Assam (N. E. India). *Hydrobiol.* 446/447: 305-313.

Sladecek, V. 1983. Rotifers as indicators of water quality. *Hydrobiol.* 100: 169-201.

Starkweather, P. L. 1980. Aspects of feeding behaviour and trophic ecology of suspension-feeding rotifers. *Hydrobiol.* 73: 63-72.

Chapter 17

Finfish Diversity of Tungabhadra Reservoir, Karnataka

☆ *B. Vasanthkumar, S.V. Roopa*
and B.K. Gangadhar

ABSTRACT

A study was undertaken to access the status of available fin fishes in Tungabhadra dam. To study the diversity of fishes total five stations were selected. During the present study 36 fin fishes were identified from station I to V. The present study was carried out for a period of 12 months from May 2007 to April 2008. The important fishes available were *Catla catla, Wallago attu, Cirrhinus reba, Labeo fimbriatus* etc.

Keywords: *Finfish diversity, Tungabhadra dam, Karnataka.*

Introduction

Freshwater resources of the country are vast and varied such a reservoirs, tributaries, rivulets, streams, reservoirs, lakes, canals, tanks and ponds: The rivers of East coast system of peninsular India consists of Mahanadi, Godavari, Krishna, Tungabadra and Cauvery. The rivers of the east and west coast systems are rain fed. The river Ganga is the largest river in India. The total length of reservoirs in India is 1, 64,121 kilometers (Dixitulu and Paparao, 1994).Damming of streams has been in vogue since time immemorial. Imitating beavers, prehistoric man dammed streams by putting sticks across a stream to trap fish.

Tungabhadra river derives its name from the confluence of two streams, the Tunga and the Bhadra, both of which rise in the wooded eastern slopes of the western Ghats in the state of Karnataka and flow eastward. After confluence of these two streams at Kudali about 531 km and joins the river Krishna at Sangamaheshwarm near Kurnool in the state of Andhra Pradesh. The river runs for 382 km in Karnataka, thereafter forms the boundary between Karnataka and Andhra Pradesh for 58 km and flows for the remaining 91 km in Andhra Pradesh. Tungabhadra sub-basin of 69,552km^2 of this the catchment area of the Tungabhadra reservoir is 28,117 krn". The river basin is influenced by the south-west monsoon, with dwindles to few cumecs in summer months.

Materials and Methods

In the present investigation five stations were selected namely Station1- T.B.Dam, Station 2 – Munirabad, Station: 3 – Vysanakere, Station: 4 - Narayana Devar kere, Station: 5 – Hosalli. The present study was carried out for a period of 12 months from May 2007 to April 2008. During the present investigation an attempt was made to study the biotic *i.e.* with reference to fish species diversity was observed from five study locales. During the present observation 36 varieties of fishes were noticed at station I to V.

For identification of fishes Day's Volume-I (1978), Talwar and Jhingran (1991) Beavan (1990) and F.A.O sheets were made use of. Identification of pelagic, mid-water and benthic fishes was carried out simultaneously for their systemic study.

Figure 17.1: Overview of Tungabhadra Dam.

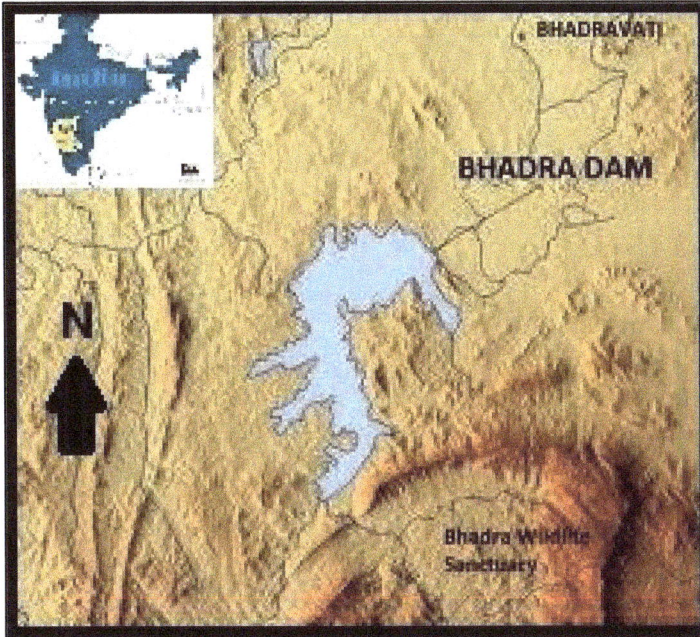

Figure 17.2: Location of Tungabhadra Dam.

Results and Discussion

The fish catch of Tungabhadra reservoir during the mid-1960s was dominated by predatory catfishes, an undesirable component of fisheries by any standards. The plankton, benthos and detritus resources were not directly utilized by any of the major commercial species. More than 75 per cent of the total catch comprised *Aorichthys seenghala, Wallago attu, Silonia childreni* and *Pseudeutropius taakree,* all living on a long-food chain. This is clearly the result of management failure to induct fast-growing, short food chain fishes into the system, during the early years of the reservoir.

Tungabhadra reservoir, when it was impounded, had a population of *Puntius kolus,* which contributed up to a third of its total catch. Other species of *Puntius* (*P. dubius, P. sarana and P. pulchellus*), *Tor tor, Labeo fimbriatus, L. calbasu, L. porcellus, L. potail* and *L. pangusia* formed the other indigenous forms (Krishnamoorthy, 1979). Most of the native species found the changed environment after impoundment hard to cope with and started declining, the main reasons being destruction of breeding grounds, absence of fluviatile environment, and the changed trophic structure. Their share in the total fish catch has declined drastically from 74.89 per cent in 1958 to 28.91 per cent in 1965. Unlike the case of reservoirs in Tamil Nadu, no serious attempts were made in Tungabhadra to introduce Indo-Gangetic major carps to fill the vacant niches created by the receding population of *Puntius* and *Labeo* species. As a result, the carp minnows and minor weed fishes took advantage of the new spurt in plankton and benthic communities and these fishes, in turn, provided good forage to predatory catfishes.

Table 17.1: Checklist of Finfishes

Sl.No.	Scientific Name of the Fish	Common Name of the Fish
1.	*Notopterus notopterus*	–
2.	*Salmostoma clupeodes*	Bloch razorbelly minnow
3.	*Cirrhinus reba*	–
4.	*Labeo dero*	Kalabans
5.	*Labeo fimbriatus*	Frig lipped peninsula carp
6.	*Labeo kontius*	Pig mouth carp
7.	*Labeo pocellus*	Bombay labeo
8.	*Catla catla*	Catia
9.	*Ostebrama cotio peninsularis*	Peninsular osteobrama
10.	*Osteobrama vigorsii*	Bheema osteobrama
11.	*Rohtee ogilbii sykes*	Vatanirohtee
12.	*Osteohilus thomassi*	Kontibarb
13.	*Puntius sophora*	–
14.	*Puntius corichonius*	–
15.	*Puntius Sarana Sarana*	–
16.	*Puntius Sarana Sabnasutus*	–
17.	*Puntius jordoni*	Peninsularolive barb
18.	*Hypselobarus kolus*	Jedons carp
19.	*Tor khudree*	Kolus
20.	*Tor Mussullab*	Yellow mahaseer
21.	*Amblypharygodon mola*	Mola carp let
22.	*Aorichthys aor*	Long whiskered cat fish
23.	*Mystus krishnensis Ramakrishaiah*	Krishna Mystus
24.	*Mystus cavasius*	Gangetic Mystus
25.	*Retagogrea*	–
26.	*Reta Pavimentatus*	Gogra rit
27.	*Proeutropiichthys taakree taakree*	Indian taa kree
28.	*Ompak bimaculatus*	–
29.	*Wallago attu*	–
30.	*Xenenteodon Cancila*	–
31.	*Parambasis ranga*	Glass fish
32.	*Chanda nama*	–
33.	*Glossogobius*	Tank goby
34.	*Channa striatus*	–
35.	*Rhinomugil corsula*	Corsula mullet spinyed
36.	*Mastacembelus*	Spinyeel

Summary

1. Tungabhadra river provides an interesting example of fluvial geomorphology in fact the course of its passage from the origin to confluence it exhibits all stages of life,youth maturity and old age.

2. The self purification capacity of the river is good.

3. Fishing practices, gear and tackle used are of conventional type. Destructive fishing are practices are prevalent,as dynamiting, lack of mesh regulations and uniform policy for the riverine system.

4. It is also suggested. to follow the habitat improvement programme for recovery of endangered population should be initiated.

5. The process in the direction of culture fishery development needs as integrated approach.Production capacities of seed centers should be enhanced and quality seed supply be ensured.

6. *Tor khudree. Mystus sps* should be transplanted into other suitable places and steps to establish their populations undertaken.

References

Beavan, C. 1990. Hand Book of The Freshwater Fishes of India.

Day, F. 1889. The fauna of British India including Ceylon and Burma Fisheries, Vol. I and II.

Dixitulu, J. V. H. and Paparao, G. 1994. Hand Book on Fisheries. Global Fishing Chimes Private Limited, Vishakapatannam. pp. 42-368.

Krishnamurthi. C. R., Bilgrami. K. S., Das. T. M. Mathur. R. P. (Eds). 1991. The Ganga- A scientific study. New Delhi. Northern Book Center, xxvi + 246pp. 8 pls.

Talwar and Jhingran, A. G. 1991. A Text book - Fish and Fisheries of India.

Chapter 18

Fish Diversity of Yeldari Reservoir, Maharashtra

☆ *S.D. Niture*

ABSTRACT

The Yeldari reservoir is a large sized reservoir of about 6,272 ha area, constructed on Purna river at Yeldari camp of Parbhani district. The present study deals with the fish diversity of Yeldari reservoir.During the present investigation, 41 species of fishes belonging to 8 orders and 15 families have been obtained from the Yeldari reservoir.

Keywords: *Fish fauna, Ichthyofauna, Fish diversity, Biodiversity, Yeldari reservoir.*

Introduction

The Yeldari reservoir is a large sized reservoir of about 6,272 ha area. It was constructed on river Purna at Yeldari camp in Parbhani district of Maharashtra. The reservoir lies in between north latitude 19°-43'-00" and East longitude 76°- 45'- 00". It is situated in hilly region on its both sides. Reservoir area is included in the survey of India toposheet map No. 56A/10. It was the first major project in Marathwada region for irrigation and power generation to initiate the process of economic development of Marathwada region. The catchment area of reservoir is 7,330 km². It is mainly hydroelectric project. The present work was mainly undertaken to investigate the fish diversity from Yeldari reservoir for a period of two years.

Materials and Method

For the present investigation fishes were collected from fish collection stations from fisher communities living around the reservoir.For the identification of the fish species, Day volumes and taxonomic records from Talwar and Jhingran (1991) were used. Johal and Jha (2007) suggested that periodic fish faunal surveys must be undertaken, so that the loss or gain of fish diversity can be evaluated. During present investigation fish diversity, their economic importance and abundance was studied.

Results and Discussion

During the present investigation, 41 species of fishes belonging to 8 orders and 15 families have been recorded from the Yeldari reservoir. Earlier Sakhare (2002) reported the occurrence of 29 fish species from this reservoir, however during present investigation the occurrence of 09 different fish species were recorded as new report from this reservoir, For this new reports we got the collection of these species from various fishing stations, village markets, collection from fisher folks and tribes living around the peripheral region of the reservoir. From Table 18.1, number 3, 4, 8, 9, 22, 24, 31, 32, 40 these species are newly reported species during this investigation.

Table 18.1: Fish Fauna of Yeldari Reservoir, Maharashtra

Phylum: Chordata
Sub-Phylum: Vertebrata
Class: Osteichthyes
Sub-class: Teleostei
Order: Cypriniformes
Family: Cyprinidae

1. *Labeo rohita* (Gunt)
2. *Labeo calbasu* (Gunt)
3. *Labeo boggut* (Cuv)
4. *Cirrhina reba* (Cuv-val)
5. *Cirrhina mrigal* (Cuv-val)
6. *Cyprinus carpio communis* (Linn)
7. *Catla catla* (Cuv-val)
8. *Rohtee cotio* (Sykes)
9. *Rohtee vigorsi* (Sykes)
10. *Rohtee ogilbi* (sykes)
11. *Chela phulo* (Ham-Buchanan)
12. *Chela bacila* (Ham-Buchanan)
13. *Puntius Sarana* (Ham)
14. *Puntius ticto* (Ham)
15. *Puntius hexacticus kolus* (Sykes)
16. *Puntius sophore* (Ham)
17. *Ctenopharyngodon idella*
18. *Hypophthalamichthys molitrix* (val)

Family: Poeciidae

19. *Gambusia affinis* (Bird and Gira)

Contd...

Table 18.1–*Contd...*

Order: Channiforms
Family: Channidal

20. *Channa marulius* (Ham)
21. *Channa striatus* (Ham)
22. *Channa punctatus* (Ham)
23. *Channa gachua* (Ham)

Family: Cobitida

24. *Amblypharyngodon mola* (Ham-Buch)

Order: Siluriformes
Family: Siluridae

25. *Wallago attu* (schne)
26. *Ompok bimaculatus* (Bloch)

Family: Bagridae

27. *Mystus seenghala* (Sykes)
28. *Mystus cavassius* (Ham)

Family: Heteropneustidae

29. *Heteropneustes fossilis* (Bloch)

Family: Clairidae

30. *Clarius batrachus* (Cuv-val)

Family: Scombersocidae

31. *Belone concila* (Cuv-val)

Order: Clupeiformes
Family: Notopteridae

32. *Notopterus kapirat* (Lacep)
33. *Notopterus chitala* (Gunt)

Order: Perciformes
Family: Percidae

34. *Ambassis nama* (Cuv-val)
35. *Ambassis ranga* (Cuv-val)

Order: Mastacembeliformes
Family: Mastacembelidae

36. *Mastacembelus armatus* (Cuv-val)
37. *Mastacembelus aculeatus* (Cuv-val)

Order: Cypriformes
Family: Cobitidae

38. *Nemacheilus botia* (Gunt)
39. *Rasbora daniconis* (Bleek)

Order: Discognathiformes
Family: Discognathidae

40. *Discognathus modestus* (Hackel)

Order: Gobiformes
Family: Gobiidae

41. *Gobius giuris* (Ham-Buch)

Table 18.2: Economic Importance of Fishfauna of Yeldari Reservoir, Maharashtra

Sl.No.	Name of Fishes	Abun-dance	Comm-ercial Food	Fine-food	Corse-food	Aquarium Fish	Other
1.	*Ambassis nama* (Cuv-val)	–	–	–	–	–	LV
2.	*Ambassis ranga* (Cuv-val)	–	–	–	–	–	LV
3.	*Amblypharyngodon mola*(Sykes)	++	–	–	√	√	LV
4.	*Belone concila* (Cuv)	–	–	–	√	–	MD
5.	*Catla catla* (Cuv-val)	+++	√	√	–	–	–
6.	*Channa punctatus* (Ham)	++	√	√	–	–	MD
7.	*Channa gachua* (Bloch)	+	√	√	–	–	MD
8.	*Channa marulius* (Bloch)	+	√	√	–	–	MD
9.	*Channa straitus* (Bloch)	+	√	√	–	–	MD
10.	*Chela phulo* (Ham-Buch)	_	–	–	√	–	–
11.	*Chela bacila* (Ham-Buch)	_	–	–	√	–	–
12.	*Cirrhina mrigal* (Cuv-val)	+++	√	√	–	–	–
13.	*Cirrhina reba* (Cuv-val)	++	√	√	–	–	–
14.	*Clarius batrachus* (Cuv-val)	++	√	√	–	–	–
15.	*Ctenopharyngodon idella* (Val)	+	√	√	–	–	–
16.	*Cyprinus carpio communis* (Linn)	+++	√	√	–	–	–
17.	*Discognathus modestus* (Hackel)	+	–	–	√	–	–
18.	*Gambusia affinis* (Bird and Gira)	+	–	–	√	–	LV
19.	*Gobius guiris* (Ham-Buch)	+	–	–	–	–	–
20.	*Heteropneustes fossilis* (Bleck)	++	√	√	–	–	MD
21.	*Hypophthalamichthys molitrix* (val)	++	√	√	–	–	–
22.	*Labeo boggut* (Cuv)	–	√	–	–	–	–
23.	*Labeo calbasu* (Gunt)	+	√	√	–	–	–
24.	*Labeo rohita* (Gunt)	+++	√	√	–	–	–
25.	*Mastacembelus aculeatus*(Cuv-val)	+	–	√	–	–	MD
26.	*Mastacembelus armatus*(Cuv-val)	++	–	√	–	–	MD
27.	*Mystus Cavassius* (Ham)	+	–	√	–	–	–
28.	*Mystus Seenghala* (Ham)	++	–	√	–	–	–
29.	*Nemacheilus botia* (Gunt)	–	–	–	√	√	–
30.	*Notopterus chitala* (Gunt)	+	–	–	√	–	MD
31.	*Notopterus kapirat* (Gunt)	+	–	–	√	–	MD
32.	*Ompok bimaculatus* (Bloch)	–	–	–	√	–	MD
33.	*Puntius hexacticus* (Sykes)	+	–	–	√	√	BT, LVMD
34.	*Puntius Sarana* (Ham)	+	–	–	√	√	BT,LV

Contd...

Table 18.2–*Contd...*

Sl.No.	Name of Fishes	Abun-dance	Comm-ercial Food	Fine-food	Corse-food	Aquarium Fish	Other
35.	*Puntius sophore* (Ham)	+	–	–	√	–	BT
36.	*Puntius ticto* (Ham)	++	– –	–	√	√	BT,LV
37.	*Rasbora daniconis* (Bleek)	+	–	–	√	– –	–
38.	*Rohtee cotio* (Sykes)	–	–	–	√	–	–
39.	*Rohtee vigorsi* (schne)	–	–	–	√	–	–
40.	*Rohtee ogilbii* (schne)	–	–	–	√	–	MD
41.	*Wallago attu* (schne)	++	–	√	–	–	–

LV: Larvivorus fish; BT: Bait; MD: Medicinal value.

The Yeldari reservoir fishery has been studied up to some extent by Sakhare (2002) especially for the physico-chemical characters of reservoir water and discussed about some aspects of fishery management and concluded that the Yeldari reservoir has good fishery potential. Sakhare (2002) also suggested regular studies on various aspects for the fisheries development and reported the occurrence of 29 fish species from Yeldari reservoir.

Niture (2009) suggested that if the fish co-operative society or companies with some other new successful management plans and strategies are permitted in the Yeldari reservoir fishery then there is a better chance of Yeldari reservoir improvement in terms of increase in fish production, socio-economic upliftment of fisher communities, employment generation to trap the untamed huge fishery potential of Yeldari reservoir in future, which will help to improve the inland fishery production of this region. In the development of the reservoir fisheries sector of this region Yeldari reservoir fishery development is the best option for application of pen-culture,cage culture methods, Yeldari reservoir was characteristically loaded with variety of weeds, located in all corners of the reservoir. The weed species mainly includes, *Hydrilla, Ipomea* and *Vallisneria*. There was a numerous quantity of the filamentous algae on the bottom and the coastal rocks and stone pieces in the reservoir. The kind of aquatic phyta support the periphyton mainly, hence the reservoir has good population of weed fishes.

Kumar (1990) reported 51 fish species belonging to 9 families from Govindsagar reservoir of Himachal Pradesh, out of which 12 fish species were commercially important. In Yeldari reservoir 12 commercially important fish species. Were recorded.

Devi (1997) studied the ichthyofauna of Ibrahimbagh and Shatamrai reservoirs of Hyderabad and reported o dominance of order cypriniformes, followed by order siluriformis, channiformis and perciformis. The present study also shows similar findings.Jain (1998) reported of 53 fish fauna and was grouped into seven categories in Rajasthan state. Kadam (2005) reported 45 different fish species from a small sized Masoli reservoir of district Parbhani, Maharashtra and Niture (2009) reported only 37 fish species from same Masoli reservoir. Sukumaran and Rahaman (1998) stated that majority of reservoirs of Karnataka state have a large population of predatory fish. A total of 27 species belonging to six families have been encountered in pong reservoir (Singh, 2001). Sakhare and Joshi (2003) reported 20 species of fishes from Bori reservoir of Maharashtra where order Cypriniformes dominated with seven species.Suresh (2003) reported 54 fish species in Loktak Lake in Manipur and reported 15 commercially important fish species. Mahapatra (2003) recorded abundance of cat fishes in Hirakud reservoir (Orissa) and reported 43 species,out of which 18 species were economically important. Sakhare and Joshi (2003) reported 34 species of fishes from reservoirs of Parbhani District of Maharashtra. Dhere (2002) reported presence of only 18 fish species from Karpara reservoir and Niture (2009) reported 27 fish species from same Karpara reservoir.

Acknowledgements

Author is thankful to Dr V. B. Garad, Head, Department of Zoology and Fishery Science, D.S.M. College, Parbhani for providing Laboratory facilities and research supervisor Dr. S.P Chavan, School of Life sciences, S.R.T.M. University, Nanded.

References

Ahirao, S. D and Mane, A. S. 2000. The diversity of ichthyofauna, taxanomy and fisheries from freshwater of Parbhani district, Maharashtra State. *J. Aqua. Biol.* 15 (1 and 2): 40-43.

Desai, S. S. 1980. Fisheries of Nathsagar reservoir. *India Today and Tomorrow*. 8 (4): 161

Biswas, K. P. 1990. A Text Book of Fish, Fisheries and Technology, Narendra Publishing House, Delhi, p143-164.

Day, Francis. 1889. The fauna of British India including Ceylon and Burma fishes.

Devi, B. S. 1997. Present status Potentialities, management and economics of fisheries of two minor reservoirs of Hyderabad.

Dhere, R. M. 2002. Hydrobiological study of Karpara reservoir in relation to fish production. Ph. D. Thesis, Swami Ramanand Teerth Marathwada University, Nanded.

Gopinath, P. and Jayakrishnan, T. N. 1984. A study on the pisicifauna of the Idukki Reservoir and catchment area. *Fish. Tech.* pp. 131 – 136.

Jain, A. K. 1998. Fisheries Resource Management in Rajasthan. An overview of present status and Future scope. *Fishing Chimes*, 17(11): 9-15.

Jayaram, K. C. 1981. The Freshwater Fishes of India, Pakistan, Bangladesh, Burma and Sri Lanka – A Hand Book, SZI, Calcutta, India.

Jayaram, K. C. 2002. Fundamental of fish taxonomy, Narendra Publications, Delhi.

Jhingran, A. G. 1980. Riverine fishery resources of India and their socio-cultural impact. *Tropical Ecology and Development*. pp. 747-756.

Jhingran, A. G. 1988. Reservoir fisheries in India. *Jr. of the Indian Fisheries Association*. 18: 261-273.

Jhingran, V. G. 1985. Fish and fisheries of India. Hindustan publishing corporation India, New Delhi. p. 106, 171-191.

Johal, M. S. and Jha, S. K. 2007. Fish Diversity of Haryana State and its Conservation Status. *Fishing Chimes*. 27(1): 107-108.

Kadam, M. S. 2005. Studies on biodiversity and fisheries management at Masoli reservoir in Parbhani Dist. M. S. India. Ph. D. Thesis, Swami Ramanand Teerth Marathwada University, Nanded.

Kumar, K. 1990. Management and Development of Govindsagar reservoir, A case study. Proc. Nat. Workshop Reservoir Fish. 13-20.

Niture, S. D. and Chavan, S. P. Fisheries Management of Yeldari Reservoir, Maharashtra Advances in Aquatic Ecology (volume 4) Ed. V. B. Sakhare). Vol. 4: 152-172.

Niture S. D 2009 Prospectivies of Inland fisheries development in Parbhani and Hingoli districts of Maharashtra. India. Ph. D. Thesis, Swami Ramanand Teerth Marathwada University, Nanded.

Pandey K. C. 1998. Concepts of Indian Fisheries. Shree Publishing House, New Delhi.

Parihar, R. P. 1999. A Text Book of Fish Biology and Indian Fisheries. Central Publishing House, Allahabad. pp. 310-313.

Sakhare, V. B. 1999. Fisheries of Yeldari reservoir. Maharashtra. *Fishing Chimes*. 19(8): 45-47.

Sakhare, V. B. and Joshi, P. K. 2002. Ecology of Palas-Nilegaon Reservoir in Osmanabad District, Maharashtra. *J. Aqua Bio*. 18. (2): 17-22.

Sakhare, V. B. 2002. Studies on Some Aspects of Fisheries Management of Yeldari Reservoir. Ph. D. Thesis, Swami Ramanand Teerth Marathwada University, Nanded.

Sakhare, V. B. and Joshi, P. K. 2002, Ecology and ichthyofauna of Bori reservoir in Maharashtra. *Fishing Chimes*. 22(4): 40-41.

Sakhare, V. B. and Joshi, P. K. 2003. Reservoir fishery potential of Parbhani district of Maharashtra. *Fishing Chimes*. 23(5) 13-16.

Sakhare, V. B. 2007. Reservoir Fisheries and Limnology. Narendra Publishing House, Delhi.

Srinivas, 2007. Fisheries of Edulabad Reservoir in A. P. *Fishing Chimes*. 26(10): 105-107.

Talwar, P. K. and Jhingran, A. G. 1991. Inland fishes of India and adjacent countries, Oxford and IBH Publishers, New Delhi.

.

Chapter 19

Studies on Water Quality of Godavari River at Ramkund, Nashik, Maharashtra

☆ *Rekha S. Bhadane*

ABSTRACT

The present investigation deals with the study of physico chemical nature of water from Ramkund of Godavari river. It is believed that Prabhu Ramchandra used to take bath in Godavari at Ramkund. Water quality of Ramkund is assessed during April 2012 to March 2013. The study includes water Temperature, transparency, total dissolved solids, pH, chlorides, dissolved oxygen, biochemical oxygen demand.The study revealed that the water is not suitable for bathing and drinking purposes.

Keywords: *Ramkund, Godavari, Water quality.*

Introduction

The river Godavari is the main source of water supply for Nashik city. After every 12 years, Kumbhamela is held in this holy city where people from all parts of the country come to Ramkund for various religious purposes, and also to take a holy dip in the water of Ramkund.

The work on physico-chemical parameters of Ramkund water can provide valuable information to organisers and planners, scientists and other authorities to examine the current status of water quality at Ramkund.

Ramkund is situated about 35 kms away from the origin of river Godavari in Nashik city of Maharashtra state. It being a famous pilgrim spot, thousands of people take holy dip on daily basis. It is believed that the river takes a 90 degree turn at this place. Now -a-days, due to various human activities, this place is heavily polluted, causing the need to analyse the water quality here.

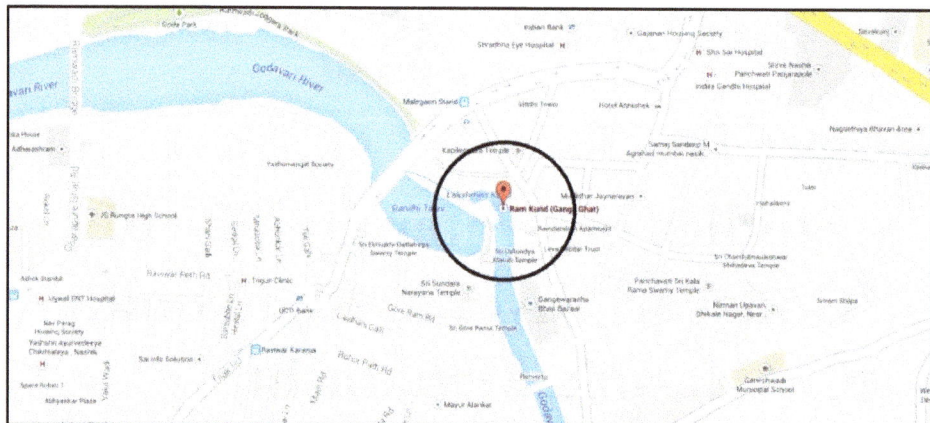

Figure 19.1: Map Location of Godavari River at Ramkund.

Materials and Methods

Monthly water samples were collected from the three stations for a period of one year from April 2012 to March 2013. Water samples were collected in one litre plastic containers and brought to the laboratory for analysis. Parameters like temperature, transparency, total dissolved solids, pH, chlorides, total hardness, dissolved oxygen and biochemical oxygen demand were analysed in the laboratory.

Table 19.1: Physico-chemical Profile of Ramkund of River Godavari at Nashik

Sl.No.	Parameters	Standard Limits (Drinking Water)	Range
1.	Water Temperature (ºC)	–	21-27
2.	pH	7.5	7.3-8.6
3.	Transparency (cm)	75	26-60
4.	Total Dissolved Solids (mg/l)	118	120-338
5.	Total Hardness(mg/l)	92	116-304
6.	Chlorides (mg/l)	10.8	36.9-48.4
7.	Dissolved Oxygen (mg/l)	–	6.4-15.2
8.	Biochemical Oxygen Demand (mg/l)	–	10-200

Results and Discussion

Temperature

In the present investigation the temperature values of water ranged from 21°C to

27°C. The water temperature showed an increase during summer and decrease during winter.

pH

The pH of the water is affected due to the reaction of CO_2 and also due to organic and inorganic solutes present in water. The pH is associated with high photosynthetic activity. pH values ranged from 7.3 to 8.6. Low pH values were recorded in monsoon; this mightbe due to high turbidity in the water. Klein (1973) pointed out that the pH values below 5 and above 8.8 are detrimental.

Transparency

The transparency ranged between 26 cms to 60 cms. The water was more transparent during summer as compared to monsoon and winter.

Total Dissolved Solids

The total dissolved solids were variable and ranged between 120 to 338 mg/lit. The amount of total dissolved solids increases due to the release of decaying matter from aquatic vegetation. The results are coinciding with earlier work carried out by Verma *et al.* (1998) and Salodia (1996).

Total Hardness

The term 'Hardness' is generally used as the assessment of the water quality. Swingle (1967) suggested that a total hardness of 50 mg/lit $CaCO_3$ equivalent is dividing line between soft and hard waters. During present investigation the total hardness ranged between 116-304 mg/lit. Higher values were recorded during summer and lower values were recorded during winter.

Chlorides

These are generally present in natural waters. It's an important factor indicating stress in a system depending on the pollution status of the system. Chloride values ranged from 36.9 to 48.4 mg/lit.

Dissolved Oxygen

The Dissolved Oxygen is of great importance to all living organisms. The dissolved oxygen values ranged from 6.4-15.2 mg/lit. Das (2002) stated that dissolved oxygen along with turbidity could provide information about the nature of an ecosystem better than any other chemical parameters.

Biochemical Oxygen Demand

This is an indicator parameter to know the presence of biodegradable matter in waste and express degree of contamination. Biochemical oxygen demand values ranged from 10-200 mg/lit., the values were more during monsoon than in winter and summer.

Conclusion

Thus the present study concludes that water from Ramkund is heavily polluted, which would possibly result in various health problems to people taking a holy dip in it. The results further indicate that pH, chloride, total hardness, total dissolved solids is not within the permissible limits. So to conclude, it can be said that the river water of Ramkund is not fit for domestic and drinking purposes and it definitely needs treatment to minimize the contamination.

Acknowledgement

The author is thankful to the Coordinator, M.G. Vidyamandir and Principal, L.V.H. Mahavidyalaya for providing necessary research laboratory facilities.

References

APHA. 1995. Standard Methods for Examination of Water and Wastewater, American Public Health Association, Washington DC, 19th edition.

Das, A. K. 2002. Evaluation of production potential in a peninsular reservoir (Yerrakalva) *J. Inland Fish Soc. India*, 34(2).

Klein, L. 1973. River Pollution, II: Causes and effects (5th Imp) Butter worth and Co. Ltd.

Salodia, P. K. 1996. Freshwater Biology – An ecological approach. pp. 64-68.

Swingle, H. S. 1967. Biological means of increasing productivity in ponds, FAO Fish. Rep., 44(3): 416-23.

Verma, S. R., Tyagi, A. K. and Dalela R. C. 1978. Pollution studies with reference to the biological indices, *Proc. Ind. Acad. Sci.* 87-113; 123-131.

WHO. 1984. WHO Guidelines for drinking water, Geneva, Switzerland, Vol. 1.

Chapter 20

Potability Studies of Drinking Water in Ambajogai, Maharashtra

☆ *V.S. Hamde*

ABSTRACT

A longitudinal study of the bacteriological quality of rural water supplies was undertaken for a movement towards self-help against diseases, such as diarrhoea, and improved water management through increased community participation. Twenty five water samples from different sources, such as well, borewells, hand pumps, and from households were collected from Ambajogai town, India. Overall, 49.8 per cent of total samples were polluted, whereas 45.9 per cent of the samples from piped water supply were polluted. The quality of groundwater was generally good compared to open wells. Irregular and/or inadequate treatment of water, lack of drainage systems, and domestic washing near the wells led to deterioration in the quality of water. No major diarrhoeal epidemics were recorded during the study, although a few sporadic cases were noted during the rainy season. As a result of a continuous feedback of bacteriological findings to the community, perceptions of the people changed with time. An increased awareness was observed through active participation of the people cutting across age-groups and different socioeconomic strata of the society.

Keywords*: Community participation, Diarrhoea, Drinking-water, MPN.*

Introduction

Water is the basic element of social and economic infrastructure, which is essential for healthy society and sustainable development. When 70 per cent of the earth's crust is water, only 2.5 per cent is freshwater but even here, including the frozen extremities of the globe and mountainous heights only 0.0085 per cent is stated to be available for the vast humanity. It is common knowledge that the management of water resources dates back to 5ᵗʰ century B.C. When,' Treatise on Air, Water and Place' became part of the Hippocratic corpus. Yet it is wonder that humanity realized only in the 19th Century A.D. that water is one of the important determinants of health and disease. Even of the important determinants of health and disease. Even in this computer Era of phenomenal scientific and technological developments, it is rather disconcerting that there are still constraints in achieving water-quality assurance especially in the developing countries. However poor water quality continues to pose a major threat to human. Village water supply contributed to reduces the mortality rate of children and to increasing life expectancy. It reduces the suffering and hardship caused by water related diseases and results in significant benefit to individuals and to society. Such as saving in Medical treatment including cost of medicines, workdays and income due to reduction in sickness, Travel, costs and time required obtaining health care, increased productivity and extended life span.

☆ It helps is to draw maximum economic benefits as it helps in considerably in saving time and energy both, in the following ways;

☆ It improves opportunities for keeping livestock or growing subsistence crops.

☆ Communities with adequate water supplies attract small business and may reduce out migration.

☆ The development process for water supply may be extended to community projects.

☆ In case of larger communities with buildings and other valuable properties, water supply may be designed to periodic improved fire fighting capacity.

☆ Easier access to safe-water can improve family and social development. When women are free from water bearing, they have more time not only for income producing work but also for childcare.

☆ Least incidence of the Water-borne diseases and improved health and hygiene particularly of the poor, resulting into reduced infant and maternal mortality.

☆ Development of industries and industrial production due to availability of additional water which will not only give boost to the economic development and increased productivity but will also in a decade's time, increase the income level and affordability for the purpose of full cost recovery.

Dhindsa *et al.* (1997) analyzed that drinking water sample was collected from the Ghat gate and Gandhi Nagar areas at selected locations of Jaipur city indicates that all the 15 samples collected by them from different points in both the localities were safe for drinking, as they did not contain any biological life and coliform organisms. Mirchandani (1998) made an attention to certain important factors that need to be considered to make a water treatment plant achieve certain important factors that need to be considered to make a water treatment plant achieve certain objectives as well as to make the operation and maintenance of the plant simple, safe and economical. Sharma *et al.* (2002) indicated that there is no major pollution hazard in the spring water of Bilaspur area as the MPN count of coliforms in the water samples has been found to be zero. The water in the area is highly alkaline and is very hard. The chloride, Iron and fluoride contents in the spring water of the area are low. The quality of ground water in most of the areas is suitable for drinking purposes. Kelkar *et al.* (2001) collected samples from various city of India [Nagpur, Panaji, Ghaziabad and Jaipur] after the start of water supply (1st flush) and after half an hour indicates during continuous water supply, samples negative for fecal coliforms were more than 90 per cent. In case of Panaji, 100 per cent samples were free from both coliforms and fecal coliforms whereas during IWS it ranged from 24 to 73 per cent. In Ghaziabad, Positive samples for bacteriological quality a CWS were lower as compared to the IWS, similar observation were recorded for Nagpur and Jaipur cities. Thus, during IWS, the percentage of hygienically safe water samples was much less as compared to that during CWS. In Ghaziabad, low percent of samples negative for fecal coliforms during both modes of operation is attributes to direct supply from tube well water without chlorination. Wherever disinfection of water by chlorination was effective as indicated by the presence of adequate residual chlorine, the quality of water was safe irrespective of the mode of supply. Thus, potential health-risk involved in IWS is of significantly higher magnitude. Bhave (1981) proposed a detailed design of a water supply system done for a design period of 30 years for the SCWSS (single conventional water supply system) and DPWSS (Dual purpose water supply system), for a suburb adjoining Mumbai from Thane District, Ambarnath. From the results it can be concluded that it is economically feasible to construct DPWSS with advanced water treatment plant for achieving high quality water for drinking and cooking purposes leading to saving of capital cost of WTP in SCWSS. Indirabai and George (2002), Drinking water quality of tap water and bore well water involving physical, chemical and biological parameters showed considerable variation in the different characteristic studied. The water quality index calculated for the various parameters tested over a period of three months ranged between 63.13 and 69.50 for tap water and 65.58 and 73.52 for bore well water. It is axiomatic that all the samples tested were of medium (poor) quality for drinking except for two. Hence it is pertinent that water needs purification prior to its utilization, so as to control the outbreak of water borne health hazards. Necessary steps should be taken to ensure good quality and quality of drinking water. They made an attempt to ascertain the ascertain the quality of the Pondicherry coastal water. Four sampling stations-2 on the 1-fathom line and two on the 5-fathom line were fixed. The results revealed that the total coliforms bacterial population showed seasonal variation *viz.* between post- and pre-monsoon periods. Secondly, the stations near the coast showed very low counts of bacterial population when compared to the 5-fathom line *i.e.* about kilometer away from the coast on the sea. However, such incidence level is well below the admissible standard values indicated by W.H.O. Asati (2009) conducted the experimental study which found the water quality of ground water was acceptable from physico-chemical analysis while as far as the bacteriological standards was not at all fit for drinking purposes, it can be used after using disinfection system. Water is the major constituent of all living things and needed by them for various purposes. The demand for quality drinking water had changed considerably with the development in olden days, the only requirement of drinking water was that it should be free flowing and non turbid. The need for better environment and health cannot be over emphasized. With increasing industrialization, urbanization, and growth of population, India's environment has become fragile and has been causing concern (Padolkar, 2004). Urbanization has direct impact on water bodies as the settlement takes place around the vicinity of water bodies and due to lack of space people have tendency to encroach upon the lake (Fukmare, 2002).

Ground water is the chief source of drinking water in India and this is only 0.61 per cent of the total available water on the Earth (Singh and Malik,2011). Reported that only 4 per cent of world's freshwater resources are available in India while India inhabitants 14 per cent of the world population. Indian population the per capita availability of water is steadily reducing; and when this drops below $1700\,m^3$/person/year, India will be water stressed. Water is a good solvent and picks up impurities easily. Pure water is tasteless, colorless, and odorless is often called the universal solvent. When water is combined with carbon dioxide to form very weak carbonic acid, an even better solvent results. As water moves through soil and rock, it dissolves very small amounts of minerals and holds them in solution. Calcium and magnesium dissolved in water are the two most common minerals that make water "hard." The degree of hardness becomes greater as the calcium and magnesium content increases and is related to the concentration of multivalent cations dissolved in the water. Excessive groundwater exploitation has resulted in lowering of water table in rural and urban areas of India. The water quality parameters decide the portability of water (WHO, 1971). This paper evaluates the quality of drinking water from three different sources *viz.,* open wells, bore wells, and corporate water supply in the Ambajogai city.

Ambajogai town is a Tehsil of Beed District (MS). It is about 98 km away from its district head quarters by road. The water quality variables are broadly classified into three categories:

1. *Physical*: The physical parameter includes general appearance, temperature, turbidity, color, taste and odor, etc.

2. *Chemical*: Chemical parameter include all possible inorganic and organic substances such as pH, acidity, alkalinity, hardness, conductivity, chlorides, sulphates, nitrates, BOD, COD, DO, etc.

3. *Bacteriological*: The bacteriological parameters includes coliform, MPN, Total plate count(TPC), bioassay counts or species diversity etc. The water samples collected from study area depend upon the method of sampling. All samples of water were properly labeled and are accompanied by complete and accurate identifying and descriptive data. Data collected include date and time of collection, type of source of the sample and temperature of water at the time of collection. When samples collected from the same sampling point for different analysis, it is essential that the sample for bacteriological examination be taken first. Samples were collected in

polythene material based plastic bottle. Sample bottles were carefully cleaned before use and rinsed with a chemical-acid cleaning mixture by adding one litres of concentrated sulphuric-acid slowly with stirring of 35ml saturated sodium dichromate solution. After having been cleaned, bottles are rinsed thoroughly with tap water and then with distilled water. About 2 to 2.5 litres of the samples was required for analysis. Prior to filing the sample bottle was rinsed out two or three times with water to be collected. Proper care taken to obtain a sample from of existing conditions and to handle it in such a way that it does not determinate or become contaminated before it reached the laboratory. The samples were reached the place of analysis as quickly as possible within 2 hrs of collections. The time elapsed between collection and analyses were recorded in the laboratory report. Parameters like temperature were carryout only at spot because it may change their characteristics significantly during transport. For physico-chemical and biological examine of water a proper sampling procedure (grab sample) was adopted. The frequency of collection of samples for chemical analysis depends on the variability of the quality of tested water, the type of treatment processes used and other local factors. Samples for chemical examination should be collected at least once every three months in supplies serving more than 50,000 inhabitants and at least twice a year or supplies nearly 50,000 inhabitants. More frequent sampling for chemical examination may be required for the control of water treatment processes.

It is necessary to collect samples of both raw and treated water for examination of toxic substances at least every three months and more frequently when sub tolerance-levels of toxic substances are known to be generally present in the source of supply or where such potential pollution exists. For bacteriological sampling which controls the safety of supply to the consumer, the frequency of sampling and the location of sampling points of pumping stations, treatment plants, reservoirs as well as the distribution system, should be such as to enable a proper evaluation of the bacteriological quality of entire water-supply. Total twenty five numbers of water samples were collected from different area. The distance between two consecutive locations is about 500 meters. The water samples were analyzed for pH, turbidity, total hardness (calcium and magnesium), chloride, alkalinity and coliforms. All these parameter were carried out at laboratory by suitable method of analyzed.

Microbiological Analysis

The microbial study was performed on randomly selected 25 samples. Of the 25 samples, 20 samples were found to have microbial numbers within the maximum permissible limit of 1 coliform per 100 ml as per the safe drinking water act. An open well sample at some places (MPN: 100,/100 ml), a sample of one location is (MPN: 60/100 ml) and samples from three bore well sources (MPN: 04/100 ml) have registered positive presumptive test results.

This study has presented the physiochemical and microbiological analysis of water samples taken from different area from Ambajogai. The WHO Guidelines for drinking water quality states that the pH range of drinking water should fall between 6.5 and 8.0.

Materials and Methods

Water sampling: The samples were collected in pre cleaned sterilized bottles and stored in an icebox. A total of 25 water samples were collected from different sources, such as shallow.

Hand pumps, dug wells and public health water supply taps. All the analyses were carried out in triplicate. The results were reproducible within ±3 per cent error limit.The pH and electrical conductivity of the water were determined on site using pH and EC scan meter. The TDS were calculated using a formula from the Manual on Water supply and Treatment(1991). Na, K and Ca concentrations were determined using ELICO CL-220 Flame photometer. Total alkalinity and total hardness were measured by titrimetric methods using standard sulfuric acid and standard EDTA solutions, respectively.

Microbiological Analysis

The microbial study was performed on randomly selected 25 samples. The samples were tested for the presence of the coliforms by the Most Probable Number (MPN) technique. Samples with 0 coliforms per 100 ml of original water were taken as excellent, with 1-10 coliforms as acceptable and above 10 coliforms as polluted (WHO, 1990).

Bacterial Culture

Besides the MPN test, 50 ml of each sample was centrifuged in a sterile plastic centrifuge tube at a speed of 2000 rpm for 10 minutes under cold conditions ($15^{\circ}C \pm 5^{\circ}C$). The pellets were then streaked onto sterile MacConkey agar, and sterile Brain Heart Infusion (BHI) agar plates incubated at $37^{\circ}C/18$-24 hours followed by propagation into BHI broth and identification by standard biochemical testing (Kreig, 1986; McFaddin, 1980; Sneath *et al.*, 1986).

Results and Discussion

The samples of Ambajogai region was tested for physical, chemical and bacteriological parameters and their results were compared with W.H.O. drinking water standards. After performing Physical, Chemical and Bacteriological tests on water sample taken at the faucet-points of water supply, the calcium hardness, Magnesium hardness, Total hardness and pH were found nearer to intolerable concentration are found to be safe for domestic and drinking purposes. Chloride, alkalinity and turbidity were found within the permissible limit.

The data obtained for iron, pH,hardness and MPN of water samples collected from different sources are presented in Table 20.1. It was observed that 11 samples shows coliform exceeding the water quality standard *i.e.* one per 100ml of sample as per safe drinking water act. However occurrence of fecal coliforms in water sample

Table 20.1: Physico-chemical and Microbiology Analysis Bore Well Water

Place of Sample Collection	pH	Iron mg/L	Hardness mg/L	Total Solids gm/L	Total Dissolved Solids gm/L	Bacterial Count MPN/ 100ml	Total Pathogen	Hetero- trophs X 10⁴
Adarsh colony	8.2	0.31	350	0.1	0.04	60	+	0.7
Adarsh colony	8.1	.31	400	0.2	0.04	60	+	0.7
Adarsh colony	8.0	0.32	300	0.2	0.01	100	+	0.2
Adarsh colony	8.2	0.32	280	0.2	0.01	10	+	0.2
Adarsh colony	8.2	0.32	320	0.1	0.01	4	+	0.3
Saraswat Colony	7.9	0.39	550	0.2	0.01	4	+	0.3
Saraswat Colony	8.0	0.38	600	0.1	0.04	–	–	0.2
Saraswat Colony	8.0	0.32	600	0.3	0.03	–	–	0.2
Saraswat Colony	8.0	0.36	300	0.3	0.03	4	–	0.2
Saraswat Colony	7.9	0.31	310	0.3	0.01	4	–	1.0
RamanandColony	7.9	0.31	330	0.1	0.01	–	–	0.1
RamanandColony	8.0	0.30	360	0.2	0.01	–	–	0.2
RamanandColony	8.0	0.32	360	0.2	0.01	–	–	0.2
Shrinagar colony	8.1	0.35	360	0.2	0.01	4	+	0.2
Shrinagar colony	8.0	0.37	360	0.2	0.01	10	+	0.9
Shrinagar colony	8.0	0.31	370	0.2	0.01	2	–	0.1
Shrinagar colony	8.0	0.31	380	0.2	0.01	–	–	0.1
Shrinagar colony	8.0	0.22	400	0.1	0.01	–	–	0.1
Shrinagar colony	8.1	0.21	410	0.2	0.01	–	–	0.1
Shrinagar colony	8.1	0.21	410	0.2	0.01	–	–	0.1
Housing Society	8.1	0.21	410	0.2	0.01	–	–	0.1
Housing Society	8.0	0.21	500	0.2	0.01	–	–	0.1
Housing Society	8.0	0.20	420	0.2	0.01	–	–	0.1
Boys Hostel YM	8.0	0.21	430	0.2	0.02	–	–	0.1
Boys Hostel YM	8.0	0.21	330	0.2	0.02	–	–	0.1

and occurrence of *Salmanella, Shigella* in 08 water samples highlights more severe environmental problem in Ambajogai region.

The level of iron was formed to be higher in 17 samples. Some microorganisms like *Leptothrix and Gallionella* are able to utilize dissolved iron as energy source. Although iron has got little concern as a health hazard. It is still consider as a nuisance in excessive quantity. The limit of iron in water is based on aesthetic and taste consideration rather than its physiological efforts.

Water having hardness below 300mg/L-is considered potable. The present study shows that all samples have hardness above the permissible limits. Hardness beyond the permissible limits may produce gastrointestinal irritation (ICMR 1975).Hard water is also unsuitable for domestic uses like washing, cleaning and laundering. Geldreich (1992) reported that presence of bacteria in well water degrading complex organic compounds. However our results indicate that majority of samples contains heterotrophic microorganisms. This may be due to percolation of domestic waste in that area. It is therefore suggested that the bore well water should be treated properly before consumption and awareness must be created among peoples

Excessive hardness in water does not cause any health hazards, however water should contain some amount of hardness because calcium salts are required for the growth of children. The soft water dissolves lead much more readily than the hard-water and hence there is less chance of lead poisoning with hard-water than with soft water. Presence of coliforms in water sample is found hazardous for human health; it may because water-Borne diseases such as cholera, dysentery, Para-Typhoid, Polio and Jaundice are viral-diseases due to untreated drinking water, which may be contaminated due to fecal matters of waste product of affected persons. And there is also a need to control Ca and Mg hardness in the water as they are in intolerable level. From the above observation (Table 20.1), it can be concluded that the water supplied to community is not good for the health of consumers and hence its proper and immediate treatment. Table 20.1 illustrates the physico-chemical parameters of ground water Ambajogai. The analytical results of physical and chemical parameters of ground water were compared with the standard guideline values as recommended by the WHO, with the statistical parameters of different chemical constituents of ground water analyzed in twenty five locations are shown in Table 20.1.

Colour

Colour in water is an important constituent in terms of aesthetic considerations. To be aesthetically pleasing, water should be virtually free from substances introduced by manmade activities which produce objectionable color.

Turbidity

Higher value of turbidity can make water unfit for drinking purposes. The average value of turbidity was 0.38g/L and it was ranging from 0.1 to 0.5g/L. The WHO standards recommendation for the turbidity is 0.5 g/L. All collected samples were well agreed within the limits.

Total Dissolved Solids (TDS)

The total dissolved solids range from 0.718 to 7.952 g/l with a mean of 4.242 g/l. According to WHO specification, TDS up to 5g/l is highest desirable and up to 1g/l is maximum permissible category. None of the samples are fall under highest desirable limit. No samples were exceeds the permissible limits. Based on the concentration of TDS ground water can be classified as follows: Up to 5g/l as desirable for drinking; up to0.1g/l as permissible for drinking and up to 3g/l as useful for irrigation. Based on the above classification, it was observed that out of 25 samples, all samples are permissible for drinking and irrigation purposes.

pH

The value of pH in the raw water used for public water supplies is more important because without adjustment to a suitable level, such waters may be corrosive and adversely affect treatment processes, including coagulation and chlorination. The pH values in the ground water samples of the study range from 7.9 to 8.2with the mean value of 7.95. The ground water samples are alkaline in nature pH. According to WHO specification, all the samples are fall under desirable limit of pH (6.5-8.5).

Total Alkalinity (TA)

In most of the natural water contains substantial amounts of dissolved carbon dioxide, which is the main source of alkalinity. In the present study, alkalinity ranges between 180 to 1020mg/l with a mean value of 603mg/l. All water samples belong to permissible limit.

Total Hardness (TH)

The determination of hardness in raw waters subsequently treated and used for domestic water supplies is useful as a parameter to characterize the TDS present and for calculating chemical dosages, where lime soda softening is practiced. Because hardness concentrations in water have not been proven health related, the final level achieved principally is a function of economics. In the present study, the total hardness of the water samples is optimum and within the permissible limit.

Calcium (Ca)

Calcium is one of the most abundant substances of the natural water. Being present in higher quantities in rocks, it is leached from these to contaminate water. Disposal of sewage and industrial wastewater is also an important source of calcium. In the present study, the calcium range from 70 to 480mg/l with a mean value of 301mg/l. All water samples are exceeds the standard given by BIS. But it has no hazardous effect on human health (data not shown).

Sulphate and Phosphate

Sulphate is unstable if it exceeds the maximum allowable limit of 400mg/l guideline given by WHO and it causes a laxative effect on human system with the excess magnesium in groundwater. In the present study, the sulphate is range from 60 to 70mg/l. According to WHO classification, all samples are fall under desirable limit. Amount of phosphate present in the water samples are negligible (Data not shown).

Conclusion

Groundwater is an important natural resource, which plays a significant role in human development. Based on the present study of groundwater samples collected from study area, the following conclusions have been drawn,

1. More than 50 per cent of water samples are not suitable for drinking and agricultural purposes in regards to the high concentration of TDS present in the water samples.

2. High value of total hardness present in the 16 locations of water samples and it leads to scaling problem.

3. All the values of pH in water samples is well agreed with guidelines given by WHO, BIS.

References

Asati S R. 2009. Report on the Study Conducted in the Campus of M Patel Institute of Engg. and Technology In Gondia Distt. of MS on Physico-Chemical and Bacteriological Parameters of Water Samples From Different Sources. *Journal of IPHE, India*.

Bhatia K. K. S. 2007. Water Quality and Water Quality Modeling. *Journal of All India Council of Technical Education*, New Delhi pp. 01-28.

Bhave, P. R. 1981. Node flow analysis of water distribution systems. *Journal of Transportation Engineering,* 107(4): 457-467.

Dixit, A J. 2003. Evaluation of Suitability of Ambanala Water Amravati, for Irrigation. *Journal of IWWA*, pp. 230-247.

Duttam, A K. 1994. Water-Borne Diseases and Drinking Water. *Journal of IWWA*, pp. 39-46.

Dhindsa, S S. 1997. Biological and Bacteriological Investigation on Contamination of Drinking Water in Jaipur: A Case Study', *Journal of IWWA,* pp. 243-249.

Elizabeth, K M. 2005. Physico- Chemical and Microbiological analysis of water from Lake Hussain-Sagar. *Journal of IWWA*, pp. 96-102.

Fukmare, A. K. 2002. Bacteriological status of drinking water in Akola City of M. S. *Asian Journal of Microbial Biotech., Environment Science*, Vol. 4: 2.

Geldreich, Edwin. E., 1996. Microbial Quality of Water Supply in Distribution Systems, Lewis Publishers.

Goyal, V. K. 1999. Water-Pollution, III Edition, Published New Age International Publication Ltd., New Delhi.

WHO. *Guidelines for Drinking Water Quality,* II Edition, WHO, Geneva (1990), pp. 97-100.

Indirabai, W. P. S. and S. George. 2002. Assessment of drinking water quality in selected areas of Tiruchirappalli town after floods. Poll. Res., vol. *21(3), 243-248.* ICMR. 1975. Manual of standards of quality for drinking water supplies. ICMR, New Delhi.

Kelkar, P. S. 2001. Water Quality Assessment in Distribution System Under Intermittent and Continuous Modes of Water Supply, *Journal of IWWA,* pp. 39-45.

Krieg, N. R. 1986. Enrichment and isolation. In *Manual of Methods of Bacteriology* eds. Gerhardt, P., Murray, R. G. E., Costilow, R. N., Nester, E. W., Wood, W. A., Krieg, N. R. and Briggs Phillips, G. pp. 112–142. Washington DC: ASM Press

McFaddin J. F. 1980. Biochemical tests for the identification of medical bacteria. New York. Wlliams and Wilkins.

Mirchandani, N. W. 1998. Planning and Layout of Water-Treatment Plants", *Journal of IWWA,* pp. 255-264.

Manual on Water Supply and Treatment. 1991. Central Public Health and Env. Engg. Organization (CPHEEO), Ministry of Urban Development New Delhi.

Padolkar, A T. 2004. Dual Purpose of Water-Supply System", *Journal of IWWA,* pp. 281-285.

Sharma, M. R. 2004. Assessment of ground water quality of Hamirpur area in Himachal Pradesh. *Poll. Res.* 23(1): 131-134.

Singh, K. P. and Malik, A. 2011. Multivariate statistical techniques for evaluation of spatial and temporal variations in water quality of Gomati River- India. A case study. Water Research. 38. pp. 3980-3992.

Sneath, P. H., Elizabeth, S. M., Holt, J. G. (eds.) 1986. Bergey's manual of Systemic bacteriology, vol. 1 Williams and Wilkins, Baltimore, Md.

WHO, 1971. International Standards for drinking water. 3 rd edn. WHO, Geneva.

Chapter 21

Impact of Bat Guano on Degradation of Water Pollutants in Purna River Buldana District, Maharashtra

☆ *Iqbal Shaikh*

ABSTRCT

The word guano originated from the Quichua language of the Inca civilization and means "the droppings of bat". The bats forage at night for insects over a particular area, and they return to the old temples during the day to sleep and care for their young. They attach themselves to ceiling, and their excrement accumulates on the floor below. In some situations the guano can reach a depth of feet in many years and appeared as guano-hip, and it has a valuable importance.

Bat guano was collected from the temple of Lonar crater of Lonar, Buldana District, Maharashtra. The bat guano, it dissolved in water of Purna River, (10:100) concentration was prepared and kept undisturbed till 30 days and parameters was noted at an interval of 2 hour and thereafter 5 days for about 24 hours and 30 days respectively. Resulted into increasing in pH and decline in chloride, nitrate, phosphate and sulphate content of industrial effluent after the addition of bat guano. Our investigation results indicated that bat guano used for degradation of water pollutants and bioremediation of aquatic ecosystem.

Keywords: *Bioremediation, Bat guano, Industrial effluents, Water pollutants.*

Introduction

Lonar crater is situated in village Lonar in the Buldhana District of Maharashtra, India. It has an almost perfectly circular shape and accumulated with water in the deeper parts of basin. Rocks in the crater reveal many characteristic features of the moon rocks. There are many old temples on the peripheral boundary of the crater which have now become roosting places for bats. Ramgaya Temple has become the source of sweet drinking water, as this is the only sweet water stream available in the crater; rest of the crater water is highly saline. Kamalja Devi temple is situated at the southern base of the crater. Morache temple (Peafowl's temple) is now famous for existence of thousands of bats and peacocks. Waghache temple (Leopards temple) is also famous for bats and people have seen leopard found in it many times.

Bat Guano

The word guano originated from the Quichua language of the Inca civilization and means "the droppings of bat". The bats forage at night for insects over a particular area, and they return to the old temples during the day to sleep and care for their young. They attach themselves to ceiling, and their excrement accumulates on the floor below. In some situations the guano can reach a depth of feet in many years and appeared as guano-hip, and it has a valuable importance.

Bioremedation and Bat Guano

One of the most serious universal, international problems facing us today is the removal of harmful compounds from industrial and municipal waste. If it is discharged into lakes and rivers, a process called eutrophication occurs (Prince, 2003).

Environmental contamination whether it is from industrial or municipal toxic waste that degrades the various environments is a vital concern to the public. Thus it is crucial to develop and implement accurate means to clean and preserve our precious and deteriorating environment. Although there are many techniques in cleaning environmental contaminations, one process has the most potential, namely bioremediation. Bioremediation, or commonly referred to as biodegradation, is a process in which microbes such as bacteria, fungi, yeast, or micro algae are involved in degrading toxic wastes (Pace, 1997 and Knezevich, 2006).

A marvelous symbiosis exits between the microorganisms and bat guano. Bacteria in the mammalian intestinal tract aid in the breakdown of food during digestion. These organisms synthesize enzymes capable of degrading a vast array of substances. Innumerable microbes are regularly excreted along with waste products and together with other organisms; they constitute the microbial population of a bat guano deposit (Steele, 1989).

Large populations of bat deposit thousands of kilograms of dropping annually. An ounce of bat guano contains billions of bacteria, and a single guano deposit may contain thousands of bacterial species. Guano being rich in bioremediation microbes cleans up toxic substances, (Barry *et al.,* 1997). At present we do not know these species.

Materials and Methods

To study the impact of bat guano on water, 10 mg bat guano was dissolved in 100 ml of Purna River water (10:100 proportions) for both times. After addition of bat guano in water, then the water was analyzed for the change in its pH, chloride, nitrate (NO_2), phosphate (PO_4) and sulphate (SO_4) contents. The change in water parameters were noted after every two hour upto 24 hours. Thereafter, the samples were kept undisturbed and analyses were carried out for 30 days at an interval of 5 days. The water was analyzed by using standard methods for water analysis suggested by APHA (1998).

Results

When bat guano was dissolved in river water with pH 5.00. After 2 hours the pH was found to be changed to 6.15 and after 4 hours increased gradually and it reached to 7.25 after 24 hours (Table 21.1). The river water was kept undisturbed till 30 days and the pH was noted after every 5 days upto 30 days. After 5 days the pH was seen to be increased upto 20 days and then it remained constant during 25 to 30 days of observations (Table 21.2).

When bat guano was dissolved in river water with chloride (201); nitrate (56.5); phosphate (57.5) and sulphate (46.0), after 2 hours the parameters was found to be changed to chloride (187), nitrate (52.4), phosphate (56.5) and sulphate (46.0) and after 4 hours decreased gradually to chloride (91), nitrate (24.4), phosphate (29.5) and sulphate (29.6) upto 24 hours (Table 21.1). The river water was kept undisturbed till 30 days and the chloride, nitrate, phosphate and sulphate was noted after every 5 days upto 30 days. After 5 days the parameters was seen to be decreased upto 20 days and then it remained constant during 25 to 30 days of observations (Table 21.2).

Discussion

Tilak *et al.* (2005) reported a number of bacterial species associated with the bat guano belonging to genera, *Azospirillum, Alcaligens, Arthrobacter, Acinetobacter, Bacillus, Burkholderia, Enterobacter, Erwinia, Flavobacterium, Pseudomonas, Rhizobium* and *Serratia.* He also suggested that this bacterium has high bioremediation capacity. Hutchens *et al.* (2004) had demonstrated aerobic methane oxidizing bacteria, *Methylomonas* and *Methylococcus* in bat guano.

The bacterial enzymes capable of degrading a number of substances (Martin, 1991; Dvorak *et al.,* 1992; Edenborn *et al.,* 1992; Bechard *et al.,* 1994; White and Chang, 1996; Frank, 2000; Kaksonen, *et al.,* 2003; Vallero *et al.,* 2003; Boshoff, *et al.,* 2004; Miranda, 2005; Seena, 2005; Tilak *et al.,* 2005). Murphy (1989) demonstrated a nutritious broth formation when the bat guano was added in water and further he proved that this broth supported the growth of numerous microbes.

Alley and Mary (1996) stated that an ounce of bat guano contains billions of bacteria and thousands of bacterial species and these bacteria are important to bioremediation. Sridhar *et al.* (2006) and Pawar *et al.* (2004) examined the fungal fauna of bat guano and used for bioremediation of Lack soil.

Table 21.1: Impact of Bat Guano on Water Content of Purna River at an Interval of 2 hrs

Ps	Sg	Time (Hrs)												
		0	2	4	6	8	10	12	14	16	18	20	22	24
pH	W1	5.00	6.15	6.49	6.55	6.65	6.82	6.55	6.84	6.85	7.91	7.95	7.02	7.25
Cl	W1	201	187	173	160	155	143	132	125	115	107	98	93	91
NO2	W1	56.5	52.4	48.8	47.0	44.8	41.5	39.5	29.5	27.8	26.1	25.4	24.8	24.4
PO4	W1	57.5	56.5	55.0	45.0	44.0	43.5	39.5	35.5	32.0	31.5	30.5	29.5	29.5
SO4	W1	46.8	46.0	45.6	44.5	43.8	42.3	39.5	37.2	34.8	33.2	30.5	29.6	29.6

All values are the mean of five replicates.

Ps: Parameters; Sg: Sampling; W1: Water from Purna River.

Table 21.2: Impact of Bat Guano on Water Content of Purna River at an Interval of 5 Days

Ps	Sg	Time (Days)							
		0	1	5	10	15	20	25	30
pH	W1	5.00±0.37	7.25±0.39 (+45.00)	7.42±0.24 (+48.40)	7.38±0.30 (+47.60)	7.40±0.32 (+48.00)	7.43±0.40 (+48.60)	7.55±0.45 (+51.00)	7.55±0.40 (+51.00)
Cl	W1	201±7.60	91±8.83 (-54.73)	86±10.95 (-57.21)	84±10.09 (-47.26)	83±9.73 (-58.21)	82±9.41 (-58.71)	81±11.06 (-59.70)	81±9.68 (-59.70)
NO2	W1	56.5±2.71	24.4±1.12 (-56.81)	24.0±1.25 (-57.52)	23.6±1.49 (-58.23)	23.1±1.28 (-58.94)	22.8±1.13 (-59.29)	22.2±1.29 (-59.29)	22.2±1.40 (-59.29)
PO4	W1	57.5±3.05	29.5±1.68 (-48.70)	27.5±1.96 (-47.83)	26.4±1.45 (-54.09)	25.3±1.29 (-56.00)	23.0±1.15 (-60.00)	21.4±1.28 (-62.78)	21.4±1.20 (-62.78)
SO4	W1	46.8±2.48	29.6±1.68 (-37.18)	28.9±1.59 (-38.25)	28.1±1.43 (-39.96)	27.8±1.39 (-40.60)	26.9±1.61 (-42.52)	26.1±1.46 (-44.23)	26.1±1.70 (-44.23)

All values are the mean ±SE of five replicates; Figures in parenthesis indicate percent change over the result on 0 day.

Ps: Parameters; Sg: Sampling; W1: Water from Purna River.

Conclusion

Anthropogenic activities, municipalities, various industries disposing their waste into the various aquatic resources.It is of utmost importance, hence, to prevent the pollution of aquatic resources by all possible means to control its quality from further deterioration. Applying microorganisms for river pollution control is an area of interest all over the world.

In the present investigation is an attempt to study the impact of bat guano with its rich microbial flora on bioremediation of aquatic resource as Purna river. The results revealed that within a period of 30 days, there was a remarkable reduction in the physico-chemical parameters of river pollutants, thus stabilizing the river pollutants, suggesting that water pollutants can be effectively treated by bat guano andis the excellent bioremediatant.

Acknowledgements

I express my sincere thanks to Dr. C. M. Bharambe, Department of Zoology, Vidnyan Mahavidyalaya, Malkapur, Dist. Buldana, for providing facilities to carry out research work.

References

Aaranson, S. 1970. Experimental Microbial Ecology. Academic Press, New York. pp. 236.

APHA, 1998. Standard methods for the examination of water and wastewater, 20th ed. APHA, AWWA and WEF New York, Washington DC.

Boyd, S. A. and E. G. Patricia. 2005. An Approach to Evaluation of the Effect of Bioremediation on Biological Activity of Environmental Contaminants: Dechlorination of Polychlorinated Biphenyls. *Environmental Health Perspectives*, 113(2): 180-185.

Chapelle, F. H. : Bioremediation: Nature's Way to a Cleaner Environment. U. S. Geological Survey. URL: http: //water. usgs. gov/wid/html/bioremed. html.

Conde-Costas, C. 1991. The effect of bat guano on the water quality of the Cueva EL Convento stream in Gauayanilla, Puerto Rico. *Nss. Bull.* 53(1): 15.

Dash, M. C., Mishra P. C., Kar G. K. and Das R. C. 1986. Hydrobiology of Hirakund Dam Reservoir. In: Ecology and pollution of Indian Lakes and Reservoirs. Mishra Publishing House, New Delhi, p. 317-337.

Dilip, K. M. and N. R. Markandey. 2002. Microorganisms in Bioremediation/edited by. New Delhi, Capital Pub., viii, 190 p., tables, figs., ISBN 81-85589-08-9.

Dvorak, D. H., Hedin, R. S. and McIntire, P. E. 1992. Treatment of metal contaminated water using bacterial sulphate reduction: results from a pilot-scale reactor. *Biotechnol. Bioeng.* 40: 609-616.

Edenborn, D. H., R. S. Hedin. 1992. Treatment of water by using sulphate reducing bacteria. *Biotech. Bioeng.* 30: 512-516.

Everett, J. W. ; J. Gonzales; L, Kennedy. 2004. Aqueous and Mineral Intrinsic Bioremediation Assessment: Natural Attenuation. *Journal of Environmental Engineering*, 130(9): 942-950.

Faison, B. D; and Knapp, R. B. 1997. A bioengineering system for in situ bioremediation of contaminated groundwater. *Journal of Industrial Microbiology and Biotechnology*, 18 (2-3): 189-197.

Keleher, S. 1996. Guano: Bats' Gifts to Gardeners. 14(1): 15-17.

Knezevich, V. ; O. Koren; E. Z. Ron; E. Rosenberg. 2006. Petroleum Bioremediation in Seawater Using Guano. Bioremediation Journal, Vol. 10th Issue 3, p. 83-91, 9p.

Pace, N. R. 1997. A molecular view of microbial diversity and the biosphere. *Science.* 276: 734-740

Pawar, K. V. and Deshmukh, S. S. 2004. Bioremedition of Lack soil using bat guano. *Indian J. Environ and Ecoplan.* 8(3): 699-704.

Pierce, W. 1999. Speech on " Bat guano " Sept., 1999. Cassette from National Vanguard Books, P. O. Box 330, Hillsboro, WV 24946.

Prince, R. C. 2003. Bioremediation in marine environments. Prince RC. Exxon Research and Engineering, Annandale, NJ 08801. Bioremediation.

Steele, D. B. 1989. Bats, Bacteria and Biotechnology. 7(1): 3-4.

Tuttle, M. D. 1986. Endangered gray bats benefits from protection. *Bat*, 4(4).

Vidali, M. 2001. Bioremediation. An overview. Dipartimento di Chimica Inorganica, Metallorganica, e Analitica, Università di Padova Via Loredan, 435128 Padova, Italy. *Pure Appl. Chem.*, 73(7): 1163–1172.

Walecha, V., Vyas V. and Walecha R. 1993. Rehabilitation of the twin lakes of Bhopal. In: Ecology and pollution on Indian lakes and reservoir. Ashish Publishing House New Delhi, p. 317-337.

Chapter 22

Abundance of Avifauna in the Mangrove Ecosystem of Kali Estuary, Karwar, West Coast of India

☆ *B. Vasanthkumar, S.V. Roopa*
and B.K. Gangadhar

ABSTRACT

Studies on avian community provide effective tools for monitoring forest in general and mangrove forest in particular. Evaluating bird communities of the mangrove forests to plan for biodiversity-friendly development is gaining significance. The present study was carried out in different mangrove ecosystem of Kali estuary, Karwar (14°50′21" N and 74°10′05"E) for the period of thirteen months from January 2008 to January 2009. The study area lying within the grid of 14° 50′ 39" N and 14° 51′ 49" N and 74°07′ 44" E and74° 13′ 25" E of was selected for the distribution and abundance of avifaunal community.

Keywords: *Avifauna, Mangroves, Kali River, Karwar, Karnataka.*

Introduction

In west coast of India, especially in mangrove ecosystems harboring a fairly rich faunal and floral wealth are relatively unexplored. Only a few reports are available pertaining to the mangrove forests. Not many studies have been made in the region of

Uttara Kanada district of Karnataka. Bhagwat (2003) examined the effects of landscape modification on bird diversity in Kodagu district. Some reports have been recorded on several aspects of bird life in and around Uttara Kannada district. Other reports related to survey of birds was also carried out in Uttara Kannda district and adjoining areas and no serious studies exclusively on bird diversity across different landscapes are made in the western Ghat regions. From the review of the above literature, it was inferred that not much studies are made in this region especially on mangrove ecosystem and that studies on census, diversity and relationship with vegetation is needed. Hence, the present study was attempted on the birds of natural forests including Mangrove forests.

Ali and Ripley (1983) reported the avifauna of Indian subcontinent being represented by 2094 forms belonging to 1200 species of which 19.9 per cent (417) forms are wetland birds (Rao *et al.,* 1997). In recent past years, importance of bird diversity and its conservation has been emphasized and such studies are encouraged. Much work related to the avifaunal diversity has been done in temperate forests, while, a very limited data is available in the tropics in general and mangrove forest in particular.

Materials and Methods

In the present study, totally five study stations were selected and fixed at different locales of mangroves of Kali estuary, namely Mavinahole creek, Kanasgiri, Sunkeri backwaters, Kadwad backwaters and Kinnar respectively (Figure 22.1). The study was carried out for the period of thirteen months from January 2008 to January 2009.

The bird community on the exposed mangrove mud flat area was scanned at monthly intervals from an indigenous dugout canoe with an outboard motor engine moving at uniform speed of 10km per hour for the period of one year. Sampling began at 7.00 AM and continued upto 9.00 AM and in the evening, bird sampling was done from 4.00 PM to 6.00 PM. In each habitat, minimum of 8 censuses were conducted. About four to eight censuses were carried out per month in each study site. At each study site, all the birds that were seen were recorded by species with field binoculars and census was not conducted on rainy and heavy misty mornings. All birds that were heard while survey and were not counted because of the lack of information on the identity of species by bird call alone. The species identification was based on Ali (1996) and the nomenclature was based on Grimmett *et al.* (1998).

Foot trails were conducted into the dense mangroves not accessible by boat and the birds therein were listed. The birds sighted were indentified using relevant field guides (Woodcock, 1980; Ali and Ripley, 1983; Sonobe and Usui, 1993; Grewal, 1995, Ali, 1996; Grimett *et al.,* 1998). A combination of Total Count Method and Linear Transact count method standardized o meet the requirement of the site were employed An binocular and 15-45x60 spotting scope was used for the this purpose.

Results and Discussion

The climate of Karwar is strongly influenced by the southwest monsoon occurring normally from June to September. However, since the present study is confined to water birds ecology the seasons have been categorized as summer, extending from

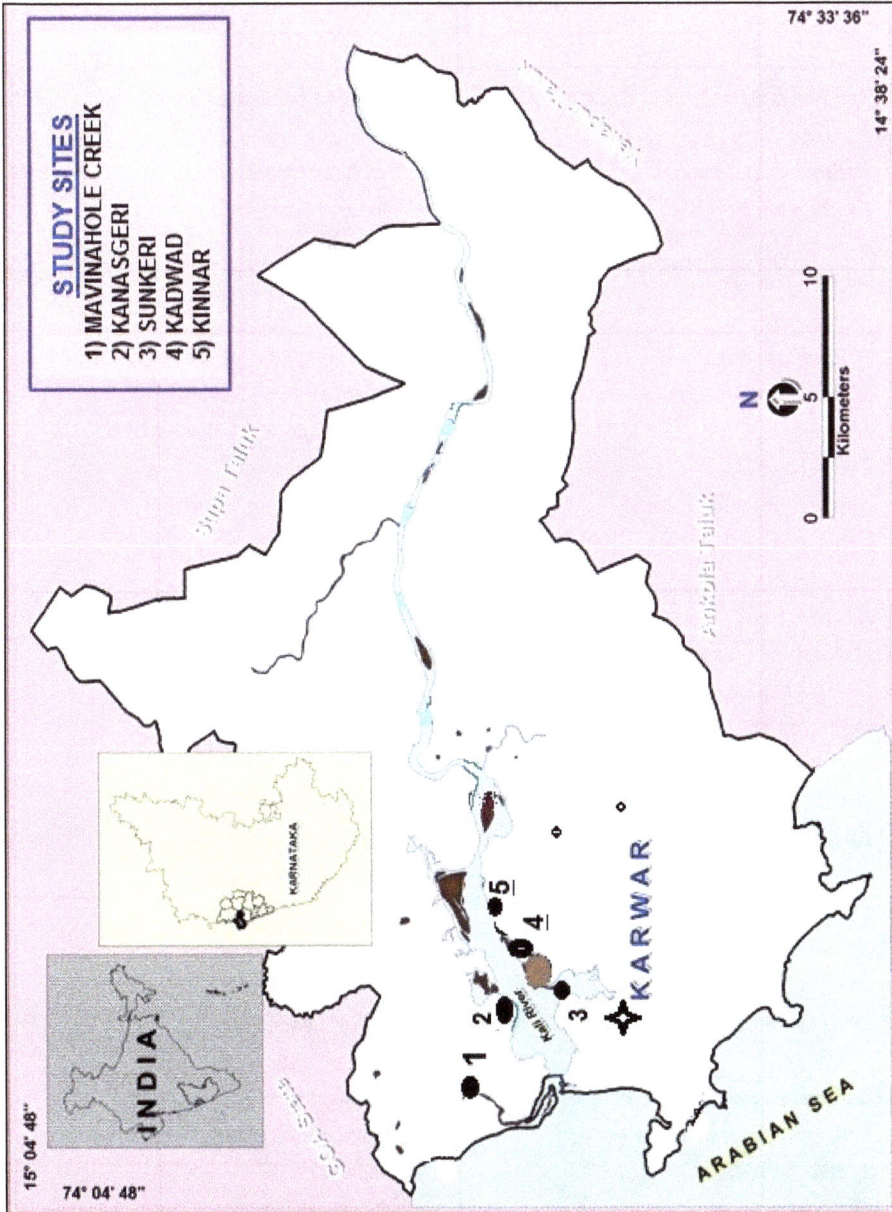

Figure 22.1: Map of Kali River Showing the Positin of Study Sites.

March to May; monsoon from June to September; post monsoon from October to November and winter from December to February. A check-list of avian fauna in the mangrove ecosystem of Kali estuary is depicted Table 22.1.

Table 22.1: List of Avian Fauna Recorded in the Mangrove Forests of Kali Estuary, Karwar

Scientific Name	Common Name
Spilornis cheela	Crested serpent eagle
Heliastur Indus	Brahminy kite
Heliaeetus leucogaster	White bellied sea eagle
Sterna aurantia	Indian River tern
S. hirunda	Common tern
Aegithina tiphia	Common Iora
Alcedo atthis	Common Kingfisher
Halcyon capensis	Stork-billed Kingfisher
H. smyrnensis	White-throated Kingfisher
H. pileata	Black-capped Kingfisher
Ceryle rudis	Pied Kingfisher
Amaurornis phoenicurus	White-breasted waterhen
Ardea alba	Large Egret
A. Cinerea	Grey heron
Ardeola grayii	Pond heron or paddy bird
Phalacrocorax niger	Little Cormorant
Pluvialis fulva	Pacific golden plover
Bubulcus ibis	Cattle Egret
Butorides striatus	Little heron
Casmerodius albus	Great Egret
Centropus sinensis	Crow-pheasant
Ceryle rudis	Lesser pied Kingfisher
Charadrius dubius	Little ringed plover
C. leschenaultia	Large sand plover
C. mongolus	Lesser sand plover
Larus brunnicephalus	Brown headed gull
L. fuscus	Lesser black backed gull
Chlidonius hybridus	Whisked tern
Ephippirhynctus asiaticus	Black-necked stork
Egretta garzetta	Little Egret
Gelocheidon nilotica	Gull-billed tern
Haematopus ostralegus	Eurasian ovstercatcher

Contd...

Table 22.1–*Contd...*

Scientific Name	Common Name
Halcyon capensis	Stork billed Kingfisher
H. chloris	White collared Kingfisher
H. smyrensis	White breasted Kingfisher
Lanius schach	Rufousbacked shrike
Merops orientslis	Green bee-eater
Nectarinia asiatica	Purple sunbird
Orilus xanthornus	Black hooded oriole
Orthotomus sutorius	Tailor bird
Passer domesticus	House sparrow
Pericrocotus cinnamomeus	Small minivet
Psitacula krameri	Roseringed parakeet
Pycnonotus cafer	Redevented bulbul
P. jocosus	Redwhiskered bulbul
Streptopelia chinensis	Spotted dove
Turdoides striatus	Jungle babbler
Zosterops palpebrosa	White eye
Acridotheris fuscus	Jungle myna
Acridotheris tristis	Common myna
Hierococcyx barius	Common hawk cuckoo
Hirundo rustica	Swallow
Eudynamys scolopacea	Asian koel
Chrysocolaptes lucidus	Greater flameback
Calmator jacobinus	Pied cuckoo
Copsychus saularis	Magpie-robin or dovel
Coriacias benghalensis	Indian roller
Coracina macei	Large cuckooshrike
Corvus splendens	House crow
Cuculus micropterus	Indian cuckoo
Cypsiurus parvus	Palm swift
Dendrocitta bagabunda	Indian tree pie
Dinopium benghalense	Lesser golden backed woodpecker
Dendronanthus indicus	Forest wagtail
Dicrurus adsimilis	Black drongo
Dicrurus aenus	Bronzed drongo

The benthic fauna of mangrove ecosystem comprised of 14 taxa, belongs to four major groups namely nematodes, polychaetes, crustaceans and molluscs. Seasonal variations in the benthic macro invertebrates' densities are given in the Table 22.2.

Table 22.2: Seasonal Variation in the Density (No./m²) of Macrobenthic Invertebrates in Mangrove Ecosystem of Kali Estuary, Karwar

Macrobenthos	Summer	Monsoon	Post Monsoon	Winter
Nematodes	1121	432	815	281
Polychaetes	1833	2153	143	786
Crustaceans	301	68	123	74
Soft-bodied forms	3255	2653	1061	1141

Polychaetes were formed the most dominant group constituting 54.96 per cent of the benthic fauna in terms of densities. On a seasonal basis their densities were highest in southwest monsoon but dropped significantly low levels in the post monsoon season. Polychaete fauna was largely dominated by Nereid and spionids. Nematodes were the second most dominant component of the benthic fauna, constituted by 38.04 per cent followed by crustaceans with a mere 7.0 per cent. These three groups such as nematodes, polychaetes and crustaceans together constituted the soft-bodied forms and were easily available to the benthivorous waders. Their collective density was lowest in late post-monsoon season but increased progressively to attain peak in late summer. This sharp decline in the density of the soft-bodied fauna in post monsoon and early winter coincided with the high avian species richness owing to the influx of migrant waders.

Mangrove ecosystem harbored as 66 species of birds of which 23 species were of terrestrial birds.On the whole the cumulative biannual bird population in Kali River was 1236 individuals. Although species diversity of terrestrial bird was relatively high, they contributed a mere 25.70 per cent to the total avifaunal population. The terrestrial birds dominating in the mangrove ecosystem included Warbler; Black capped Kingfisher, Blue eared Kingfisher and Collared Kingfisher. All three species of Kingfisher were recorded exclusively inside the mangrove ecosystem indicating that although the ecosystem is estuarine wetland, which provides an ideal habitat even for terrestrial birds that feed largely on insects. Similar findings were also reported from estuarine wetland of Goa by Sonali *et al.* (2005).

The waterfowl community of Mangrove ecosystem was constituted by 21 species including both residents and migrants, the latter far exceeding the former group. In terms of population the biannual cumulative total of the water birds alone was 1254. Anatids, ciconids, aradeids, charadrids and lards were the five dominant groups. While the ciconids recorded in the region, were migrants all the ardeids were residents. Large Egret *Casmerodius albus*, little Egret, *Egretta garzetta* and Cattle Egret, *Bubulcus ibis* were observed to breed in this ecosystem. Reef Egret, *E. gularis, Anastomus oscitans*, white-necked Storks, *Ciconia episcopus* were the wading birds inhabiting in the mangrove ecosystem of wetland area. Among the Anatids the migrant pintails, *Anas acuta* were observed actively foraging on the mudflats in extremely large flocks of more than 300 individuals were encountered thereby confirming the speculations of Walia (2000).

Among the three groups, the waders constituted the largest (48.92 per cent) and most diverse chunk of the water bird community of the mangrove ecosystem. The group was largely dominated by three families namely charadridae, scolopacidae and glareolidae. Redshanks, *Tringa tetanus*, curlew sandpipers, *Cakudrus ferruginea*, common sandpipers *Actitis* sp., lesser sandplowers Charadrius mongolud, greater sandplowers, *C. leschenaultia, and C. dubius* were the regular winter visitors that formed the bulk of the migrant population. These waders used the estuarine mudflats largely as it foraging ground and were observed feeding actively in the soft sediment. Among the larids, brown-headed Gulls, *Larus brunnicephalus*, black-headed gulls, *L. fuscus*, Gull-billed Terns, *Gelochelidon nilotica*, and River-Terns, *Sterna aurantia* dominated species.

A negative correlation between waterbirds and nematodes explains the decrease in the nematode density during peak migration while a highly significant correlation between the terrestrial birds and the insectivores highlights that a majority of the birds inhabiting the mangrove canopy were insectivores. A negative correlation was observed between waterbirds and a terrestrial component was because most of the terrestrial birds were residents while the waterbirds were migrants.

In terms of seasonality a distinctly significant seasonal variation was observed among the benthivores, frugivores, herbivores comprising exclusively of pintails and piscivores as well as between the waterfowl, waders and terrestrial birds has been shown in Tables 22.2 and 22.3. In general during the one year period of study, the populations of herbivores and benthivores were significantly higher than the other groups. The avian species diversity ranged from 2.11 to 3.01 while the species richness ranged from 6.4 to 8.14.

Table 22.3: Seasonal Variation in the Community Structure of the Avifauna in Mangrove Ecosystem of Kali Estuary, Karwar

Birds	Summer	Monsoon	Post Monsoon	Winter
Waders	623	62	982	1423
Water Birds	341	51	101	1864
Terrestrial Birds	151	123	243	318

It is evident from the present study data that, the mangrove ecosystem of river Kali is home for a large number of both wetland birds as well as terrestrial birds some of which are exclusive to the mangrove habitat.This ecosystem provides an ideal foraging ground and a stopover point for numerous waterbirds particularly waders and others. Therefore an immediate step has to be taken to protect and preserve this wonderful community and an awareness campaign has to conduct to upgrade the knowledge about avian faunal community and their importance to not only to the nature but also for entire mankind.

Conclusion

During the present study, totally 66 species of birds were recorded of which 23 species were terrestrial birds. On the whole the cumulative biannual bird population observed in the mangrove ecosystem of Kali estuary was 1236 individuals. The species diversity of terrestrial bird was relatively high (25.70 per cent) to the total avifaunal populations. The dominant species were Warbler, black capped Kingfisher, blue eared Kingfisher and collared Kingfisher were some of the dominant species of terrestrial birds. The water fowl community of mangrove ecosystem was constituted by 21 species including both resident and migrants. Seasonally these avifauna showed remarked variation in their population, with highest density recrded during the winter (1864nos.) and lowest in southwest monsoon period (51 nos.) for water birds followed by waders (62 – 142 individuals) and terrestrial birds (123 – 318 (individuals)- Good correlation exists between avian fauna and macrobenthic fauna in the mangrove ecosystem.

References

Ali, S. and S. D., Ripley, 1983. A pictorial Guide to the Birds of the Indian Subcontinent.

Ali, S., 1996. The Book of Indian Birds. Bombay Natural History Society and Oxford.

Bhagwat, S., 2003. Biodiversity and conservation of a cultural landscape in the Western Ghats of India. Ph. D. Thesis, Oxford University, Oxford, U. K.

Bombay Natural History Society and Oxford University Press.

Grewal, B., 1995. Birds of the Indian Subcontinent. Guide Book Company Ltd. Hong Kong.

Grimmett, R., C. Inskipp and T. Inskipp, 1998. Birds of the Indian Subcontinent. Oxford University Press, Delhi.

Rao, V. V., V. NaguluB. S. Anjaniyulu, C. Srinivasulu and Ramana Rao, 1997.

Sonali D. Borges and A. B. Shanbhag, 2005. Avifauna of the Dr. Salim Ali bird sanctuary at Chorao-an Estuarine wetland of Goa, India.

Sonobe, K. and S. Usui (eds.), 1993. A field guide to the Waterbirds of Asia. Wildbird Societyof Japan, Tokyo.

Walia, R, 2000. Limnological studies on some freshwater bodies of Southern Tiswadi (Goa, India) with special reference to waterfowl. Ph. D. Thesis, Goa University, Goa.

Woodcock, M., 1989. Collins Handguide to the Birds of the Indian Subcontinent. Williams Collins Sons and Co. Ltd, London.

Chapter 23

A Study on Distribution of Fe and Mg in Karwar Coast, West Coast of India

☆ *B.K. Gangadhar, B. Vasanthkumar,*
J.L. Rathod and S.V. Roopa

ABSTRACT

The present study deals with the estimation of Fe and Mg along with the hydrological parameters of the inshore water of Karwar coast. Totally three stations were selected in Karwar region. Samples were collected twice in month from the 23 November 2010 to 23 March 2011. Along with the water sample collection, hydrological parameters such as water temperature, air temperature, sediment temperature, salinity, pH, dissolved oxygen were also analyzed.

Keywords: *Trace metals, Hydrology, Karwar, West coast of India.*

Introduction

Sea water is the most complex solution known to containing virtually all the naturally occurring elements including 'Trace Metals' found on earth. It has been borne out by experimental evidence that the role of heavy metal ions living system follows the pattern of natural availability and abundance of the some metals occurring in nature (Tripathi,1973, Singhal,1952). Nb organic life can develop and survive without the participation of metal ions.

The term trace metal identifies a large group of metallic elements which are present in living organism in limited amounts. Trace metals are usually divided into two sub classes; the first includes Fe, Mg, Mn, Co, Zn, and Cu which are essential for the correct functioning of biochemical processes. Cd, Hg, Cr, Pb, *etc.* belong to the second subclass which is made up of metal without any established biological function and includes the more important contaminations in the aquatic environment.

Materials and Methods

In the present study, totally three study stations were selected and fixed at different locales of Karwar coast namely Majali, Kali river mouth, Baithkhol.The study was carried out for the period of 5 months from the 23 November 2010 to 23 March 2011.

To the north of Uttar Karnataka district is Majali, which is also a fish landing centre. The coastline is about 3-4 kms. Majali having largest rocky shore area. It is situated 14° 53″46.50″ N and 74° 05′ 51.65″ E.

Kali estuary is situated 14° 50′ 21″ N and 74° 10′ 06″ E being one of the major estuary of Uttara Kannada district of Karnataka located along central west coast of India. Kali River is one of the productive rivers. The river mouth opens from Arabian Sea through a narrow mouth and tidal limit extends unto 29 kms (Neelakantan, 1976). The depth of estuary ranges from 8-12 mts at river mouth about 5 mts during high tides.

This landing centre is situated about 2 km away from Karwar at Baithkol. This is the major fish landing center of Karwar and has facilities for landing purse seines and trawlers.

Collection and Preparation of Samples

Surface water samples were collected for physico-chemical parameters in 500ml stopped glass bottles. The new acid washed 250ml plastic bottles were used for collecting trace metals samples.

Water samples for physico-chemical parameters were analyzed as soon as possible and that of trace metals possible and that of trace metals were mixed with a ml. concentrated HNo_3, immediately after collection and latter preserved in the refrigerator.

Water Sample Analysis

Water samples were first filtered through Wahtman No. 42 filter paper. 100ml filtered sample was taken in a 250ml beaker and pH was adjusted to 2-3 by adding 1N HCl.

Analysis of Sample

Concentration of metals was estimated by atomic absorption Sepctrophotometry in USIC at Karnataka University. Atomic absorption spectrophotometer (Australian) using air acetylene flame. Copper, chromium, and manganese were determined at their respective were length against suitable standards. All values were expressed in ppm dry weight (APHA, 1980; AWWA, 1998).

Results and Discussion

Environmental Parameter

The span of the study was from November – March highest water temperature at station I was 32 in February and then between 28-30 in other months. It may be due to the extreme heat or due to the time of analysis.

Salinity was highest in December, it was around 34, because due to the mineral contents, as well as there was no rain hence the salinity at this area increased. In other station it was within 25-31, Lowest were 25 in the month of January due to the dew factors and also due to run off from the land etc.

Dissolved oxygen is directly proportional to salinity hence the dissolved oxygen was highest in the month of March *i.e.*, 5.9 and lowest in the month of February at Station I due to salinity factors, light factors etc.

At Station II, highest water temperature was 30, and lowest was 29, it was not fluctuated much throughout the study season. Salinity was highest 31, and lowest was 22, it was mainly in the March due to extreme heat and time at analaysis. Dissolved oxygen was highest in December and lowest in February it is due to salinity and light factor.

At Station III, Salinity was highest in Feb *i.e.*, 32 and lowest in December. The fluctuation was generally more at Station III due to different anthropogenic and various environmental factors. Water temperature was recorded in March *i.e.*, 31 and later it fluctuated between 29.30 water temperature did not fluctuate much. Dissolved oxygen also did not fluctuate much.

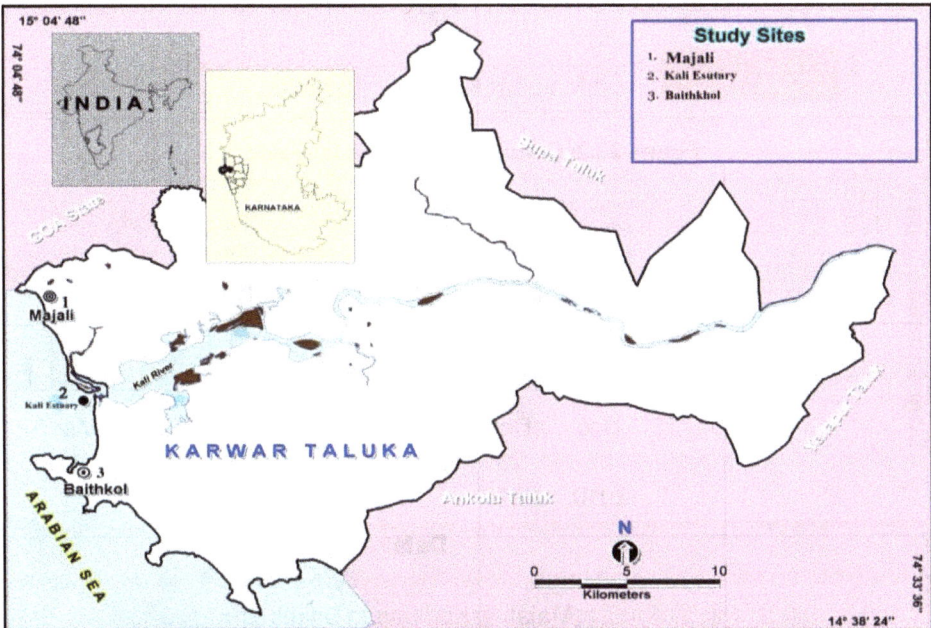

Figure 23.1: Map of Kali River Showing the Positin of Study Sites.

Trace Metal at Station I

The trace metals are usually the elements which are present in the environment in small traces and are useful in small amount but their excess presence can cause severe effects in the environment.Here in this study 2 metals were analysed Fe and Mg in water. At Station I, the highest range in Fe was determined in the month of February *i.e.*, 4.789 and lowest was 0.442 in February. But during the starting of the month here the Fe is the main component of any shipping vessel; hence the cleaning of their vessels could have been the source of this contamination and also the land runoff.

Mg was highest in December *i.e.*, 122.74 and lowest was 35.13 here the Mg content may be due to activities of human settlement or else due to the mineral content in the sediment itself.

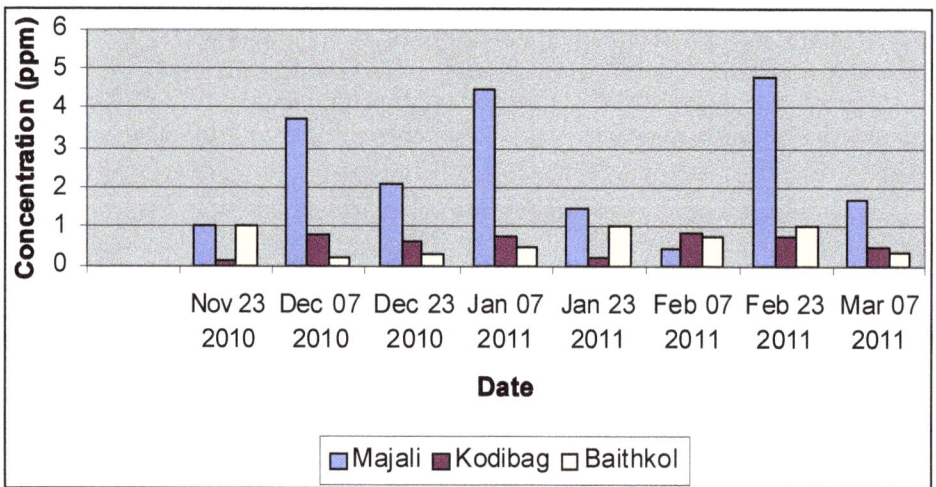

Figure 23.2: Concentration of Fe in Water.

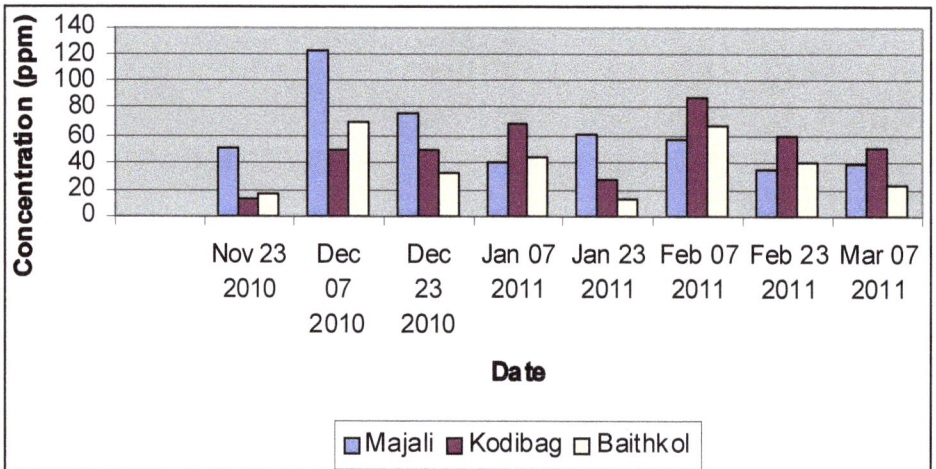

Figure 23.3: Concentration of Mg (in ppm) in Water.

Trace Metal at Station II

Highest Fe was analysed in Februry 0.821 and 0.087, it did not fluctuate more in the water sample. Mg was highest in February *i.e.*, 87.67 and lowest in 11.883. Mg was lowest throughout the season and fluctuated less.

Trace Metals at Station III

Highest Fe was recorded in the month of February *i.e.* 0.990 and lowest of 0.216. Here there is an decrease in concentration of Fe from November to March.

Highest Mg in water was seen in December and gradual decrease in January *i.e.* 12.769. Here the concentration of both Fe and Mg was more in November and December.

Conclusion

The areas selected were Majali which is an rocky shore, here the human settlement is nearby to the shore and hence the dumping of any materials and machinery in the water show a greater percentage.

The next station was Kodibagh which is generally at the mouth of the Kali estuary it bring all the contaminant from the other parts of the river. Lastly Batihkol, here the fishing activities is the major occupation, here the left over from the fish meal plant and is left in the water and also the ships are repainted and treated here and that waste too is dumped into the water. So this study is made in order to analyse the trace metals in some biotic and abiotic factors which may help us in better understanding the pollution status of that area.

References

APHA, AWA and WPCF. 1998. In: Standard Method for the examination of water and wastewater, American Public Health Association, Washington, D. C. 20th Edition, New York.

APHA. 1980. American Public Health Association, Standard Methods for Examination of Water and Wastewater, pp. 1134.

Tripathi R. M. 1973. Dietary intake of heavy metals in Mumbai city, India. The science total environment 208: 149-159.

Singhal, S. Y. S., M. D. George, R. S. Topgi and R. Noronha. 1952. The levels of certain heavy metals in Marine Organisms from Aguada. Bay (Goa*)*. *Mahahsagar,* 15 (2): 121 - 124.

Neelkantan, B. 1976. Distribution of Heavy Metals in the Northern Shrimp *Padalus borealis* from the Oslofiord. *Fish. Tech.* 8 (1): 20 –25.

Chapter 24

Impact Assessment of Mercuric Chloride on the Histhpathology of Liver of Fish *Catla catla* (Ham)

☆ *Quazi Saleem and Seema Hashmi*

Introduction

The toxicity of any pollutant is either acute or chronic. The chronic studies include both histochemistry and pathology. Although toxicant impairs the metabolic and physiological activities of the organisms, physiological studies done does not satisfy the complete understanding of pathological conditions of tissues under toxic stress, hence it is useful to analyze the histological aspects, the extent of severity of tissue damage is a consequence of the concentration of the toxicant and is time dependent also the severity of damage depends on the toxic potentiality of a particular compound or pesticide accumulated in the tissue.

Mode of action of different chemicals varies leading to varied effects on many tissues. Some toxins exert their effect locally at the portal of entry, resulting in the damage to external surface of the body, some toxins when ingested affect the different regions of gastrointestinal tract. There exists a different group of toxins that do not cause deleterious effect of the portal entry but they systematically effect the tissue in which they get accumulated. Thus various chemicals with their varied mode of action affect different tissues thereby bringing about certain architectural changes ultimately culminating in either death of the organisms or making the organisms less labile for its survival the extent of severity of tissue damage of a particular compound as toxicant depends on the toxic potentiality of it in the tissues of organisms.

Liver is the main organ responsible for detoxification of harmful substances which reach it through circulation thus the liver is the most susceptible organ to toxicants entering the body of an animal. Histopathological changes in the liver after exposure to metals includes, necrotic effect on hepatocytes, hyperplasia, rupture of cell membrane resulting into multinucleated regions and accumulation of red blood corpuscles in the severely affected condition vacuolization is also observed (Patel, 2000).

Liver plays a vital role in detoxification of toxins by breaking down substances and metabolic product, as a result of which hepatic cells exhibit more damage than cells from any other organs when an animal is exposed to a toxicant. Effects of toxicosis on liver is reported by many researchers. The pathological changes noticed in the liver might effect the physiological activity of fish such as reduction in enzyme synthesis this reduces the functional ability of liver which indirectly effects all metabolic activities of the organisms.

There is considerable variation in total mercury concentrations among different species biotic and abiotic factors may affect mercury accumulation in marine organisms among these different ecological and physiological factors the diet and the position in the trophic web are determining elements (Thorat, 1999).

The histopathological effects of different pollutants on fish have been reported by many scientists. Sastry and gupta (1997) studied the histopathological changes induced in *Channa punctatus* after chronic exposure to lead nitrate. Ramamurthy (1989) studied the lethal and sub lethal concentrations of cadmium chloride on the liver of *Tilapia mossambica*. Bhatacharya (1989) studied the effect of mercuric chloride on the ovary of *Channa punctatus.*

Materials and Methods

The fish *Catla catla* of size 6-7 cm ± 0.5 cm and 6-8 ± 0.5 grams weight were procured from the Sindhphana river in Beed district.The fish were acclimatized at 28 ± 2°C in the laboratory for 5-6 hours in aquarium. The tanks were constantly aerated and the fish were fed with blood worms as food. The food remnants were removed everyday. The fish were divided into four groups, one for control and three for experimental each group was having 5-6 fishes, forty litres of water was added in each aquarium. The experimental group was exposed to 0.1 per cent mercuric chloride then after exposure time of 8 and 16 hours the fish were sacrificed and the liver excised out and then washed thoroughly with sterile cold distilled water it was fixed in bouins fluid dehydrated in alcohol and blocks were prepared in paraffin wax (58-60°C) the sections of 7µ thickness were cut and stained with mallorys triple and mounted in DPX. Same procedure was repeated for 0.2 per cent 0.3 per cent mercuric chloride for 8, 16, hours respectively.

Results

The changes observed due to exposure to heavy metal mercuric chloride were more or less similar to the changes observed earlier also which include rupture of blood sinusoids, disorganized hepatic chords, change of shape of hepatocytes. The histological structure of normal liver lobe show that it consists of polygonal cells called hepatocytes, these hepatocytes are present in irregular lobules separated by connective tissue each hepatocyte shows clear cytoplasm and distinct central rounded nucleus, with nucleoli the hepatocytes are arranged in cords and separated by adjacent one, by blood space. The blood space is lined by connective tissue, the islets of langerhans lie scattered in hepatocytes and can be easily recognized.

Histopathology of Liver

There is a decrease in the cellular size of hepatic cells and the outline becomes indistinguishable stored materials disappear the nucleus and the nucleolie become smaller and pycnotic after exposure to 0.1 per cent of mercuric chloride as the concentration of mercuric chloride increases the nucleus become more pycnotic and eventually undergoes karyolysis, necrotic cells may appear sporadically or in clusters at 0.2 per cent concenteration, the hepatic cells undergoes congestion necrotic cells undergoes cytolysis or phagocytosis by lymphocytes or histocytes fatty degeneration i.e excessive fat in cytoplasm is due to nuclear atrophy and vascular degeneration is also observed. Changes in stroma is observed when the number of damaged cells is large e.g the connective tissue often shows a tendency to become hyper plastic i.e cirrhosis results from proliferation of connective tissue at all concentration and it advances as the exposure time increases, the desquamation of mucosa leaving submucosa or muscularis exposed to the human accompanied with hemorrhage was observed at 0.2 per cent and 0.3 per cent mercuric chloride right from eight hours exposure (Figure 24.1–24.6).

Discussion

Study of histology is of prime importance in the diagnosis, etiology and prevention of disease. Any particular type of alteration of cell may indicate diseased condition or effects of toxic substance. Jagadessan (1999) reported necrosis vacculation, breakdown of cell wall and hepatocytes after exposure to mercuric chloride, certain progressive changes like hypertrophy mitotic fusion and pycnotic nuclei were observed. *Channa gachua* exposed to mercuric chloride showed disturbance of hepatic chords and clumped cytoplasms and increased blood sinusoids.

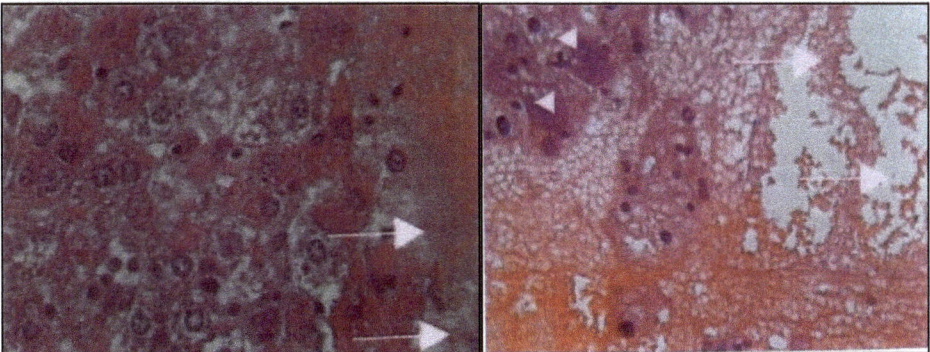

Figure 24.1: Exposure to 0.1 per cent HgCl$_2$ for 8 hrs.   Figure 24.2: Exposure to 0.1 per cent HgCl$_2$ for 16 hrs.

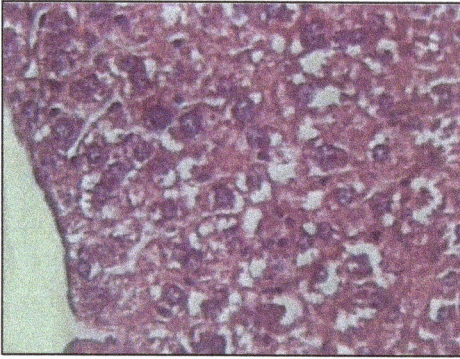

Figure 24.3: Exposure to 0.2 per cent HgCl₂ for 8 hrs.

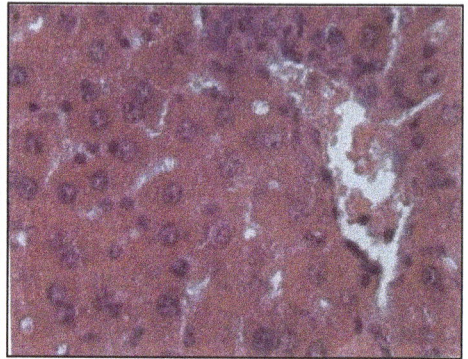

Figure 24.4: Exposure to 0.2 per cent HgCl₂ for 16 hrs.

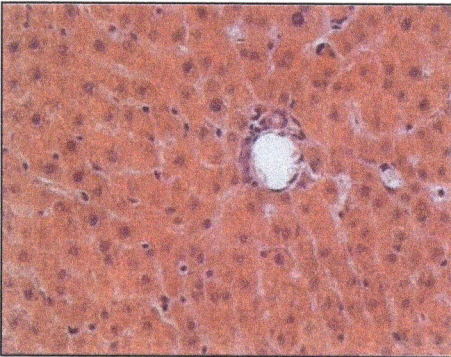

Figure 24.5: Exposure to 0.3 per cent HgCl₂ for 8 hrs.

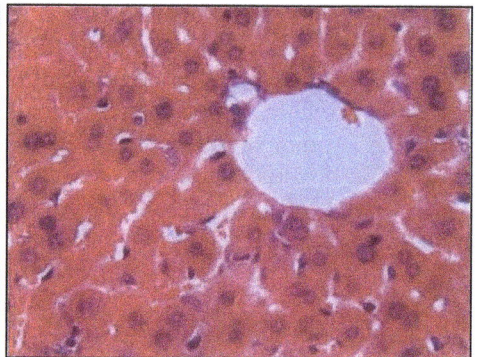

Figure 24.6: Exposure to 0.3 per cent HgCl₂ for 16hrs.

Heavy metals results in the hyperplasia of hepatocytes, aggregation of nuclei and ruptured hepatocytes with denucleated cell. Rupture of blood sinusoids vacculation and disarray of hepatic chords of *Lepidocephhalychthys thermalis* was observed after exposure to copper. Vincent *et al.* (1996) reported impact of cadmium chloride on *Catla-catla* as drastic decline in total cell count, hemoglobin, and reached anaemic state of the fish. The liver was also prominently affected by necrosis and hypertrophy of the liver cells. Enlargement of liver cells in *Anguilla anguilla* due to pentachlorophenol poisoning is also reported histopathological changes in the head and trunk of *C. punctatus* induced by chronic non lethal levels of mercuric chloride (16-17 ppb) was studied for 7, 28, 63 and 90 days of exposure the study demonstrated that mercuric chloride affected both endocrine and excretory parts of kidney (Gaikwad, 2003).

Mercuric chloride not only affects the liver but it also affects other vital organs, such as kidney, gills intestine of many fishes. The gill surfaces of *R.daniconius* exposed to 0.05 mg per litre mercuric chloride for 96 hours were damaged with fusion and dumping in the middle and distal parts of the primary lamellae, swollen deterioration and modification of arborixing ridges into more expanded surface area in the secondary lamellae. Under mercuric chloride stress the liver exhibits hypertrophied nucleus, vaccuolation in the hepatic tissue due to the shrinkage in cytoplasm, disruption of sinusoids and disappearance of cell boundaries in fish *Cirrihinus mrigala.* Liver is the main detoxifying organ and its quite reasonable to expect injury to it. The hepatotoxic action of heavy metal on fish structural and functional level is frequently encountered.

Patel *et al.* (2000) reported on the histopathological changes in the liver of *Nemicheilus botia* due to exposed to dimecron. Their observation include the hyperplasia of hepatocytes, rupture of cell boundaries, multinucleated zones and accumulation of red blood corpuscles at some places. Patil and Dhande (2000) studied the histopathological changes in the liver of *Channa punctatus* exposed exposed to $HgCl_2$, $CdCl_2$, $CuCl_2$ the degree of metal toxicity was found as mercuric chloride › cadmium chloride › copper chloride they suggested that the enzyme liberated from the hepatocytes proved to be the cause of necrosis of liver and thereafter caused necrosis in trunk kidney. Nickel toxicity was done on *Clarius batrachus* and it was found that more nickel accumulated with increase in concentration and time, nickel accumulation was recorded highest in liver.

The liver metabolizes various xenobiotics to convert them into a less toxic substance and eventually eliminates through different routes however liver is susceptible to a number of chemical substances and undergoes cellular damage. The nature of damage depends on the duration of the exposure and the biological half life values of the parent xenobiotic compound and their degraded metabolites. In the present investigation the intracellular vacuolation and nuclear swelling were observed in the hepatocytes, the cells showed loss of cell membrane and the cells were disarranged. (Figures 24.1–24.6). The swollen nuclei cytoplasmic vaccuolation gaps between hepatocytes and the loss of cell boundaries were observed in *Catla catla* treated with mercuric chloride.

Heavy metal induced hepatic changes and damage has been attributed either to the alterations in the activities of functional enzymes or to their altered synthesis. Metallothioniens in hepatocytes bind the metal ions as a protective function when the heavy metal exceeds the rate of metallotheionin like protein production, the pathological changes becomes apparent. Our observations on histopathological changes in the liver of *Catla-catla* treated with mercuric chloride were somewhat similar to those discussed above. The observations suggest that probably there is a burden of detoxification of mercuric chloride on the liver.

References

Bhattacharya. 1989. Effect of mercuric chloride on ovary of *Channa punctatus. Comp. Biochem. Physiol.* 644-693.

Gaikwad, P. T. 2003. Biology of freshwater fish in relation to pollution. Ph. D. thesis Submitted to Dr, B. A. M. U., Aurangabad.

Jagadessan G. 1999. *In vivo* recovery of gill tissue of freshwater fish *Channa gachua* after exposure to different sublethal concentration of mercury, *Poll Res.* 18(3): 289-291.

Patel, N. G., Khalid, S., Powar, L. B. 2000. Histopathological changes in the stomach and liver of Nemacheleilus botia on acute exposure to dimecron, *J. Aqua Biol.* 15(1 and 2): 105-107.

Patel, G. P., Dhande. R. R. (2000) Studies on histopathological and biochemical changes induced by metal in liver and trunk of the fish *Channa punctatus*, I nternational conference on probong in biological system, abstracts 92-111

Ramamurthi, 1989. Cadmium induced abnormalities of *Tilapia mossambica. Environ. Ecol* 15(1): 168-169.

Sastry, K. V., Gupta., P. K. 1997. The invivo effect of mercuric chloride on some digestive enzymes of freshwater fish teleost fish *Channa punctatus. Bull. Environ. Contamination and Toxicology,* May 22 (1-2): 9-16.

Thorat and suryawanshi. 1999. The effect of tannery effluent on freshwater fish *Channa gachua.* A thesis submitted to Dr, B. A. M. U. Aurangabad.

Chapter 25

Effect of Neem Leaves Induced Histopathological Alteration in Liver of the Freshwater Fish, *Garra mullya* (Sykes)

☆ *V.R. Borane*

ABSTRACT

Azadirachta indica is well known in India for more than 2000 years as one of most versatile eco-friendly, evergreen, medicinal plants having a wide spectrum of biological activity. The tree is 'village dispensary' in India. The importance of tree has been recognized by the US National Academy of Sciences in 1992 entitled 'Neem- a tree solves global problems'. It contains chemical active substance of several biological properties. Histopathological changes observed in fish liver, *Garra mullya* after exposure to conc. of aqueous extract of dried leaves for 96 hr. at sub-lethal concentration of 0.0175ppm.

Keywords: *Neem leaves, Garra mullya, liver, 96hr.*

Introduction

Neem is an evergreen, eco-friendly tree, cultivated in various part of Indian subcontinent. It has biologically active component of neem based insecticides. These natural pesticides are known to have strong antifeedant, growth regulatory and sterility effects on insects (Jacobson, 1989).Chemical investigation on the products of the neem tree was extensively undertaken in the middle of the twenth centure. Siddiqui (1942) report on the isolation of nimbin, the first bitter compound isolated from neem oil, more than 135 compounds have been isolated from different part of neem. Martinez 2002 reported that aqueous extract of neem leaves and byproduct have been used in fish farm for controlling fish parasite and predators. Most of the toxic effect in fishes due to alternation of metabolism disturbances.

Methanolic extract of neem leaf exhibits oral toxicity to mice showing ill health and discomfort, gastrointestinal spasms, apathy, hypothermia, convulsion and leading to death, also shows antifertility effect in mice and leading to death, Kanungo (1996). Ibrahim *et al.* (1992) reported Brown hisex chicks, when feed with diet containing 2 per cent and 5 per cent neem leaf from their 7'th and 35'th day after birth developed hepatonephropathy and significant change in blood parameter. Present study to assess the effect of neem leaves extract on fish liver for the period of 96 hr. at sub-lethal concentration of 0.0175ppm.

Materials and Methods

Medium sized freshwater fishes, *Garra mullya* with an average body wt. 8-9 g and length 5-6 cm were collected from shiven river, Nandurbar. They were brought to the laboratory condition and were acclimatized in well aerated for about 7-8 days. The physico-chemical of the water by, APHA, (2005). The fishes were divided into two groups A and B. Group A maintained as a control. The Group B fishes were exposed to $LC_{50/10}$ dose of fresh neem extract 0.0175ppm for 96 hr. Leaves of neem were dried and finally copped then dissolved in tap water at concentration of 500 g. of dried leaves per litre of water for 24 hr. at room temperature by Cruz *et al.* (2004). The mixture was filter and the fresh extract was used in the experiments. For histopathology live specimens were removed from control and exposure concentration.The liver was removed from dissected animals and fixed in Bouin's fixative and processed for microtome sectioning at 5µ and stained with HE mounted with DPX, Durvy and Willington (1967).

Results and Discussion

The histopathological changes were more evident in specimens exposed to neem leaves extract and were not observed in the control fish. The liver cells in *Garra*

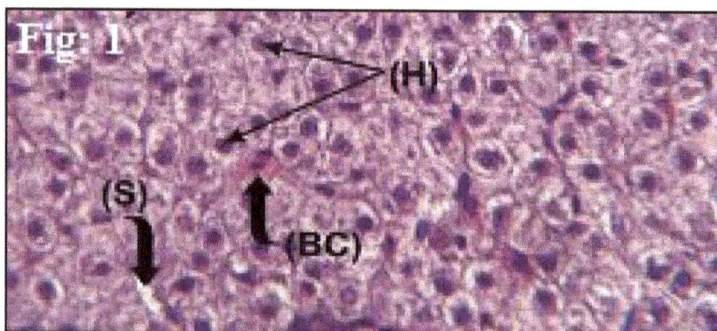

Figure 25.1: Photomicrograph of Control Liver Shows Regular Arrangement HC (Hepatic Cord), Blood Cell (BC), Hepatocyte (H), Sinus (S).

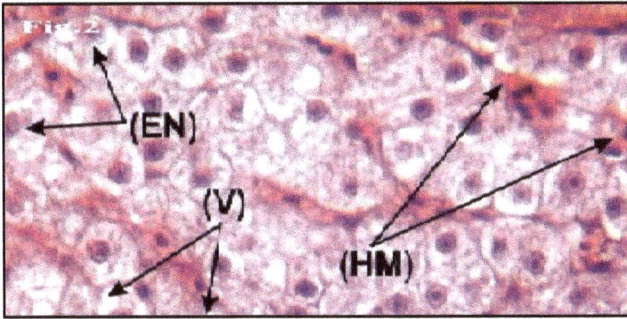

Figure 25.2: Photomicrograph of Treated Liver for 96 hr
Conspicuous Haemorrhage (HM), Eccentric Nucleus (EC), Vacuolisation (V).

mullaya are polygonal containing spherical central nucleus (Figure 25.1). After 96 hr. of exposure the hepatocytes became irregular and loose their polygonal shape. Some cells exhibited cloudy swelling, their contour becoming indistinguishable. There were many regions in the liver where cells were highly vacuolated. Many cells had exhibited pycnosis, intensive vaculation in cytoplasm and eccentric nucleus were also observed (Figure 25.2). Fish liver has the ability to detoxify pesticides but high concentrations of these compounds can alter hepatic enzyme activities that can result in damage of hepatocytes, Paris-Palacios, (2000). Several studies demonstrated that alterations in number, size and shape of the hepatocyte nucleus can be due to contaminants

According to Gingerich (1982) the vacuolization of hepatocytes might indicate an imbalance between rate of synthesis and rate of release of substance in hepatocytes. Shrunken and pycnotic nuclei indicated that cells became hypo functional and at the end, necrosis was extensive. The stagnation of bile inside the hepatocytes signifies affected metabolism Fanta *et al.* (2003).

Neem extract cause pathological changes under sublethal exposure in fish liver such as vacuole formation, cytoplasmic and nuclear pyknosis, necrosis inflammation and haemorrhage.Winkler *et al.* (2007) reported similar observation alteration in liver of fish exposed pesticides.

References

APHA 2005. Methods of water and wastewaters analysis, 21st Edn., Washington. DC. USA.

Cruz, C. Machado-Neto J. G. and Menezes M. L. 2004. Toxicidade aguda to insecticida Paration metilico e do biopesticida azardiractina de folhas de neem (*Azadirachta indica*) para alevino e juvenile de pacu(Piaractus mesopotamicus). Pesticidas: *R. Ecotoxical* e Meio Ambiente 14: 92-102

Durvy R. A. B. and Willington E. A. 1967. Caleton histological techniques 4'th edn. Oxford University Press. London.

Fanta, E., Anaa Rios F. S., Ramao, S., Vianna, A. C. C. and Freiberger, S. 2003. Histopathology of the fish, *Corydoras paleatus* contaminated with sublethal levels of organophosphorus in water and food. *Excot. Environ. Saf.*, 54(2): 119-130.

Gingerich, W. H. 1982. Hepatic toxicology of fishes. In: *Aquatic toxicology*. (Eds.: L. J. Weber). H Raven Press, NewYork. p. 55-105.

Ibrahim, I. A., Khalid S. A., Omer S. A. and Adem S. E. 1992. *J. Ethenopharmacol.*, 35: 267-273.

Jacobson, M. 1989. Focus on phytochemical pesticides, Vol. 1, The Neem Tree. CRC Press, Boca, Ratson, FL, p. 197.

Kanungo D. 1996. In neem (eds Randhawa and Parmar, B. S.) 2'nd edn., 77-110.

Martinez S. O. 2002. NIM- Azadirachta indica natureza, usos muùltoplose produção. Instituto Agronômico do Paraná (IAPARA). Londrina, P. R. 142.

Paris-Palacios S., Biagianti-Risbourg S. and Vernet G. 2000. Biochemical and (ultra) structural hepatic perturbation of *Brachydanio rerio* (Teleostei, Cyprinidae) exposed to two sublethal concentrations of copper sulphate, *Aquat. Toxicol.*, 50(1-2), 109-124.

Siddiqui, S. 1942. *Current Science*, 11: 278-279.

Winkaler, E. U. Santos T. R. M., Machado Neto J. G. and R. Martinaz, C. B. 2007. Acute lethal and sub lethal effects of neem leaf extract on the Neotropical freshwater fish, *Prochilodus lineatus Comp. Biochem. and Physiol* (Part C), 145: 236-244.

Chapter 26

Ascorbic Effect on the Cypermethrin Induced Alterations in Blood Glucose Level of the Freshwater Fish, *Channa orientalis* (Schneider)

☆ *B.R. Shinde*

ABSTRACT

The study on fishes and their diseases has importance in life of human being because it has nutritive value. Freshwater fishes, *Channa orientalis* were exposed to chronic dose of cypermethrin without and with ascorbic acid. Total count of Blood Glucose content was recorded. Remarkable decreases in Blood Glucose were observed in cypermethrin exposed fishes. Fishes were exposed to cypermethrin with L-ascorbic acid showed less present variation in the Blood Glucose. Pre-exposed fishes to pesticides showed fast recovery with ascorbic acid as compared to cured naturally. The role of ascorbic acid on exposure to cypermethrin of an experimental fish, *Channa orientalis* is discussed in the paper.

Keywords: *Cypermethrin, Ascorbic acid (50mg/l.), Blood glucose, Channa orientalis (Schneider).*

Introduction

In vertebrate main function of blood is transportation of oxygen, essential nutrients, and removal of waste products from tissues and organs systems to investigate physiological and metabolic changes. Also it acts as a medium for the translocation of pesticides from the medium to different organs or systems of an animal. The blood carries heavy metals and pesticides to different organ or system hence as blood components are directly affected of blood carries substances. Biochemically its effect is interference with heme synthesis leading to hematological damage (Awad, 1997). Decrease in hematological parameters in malathion exposed freshwater fish, *Cyprinus carpio,* (Ramesh and Mahavalaramanujam, 1992). Cypermethrin is one of the synthetic pyrethroid uses in normal practice. Cypermethrin is readily absorbed by gills and mouth even from very low concentration in water. The symptoms of poisoning in insects and animals are hypersensitivity, hyperactivity with violent burst of convulsions and finally complete prostration with convulsive movement, disturbance in the ganglia of the central nervous system rather than in the peripheral nerves.

Ascorbic acid plays an important role in distribution and excretion of toxic metals. Ascorbic acid has reversed dysfunction of cells lining blood vessels. The normalization of functioning of these cells may be link to prevention of heart diseases (Chambers, 1999). It has been realized that antioxidant can play significant role in the treatment of metal induced oxidative stress. Some antioxidants behave as efficient chelators (Gurer and Eracel, 2000). The SH group of protein is mainly responsible metal interaction or bindings L-ascorbic acid is antioxidant and may extent in protective effects by chelating the metal and removing them from the system (Tajmir Riahi, 1991). During toxicosis ascorbic acid indicate positive role in detoxification. It is necessary for the synthesis of collagen, growth and maintenance of epithelial tissue. It can acts as a hydrogen carrier it may have an essential role in the metabolism of carbohydrate or protein or both. It appears to function it maintaining strength in blood vessels.

Materials and Methods

Medium sized freshwater fishes *Channa orientalis* were collected from Shiven river area Nandurbar Dist. Nandurbar. The physico-chemical parameters of the water used for the maintenance of the fishes were analyzed as per the methods given in APHA and AWWA (2005). The fishes were divided in to three groups A, B and C. Group A fishes were maintained as a control. The Group B fishes were exposed to $LC_{50/10}$ dose of cypermethrin (0.6713 ppm) for 30 days, while group C fishes were exposed to respective chronic concentration of pesticide with 50mg/l. of ascorbic acid for 30 days. Fishes from B groups were divided into two groups after 30 days exposure to cypermethrin into D and E groups. Fishes of D groups were allowed to cure naturally while those of E groups were exposed to ascorbic acid (50 mg/l.). Blood glucose content were recorded from A, B and C group fishes after 15, and 30 days of exposure and from D and E groups after 35'th and 40'th days of recovery.

Blood was obtained by cutting the caudal peduncle dissection method (Reichenbach-Klinke, 1982; Roberts, 1978), using heparin as anticoagulant. First few drop were discarded and only the first 2ml. of blood was taken since the entry of lymph into the blood is reported (Schreck, 1975) to affect haematocrit value. The blood glucose was determined by the method (Nelson and Somoggi (1944).

Table 26.1: Physico-chemical Parameters of Water Used for Experimentation

Temperature	$25.1 \pm 3.2^{\circ}$
pH	7.60 ± 0.3
Conductivity	$140 \pm 15.7\ \mu\ mho^{-cm}$
Free CO_2	$3{\cdot}34 \pm 1{\cdot}3\ ml^{-1}$
Dissolved O_2	$6{\cdot}3 \pm 1{\cdot}1\ ml^{-1}$
Total Hardness	$204 \pm 12{\cdot}0\ mg^{-1}$
Total Alkalinity	$585.6 \pm 32.8\ mg^{-1}$
Magnesium	$31.67 \pm 2.9\ mg^{-1}$
Calcium	$30.46 \pm 3.06\ mg^{-1}$
Chloride	$107.92 \pm 16.34\ mg^{-1}$

Table 26.2: Blood Glucose in *Channa orientalis* after Chronic Exposure to Cypermethrin without and with Ascorbic Acid (Values are expressed in mg of glucose/100ml)

Group	Treatment	15d	30d	35d	40d
A	Control	89.53±0.48	88.86±0.83	—	—
B	Cypermethrin (0.6713 ppm)	83.6±0.19** (-6.62)	72.2±0.17*** (-18.74)	—	—
C	Cypermethrin (0.6713 ppm)+A A	84.9±0.25** (-5.17)	79.1 ±0.13*** (-10.98)	—	—
D	Recovery in Normal Water	—	—	71.5±0.11$^{\Delta\Delta\Delta}$ [+2.21]	76.33±0.46$^{\Delta\Delta\Delta}$ [+5.72]
E	Recovery inAA	—	—	76.7±0.65$^{\Delta\Delta\Delta}$ [+6.23]	86.0±0.58 NS [+21.74]

AA: Ascorbic acid (50 mg/l).

± indicates S.D. of three observations.

Values in ( ) indicates percent change over respective control.

Values in [ ] indicates percent change over 45 days of respective B.

* indicates significance with the respective control.

Δ indicates significance with 45 days of respective B. $p<0.05$ = * and Δ

$p<0.01$ =** and $\Delta\Delta$, $p<0.001$ = *** and $\Delta\Delta\Delta$, NS and ΔNS = Not significant.

Results and Discussion

Fishes experimentally exposed to cypermethrin for a period of 15, and 30, days in a group B and C showed significant decrease in blood glucose in cypermethrin exposed fishes. When dose of cypermethrin along with ascorbic acid was given the depletion in blood glucose was 86·0 for 40 days observation. The pre exposed fish to cypermethrin for 40 days showed fast recovery in blood glucose and significant improvement with ascorbic acid as compared to those cured naturally in normal water after 5 and 10 days.

Post stressor increase in blood glucose level may be used as indicator of the secondary phase of stress response (Leatherland and Woo, 1998). These changes include an activation of liver glycogenolysis and glycolysis as well as increased level of plasma glucose and lactate. A significant increase in conc. of the enzymes in blood plasma indicates tissue impairment caused by stress (Svoboda, 2001). Blood glucose in pesticides decrease indicates damage to gills. Borane and Zambare (2006) observed decrease in blood glucose in $PbCl_2$ as compare to $CdCl_2$ exposed fishes. When dose of $PbCl_2$ and $CdCl_2$ along with ascorbic acid was observed the depletion in blood glucose.

Rangaswamy (1984) observed that there is a continuous breakdown of glycogen reserve to meet the energy demand of the fish as a result of pesticide stress, thus increasing the blood glucose level. The hyperglycemic condition might be the hypoxic where oxygen consumption of the fish has been reduced. Srinivas *et al.* (2001) reported that the fish, *Catla catla* exposed to malathion and dichlorovos pesticide showed that the blood glucose level was increased.

The alterations in hematological parameter caused by pesticides were on the basis of chemical nature and time dependent and the toxicity of pesticides constitute certain health indices. Hypoglycemic response in the treated fish is due to the rapid utilization of blood glucose during hyper excitability, tremors, impaired liver function and convulsions which are the characteristic behavior of toxicosis in fish and mammals, (Rahman *et al.,* 1996).

References

APHA, AWWA AND WPCF 2005. Standard methods for the examination of water and wastewater. APHA (17th ed.) Inc. New York.

Awad, M. and Jr. William. 1997. Textbook of biochemistry with clinical correlations. John Wiley and Sons, INC, New York.

Chambers, J. C., Greger, M. C. and Jean Marie. 1999. Demonstration of rapid onset vascular endocrine disfunction after hyperhomocy teinemia. An effect reversible Vit. C therapy. Circulation, 99: 1156-60.

Gurer, H. N. and Eracel, C. 2000. An antioxidant beneficial in the treatment of lead poisoning? Free Rad. and Med. 29 (10): 927-945.

Leatherland, P. T. K. Woo. 1998; Fish diseases and disorders Non- infectious disorders, CABI Publishing Oxon, UK.

Nelson, N., Somoggi, R. 1944; A photometric adaptation of somogyi method for the determination of glucose. *J. Biol. Chem.* 151: 375-38.

Rahman, M. F., M. K. J. Siddique and M. Mustafa1996. Effect of repeated oral administration of Vepacide (*Azadirachta indica*) on some hematological and biochemical parameters in rats. *Indian. J. Toxicol.* 3(1): 1-8.

Ramesh M and Mahavalaramanujam, S. K. 1992. Effect of water hardness and toxicity of malathion on hematological parameters of fish, *Cyprinus carpio J. Ecotoxicol. Monit.*, 31-34.

Rangaswamy, C. D. 1984; Impact of endosulfan toxicity on some physiological properties of the blood and aspects of energy metabolism of a freshwater fish, *Tilapia mossambica* (peters). Ph. D. Thesis Sri. Venkateshwara University, Tirupati.

Reichenbach-Klinke, H. H. 1982; Enfermedades de los peces. Ed Acribia, Zaragoza, España, p507.

Roberts R. J. 1978; Fish Pathology. Bailliere Tindall. New York, USA, p. 377.

Schreck, C. B. and P., Brouna1975. Dissloved oxygen depletion in static bioassay system *Bull. Environ Contam. Toxicol.* 14: 149- 152.

Srinivas A., Venugopal, G., Piska, R. S. and Waghray, S. 2001; Some aspects of haemato-biochemistry of Indian major carp, *Catla catla* influenced by malathion and dichlorovos (ddvp) *J. Aqua. Biol.* 16 (2): 53-56.

Svobodov'a, Z., J. Màchov'a, B. Vykusov'a and V. Piaèka. 1996; Metals in ecosystem in surface waters, Metodika VuRH Vodòany Czech Republic, No. 49.

Tajmir – Riahi, H. A. 1991. Coordination chemistry of vitamin C. Part II. Interaction of L –ascorbic acid with Zn (II), Cd (II) and Mn (II) ions in the solid state and in aqueous solution. *J. Inorganic Biochem*, 42: 47 – 56.

Chapter 27

Studies on the Alteration of Blood Parameters of a Freshwater Fish *Oreochromis mossambicus* Exposed to Synthetic Detergent Aerial

☆ *S. Jeyakumar and S. Mala*

ABSTRACT

Haematological Changes of *Oreochromis mossambicus* exposed to sub – lethal concentrations of synthetic detergent Aerial for a short period was investigated. An increasing trend was observed in TEC, TLC, HBC, PCV, and MCV values on exposure to sub- lethal concentrations of Aerial. Effects of sub – lethal concentrations of Aerial in the morphology of the blood cells were also studied. The observed changes confirmed the impact of the synthetic detergent Aerial on the physiology of fish.

Keywords: *Oreochromis mossambicus, Detergent, Hematological study.*

Introduction

Now-a-days most of the water bodies are dumped with household wastes which includes synthetic detergent a new group of pollutant contaminating our aquatic system. Primarily introduced detergent powder contains alkyl benzene sulphate (ABS) type (Abel,1974). In recent years, detergents are having methylene blue active substances which are an anionic surfactant type (Jeyasurya *et al.,* 1991).Various hematological studies have been conducted to show the toxicity of non-ionic and anionic synthetic detergents on fishes (Rauthan *et al.,* 1995). In the present study, it is aimed to investigate the toxic effects of commercially available synthetic detergent (aerial) on the Haematological parameters of *Oreochromis mossambicus* as blood is a valuable diagnostic tool for investigation of diseases and physiological disorder (Xavier innocent and Martin, 2003).

Materials and Methods

The test animal *Oreochromis mossambicus* (10gm ± 0.5 g live weight) were procured from manimuthar dam (longitudes: 77.6° E, Latitude 8.5° N, Altitude 11.62 m above sea level, Tirunelveli Dt, Tamil Nadu, South India) and acclimated to laboratory conditions for a period of one month in dechlorinated tap water. During the acclimation period, the fish were fed on pelleted feed with 38 per cent protein level. After the static bio – assay of the detergent aerial, the test fishes were exposed to chosen sub-lethal levels(0.004 per cent - 0.008 per cent) of aerial which was taken in the plastic troughs of 30 litres capacity. Ten fishes were introduced into each trough and triplicates were maintained in each concentration. The test fish reared in dechlorinated tap water served as control.

Haematological studies were carried out at the 30th day of exposure. Methods described by Hesser (1960) for routine fish hematology were followed to estimate all blood parameters in this study. Red cell indices were worked out using the formula suggested by Johansson-Sjobock and Larson (1978).

Results and Discussion

Dose dependent increase in RBC count, Hb count, PCV and MCV were observed in Aerial exposed fish on the other hand MCH and MCHC showed gradual decrease in relation to detergent concentration.

Total RBC count was increased from 2.28 to2.75 X 10^6 cells/mm^3 in 0.004 per cent and 0.008 per cent detergent concentration respectively for an exposure period of 30 days. Similarly Hb content also increased from 4.0 per cent to 4.58 per cent. The significant increase in RBC and Hb could be ascribed to enhanced erythropoietin which is triggered as a typical stress response to withstand pollutant induced stress conditions(Munkittrick and Leatherland,1983).

In the present study, the increase in total RBC count, Hb per cent and Haematocrit value may be attributed to the stress response that is induced by the detergent.

Increased PCV, may be considered as a combined effect of erythrocyte swelling and compensatory mechanism of the fish to increase the O$_2$ carrying capacity of the blood as stated by Gupta *et al.* (1997). Elicited value observed in MCH increase in lymphocytes population and slow onset of macrocytic anemia condition due to pollutant stress.In aquatic organisms stress situations causes release of ACTH, which leads to increased defense mechanism, which is necessary for the adaptive value of fishes. Immediate effect was shown by RBC in the form of abnormality in shape (Nayak, Madhyastha, 1980) which elicits the PCV and MCV values.

Table 27.1: Different Blood Parameters of Control and Detergent Treated Fishes

Concentrations in per cent Parameters	0	0.004 per cent	0.005 per cent	0.006 per cent	0.007 per cent	0.008 per cent
Erythrocyte Mean X 10^6 cells/mm^3	2.28	2.394	2.485	2.576	2.667	2.758
Leucocytes cells/mm^3	31760.00	32462.00	33460.00	34458.00	35456.00	36454.00
Hb per cent	4.0	4.164	4.272	4.380	4.488	4.596
PCV per cent	10.37	11.062	12.084	13.106	14.128	15.15
MCV per cent	45.481	46.674	48.753	50.832	52.911	54.990
MCH	17.54	17.36	17,22	17.0	16.82	16.65
MCHC	38.56	37.59	35.06	33.28	31.54	30.56

**Table 27.2: The Relationship of Haematological Parameters
with the Concentration of Detergent (Aerial)**

Exposure	Y=a+bx	r-Value
TEC	2.394	
	2.485	
	2.576	0.998404
	2.667	
	2.758	
TLC	32462	
	33460	
	34458	0.999832
	35456	
	36454	
HBC	4.164	
	4.272	
	4.379	0.994748
	4.488	
	4.596	
PCV	11.062	
	12.084	
	13.106	0.995126
	14.128	
	15.15	
MCV	46.674	
	48.753	
	50.832	0.990942
	52.911	
	54.99	

Treated fishes blood cells is compared with control showing in Figures 27.1–27.4.

The total leucocytes count also increased slowly as the concentration of the detergent increases. Similar results were obtained by (Kumari *et al.,* 1989) and Dubey (2001). The increase in the total leucocytes count may be attributed to enhanced

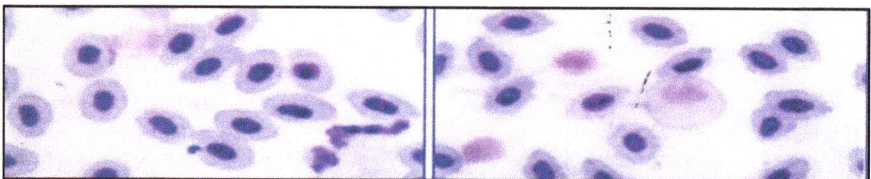

Figure 27.1: Control.   **Figure 27.2: Treated.**

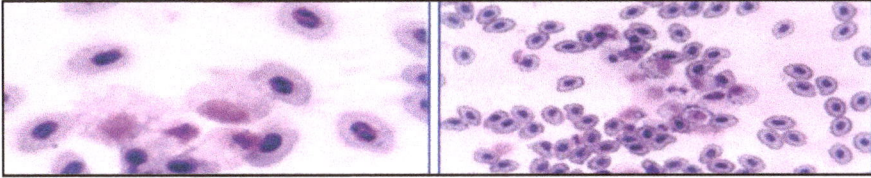

Figure 27.3: Treated.   **Figure 27.4: Treated.**

immune response of the test animals. The increased number of TLC is associated with the increased number of circulatory levels of Granulocytes, which may be probably for phagocytosis and immunological responses. to stressors (Bromage and Fuchs,1976). Thus the present study clearly reveals that the commercial detergent aerial affects the blood parameters of the test animals even at a very low concentration.

References

Abel, P. D. 1974. Toxicity of synthetic detergents to fish and aquatic invertebrates. *J. Fish Biol.*, 6: 279-298.

Bromage, N. R and Fuchs. A. 1976. Histological study of the response of internal cells of Gold Fish *Carassium auratus* to treatment with SLS. *J. Fish. Biol.* (2): 529-535.

Gupta, A. K., Kumar, P. and Rajana. 1997. Enzymological study on the effects of Aldein on a freshwater Teleost Fish, *Notopterus notopterus. J. Notion, 9: 9-12*

Hesser, E. F. 1960. Methods for routine fish hematology prog. *Fish. Cult.* 22: 164 – 171.

Jeyasurya. M. A, Subramaniam and Varadaraj. 1991. Effects of detergents on the oxygen consumption of the cat fish *Mystus vittatus. J. Ecobiol.* 3(3): 217-220.

Kumari, Munni and Kumari, G. 1989. Perils of Environmental Pollution by pesticides. An assessment through leucocytic and haemostatic response of a freshwater fish, *Clarius batrachus. Him, J. Env.* 2001. 3: 36-43.

Madhyastha, M. N., and Nayak, R. R. 1979. Preliminary studies on the effect of a detergent point on the blood cells of *Rasbora daniconius. Proc. Environ. Bio. unl. Kerala, India.*

Munkittrick, K. R. and Leatherland, J. F. 1983. Haematocrit values in Gold fish *Carassius auratus,* as indicator of the health of fish population. *J. Fish, Biol.* 23: 153-162.

Rauthan, J. U. S, Gover S. P and Jaiwal, P. 1995. Studies on some haematological changes in a hill stream fish *Borilius bendelsis, Flora and Fauna.* 2: 165–166.

Roy, D. N., Dubey, N. K. 2001. Haematological response of Trivalent chemium poisoning in India teleost cat fish *Heteropneustes fossilis. Indian J. Environ and Ecoplan* 5(3): 607 – 616.

Xavier, B and Martin. P. 2003. Haematological indices as bio-indicators pollution- A study on freshwater Crab *Paratelphusa hydrodromous* (Herbst) exposed to cyberguard. *Ind. Jr. of Env. Stu.* 7(1): 41-45.

Chapter 28

Activity of Protease Enzyme in Freshwater Fish, *Channa punctatus* (Bloch) from River Godavari, Nanded

☆ *A.R. Jagtap and R.P. Mali*

ABSTRACT

Temperature is the dominant ecological factor on all animal lives. The aquatic animals are highly sensitive to the temperature fluctuations. In present work, freshwater fish *Channa punctatus* was exposed up to 96 hours under temperature stress. The enzyme activity in stomach were observed and compared with control set. The present investigation showed depletion in the protease activity as temperature decreases. The protease activity was also found to be decreased as temperature increases in freshwater fish, *Channa punctatus*.

Keywords: *Temperature, Protease, Stomach, Channa punctatus.*

Introduction

Out of various environmental factors that influence aquatic organisms, temperature is the most all-pervasive (Brett, 1970). Temperature is a major factor of influence of enzyme activity (Kuzmina *et al.,* 1996).

Enzymes reduce the activation energy required for reaction, while temperature influences the fraction of molecules with enough energy to react (Hochachka and

Somero, 1984). A primary determinant of the inherent temperature sensitivity of any reaction is the enzyme catalytic efficiency. Enzymes which are highly efficient catalysts typically have low temperature sensitivity. There are so many factors, which can change the functioning of enzymes and thereby the temperature sensitivity of biochemical reactions that this could be considered as one of the mega problems of ectothermy. Any change in temperature may well differentially perturb a wide range of biochemical processes and integrating these effects to achieve an overall function is a huge problem for ectotherms (Hochachka, 1991).

Digestive glands are the main site of extra and intracellular digestion; they typically store large amounts of sugars, proteins and lipids. Aquatic environment contains the largest pool of diversified genetic material and, hence represents an enormous potential for different sources of enzymes. Freshwater fishes were regularly sampled for physiological measurements. Therefore the present investigation focused study on changes in composition of digestive enzyme *i.e.* protease under cold and warm temperature stress.

Materials and Methods

The freshwater fish, *Channa punctatus* was collected from the Godavari River, Nanded (Maharashtra). They were kept in glass aquarium with continuously aerated tap water. The fishes were acclimated at room temperature for 8-10 days before experiment. Fishes were feed with the small pieces of earthworms. The fishes were subjected to above and below room temperatures to carry out experiment. The four experimental sets were designed to carry out experiments (two sets at below room temperatures and two sets at above room temperatures). The animals were divided into five groups having 10 fishes in each aquarium (Fletcher, 1977). The fishes were acclimated to different temperatures for 24 hrs, 48 hrs, 72 hrs and 96 hrs. Healthy uninjured fishes ranging from 60- 70 grams were selected for present study.

The equal volume of 1 per cent peptone was added as a substrate in a 2 per cent tissue homogenate. The contents were mixed thoroughly; a thin film of toulene also added and incubated at room temperature for 24 hours. This serves as experimental extract. After boiling the extract of control set the same procedure was applied up to incubation period. After 24 hours, the toluene layer was removed from control and experimental set. 20 ml of control extract and experimental extract was used for titration. To the 20 ml extract 20 ml of 10 per cent formaldehyde was taken in conical flask and 3-4 drops of phenolphthalein was used as an indicator. The mixture titrated against

N/10 NaOH solution. The end point was recorded when the colourless solution turns pink. The same procedure applied for control extract. The difference between two readings gives the amount of protease activity. The protease activity expressed in terms of mg/gm wet wt. of tissue/hr.

Results and Discussion

The protease activity in stomach of fish, *Channa punctatus* under ambient temperature at 15 °C, 20 ± 1° C, 30 ± 1° C and 35 ± 1° C was observed and the results were compared with the observed experimental values up to 96 hrs period of exposure.

The level of protease activity in stomach subjected to cold condition was found to be decreased slightly up to 96 hrs period of exposure. The decreasing trend was more up to 96 hrs at warm temperature stress *i.e.* at $35 \pm 1°C$. The obtained values were expressed in mg/gm wet wt. of tissue/hr. The results obtained are cited with graphical representation.

Table 28.1: Effect of Temperature on Protease Activity in *Channa punctatus* at Cold and Warm Temperature Stress

Sl.No.	Temperature of Water Bath Maintained	Period of Exposure	Stomach (mg/gm Wet Wt. of Tissue/hr)
1	15° C ± 1° C	CONTROL SET (26° C ± 1° C)	1.124 ± 0.38
		24 Hrs	1.122 ±1.10
		48 Hrs	1.120 ±1.19
		72 Hrs	1.118 ±0.26
		96 Hrs	1.114 ±0.76
2	20° C ± 1° C	CONTROL SET (26° C ± 1° C)	1.123 ± 0.72
		24 Hrs	1.122 ±1.66
		48 Hrs	1.121 ±2.50
		72 Hrs	1.119 ±1.43
		96 Hrs	1.117 ±0.92
3	30° C ± 1° C	CONTROL SET (26° C ± 1° C)	1.125 ± 0.42
		24 Hrs	1.120 ±1.66
		48 Hrs	1.119 ±1.46
		72 Hrs	1.114 ±1.65
		96 Hrs	1.110 ±2.30
4	35° C ± 1° C	CONTROL SET (26° C ± 1° C)	1.124 ± 0.44
		24 Hrs	1.113 ±0.76
		48 Hrs	1.109 ±2.39
		72 Hrs	1.107 ±0.86
		96 Hrs	1.104 ±2.01

(Each Value is Mean of Six Observations ± S. D.)

The impact of climatic variations on aquatic communities has been well-documented (Beamish 1995; Bakun 1996). Exposure of animals to temperatures above or below the limits of critical temperatures leads to death if thermal acclimation is not prevalent (Sommer *et al.*, 1997). A regulatory enzyme must be capable of both efficiently catalyzing a metabolic transformation and of varying its rate of catalysis in response to changes in the cell's need for the product(s) of the pathway. Changes in the external environment of the organism, such as the ambient temperature and oxygen content, may also affect the chemistry of the cell. Therefore, modulations in the activities of regulatory enzymes must occur according to the limitations imposed by the changing external environment.

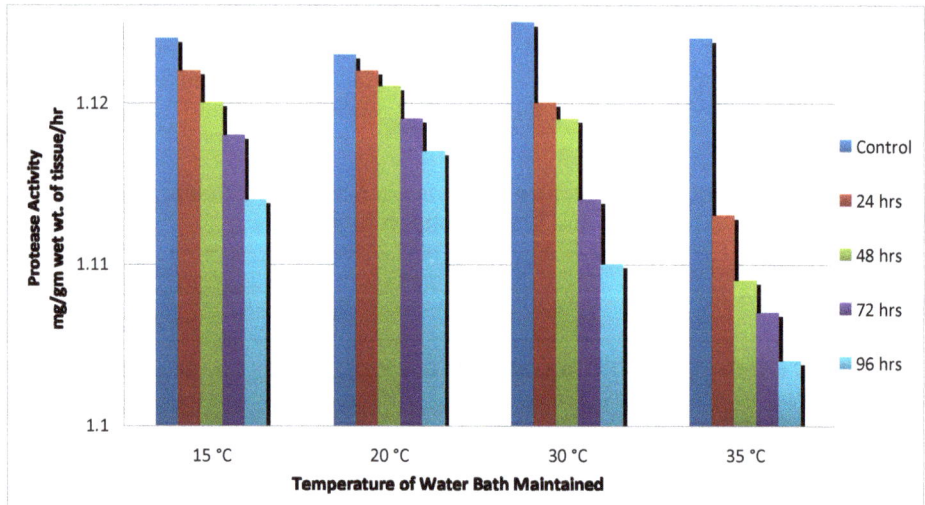

Figure 28.1: Protease Activity in Stomach of Freshwater Fish, *Channa punctatus* at cold and warm temperature stress (Each Value is Mean of Six Observations ± S. D.).

Al-Hussain (1949 b), reviewing the study on the physiology of digestion in the fishes, described the correlation between the food and the digestive enzymes. Babkin and Bowie (1928), and Dhage, (1969) showed the presence of amylase in the alimentary tract of the stomach less fishes. Dhage also reported that the concentration of amylase was much higher in the herbivorous fishes such as the major carps than in the carnivorous sh, *Ephinephalus tauvina*. Kawai and Ikada, (1971) described more amylase activity in the extracts of the intestine of the teleosts.

The enzyme proteases facilitate the breakdown of proteins. In present investigation the stomach showed decrease in protease activity at cold and warm temperature stress in *Channa punctatus*. The decreasing trend was more at warm temperature stress as compared to cold stress in stomach of freshwater fish, *Channa punctatus*.

The digestion rate and amount of enzymes produced decrease with decreasing temperature (Smith, 1980; Vonk *et al.,* 1984; Jobling, 1995). Several workers reported the information on different teleost species (Ugolev, *et al.,* 1983; Mc Leese *et al.,* 1986; Hazel, 1993) demonstrates a different effect of temperature on digestive enzyme performance of warm water and coldwater species. Several researchers report digestive enzyme optima at temperatures not encountered in nature (45-60 °C) (Pyeun, *et al.,* 1991), while other studies confirm thermal inactivation (50 per cent) of fish digestive enzymes at 35-55 °C (60min) (Dimes, *et al.,* 1994).

Temperature affects the digestion system in poikilotherms in several ways (Windell, 1978, Jobling, 1994). The need of energy and activity increases with increase in temperature. The gastric acid secretion and pepsin secretion in the stomach are influenced by temperature (Smit, 1967, Moyle and Cech, 1988).

References

Al Hussain, 1949: On the functional morphology of the alimentary tract of some fishes in relation to their feeding habits. *Qurt. J. Micro. Sci. London*. 90: 328.

Babkin B. P. and Bowie D. J. 1928. The digestive system and its function in *Fundulus heteroclitus Bio. Bull*. 54: pp. 254

Bakun, A. 1996. Patterns in the Ocean Processes and Marine Population Dynamics. California Sea Grant College System, La Jolla. pp 323.

Beamish, R. J. 1995. Climate change and northern fish populations. *Canadian Special Publication of Fisheries and Aquatic Sciences*. 121: pp. 739

Brett, J. R. 1970. Temperature-Animals-Fishes. O. Kinne, Ed., in Marine Ecology, Vol. 1. Environmental Factors, 1: 515-560.

Dhage K. P. 1969. Study of the digestive enzymes in *Epinephalus tauvina* (Forskal). Ibid. 12(2); pp. 31-37

Dimes L., Garcia-Carreno F. and Haard N. 1994. Estimation of protein digestibility - 3. Studies on the digestive enzymes from the pyloric caeca of rainbow trout and salmon. *Comparative Biochemistry and Physiology* 109A, pp. 349-360.

Fletcher, G. L. 1977. Circannual cycles of blood plasma freezing point Na^+ and Cl^- concentrations in Newfoundland winter flounder (*Pseudopleuronectes americanus*): correlation with water temperature and photoperiod. *Can. J. Zool*. 55: pp. 789-795.

Hochachka, P. W. and Somero, G. N. 1984. *Biochemical adaptation*. Princeton, New Jersey: Princeton University Press

Hochachka, P. W. 1991. Temperature: The ectothermy option. In: Biochemistry and molecular biology of fishes (Hochachka, P. W. and Ommsen, T. P., eds), 1: 313-322. *Amsterdam: Elsevier Science Publishers B. V.*

Jobhng M, Hjelrneland K. 1992. Ernering og fordsyelse. In: Dsving K, Reimers E (eds) Fiskens fysiologi. John Grieg Forlag, Bergen, pp. 234-257.

Jobling M. 1994. Fish bioenergetics. Chapman and Hall, London.

Jobling M. 1995. Digestion and absorption. In: Environmental Biology of Fish (ed. by M. Jobling), pp. 176-210. Chapman and Hall, London, UK.

Kawai and Ikada S. 1971. Studies on the digestive enzymes of sih. I. carbohydrases in digestive organs of several fishes. *Bull. Jap. Soc. Sc. Fish*. 37 (4): 101.

Kuzmina V. V., Golovanova L. L. and Izvekova G. I. 1996. Influence of temperature and season on some characteristics of intestinal mucosa carbohydrases in six freshwater fishes. *Comparative Biochemistry and Physiology* 113B, pp. 255-260.

Mc Leese J. M. and Stevens E. D. 1986. Trypsin from two strains of rainbow trout, *Salmo gairdneri*, is influenced differently by assay and acclimation temperature. *Canadian Journal of Fisheries and Aquatic Sciences 43*, pp. 1664-1667.

Moyle, P. B., Cech J. J. 1988. Fishes. An introduction to Ichthyology, 2nd edn. Prentice-Hall, Englewood Cliffs, NJ.

Pyeun, J., Cho D. and Heu, M. 1991. Comparative studies on the enzymatic properties of trypsins from cat-shark and mackerel. 1. Purifications and reaction conditions of the trypsins. *Bulletin of Korean Fisheries Society* 24, pp. 273-288.

Smit, H. 1967. Influence of temperature on the rate of gastric juice secretion in the brown bullhead, *Ictalurus nebulosus. Comp Biochem Physiol.* 21: pp. 125-132

Smith, L. S. 1980. Digestion in teleost fishes. In: Fish Feed Technology Lectures ACDP (ed. by L. S. Smith), pp. 4-18. FAO/UNDP, Rome, Italy, 395pp. (Chapter1).

Sommer, A., Klein B. and Portner, H. O. 1997. Temperature induced anaerobiosis in two populations of the polychaete worm *Arenicola marina* (L). *J. Comp. Physiol* (B). 167: pp. 25–35.

Ugolev A. M., Egorova V., Kuzmina V. V. and Grudskov A. 1983. Comparative molecular characterization of membrane digestion in fish and mammals. *Comparative Biochemistry and Physiology* 76B, pp. 627-635

Vonk H. J. and Western J. R. H. 1984. Comparative Biochemistry and *Physiology of Enzymatic Digestion. Academic Press, New York*, NY, USA, 501pp.

Windell, S. T. 1978. Digestion and the daily ration of fishes. In: Gerking S. D. (ed) Ecology of freshwater fish production. Blackwell, *Oxford*, p. 159-183.

Chapter 29

Toxic Effect of Pesticide Phosphamidon at Sub-lethal Levels on the Blood and Tissue Carbohydrate Levels in Freshwater Indian Major Carp *Labeo rohita* (Rohu)

☆ *M. Rafi Ahamed and Basha Mohidden*

ABSTRACT

An attempt has been made to study the sub lethal effects of Organic phosphate pesticide phosphamidon on the carbohydrate levels like blood glucose.Glycogen in the tissues, red muscle, white muscle and liver in Indian major carp *Labeo rohita*.During pesticide exposure the glycogen levels were studied at regular intervals like 24hours 7th day,15th day and 30th day, when compared to control levels it was revealed in the investigation that at initial levels the glycogen content(24 hours) increased this enhancement was noticed gradually decreased at 7th day,15th day and at 30th day the glycogen levels reached relatively to the normal level of the controlled fish this indicated that the fish was adopted at sub-lethal levels due to detoxifying effect.

Keywords: *Labeo rohita, Phosphamidon, Sub-lethal exposure, Glycogen.*

Introduction

In recent years industrial countries have been facing a new calamity in the shape of pollution it may be air, water and notice pollution. The pollution havac in different forms is increasing at alarming rate. The water pollution by pesticides is drawing attention from various fields like press, scientists, and industrialists in the present investigation an organo phosphorus pesticide phosphamidon was studied on freshwater major carp *Labeo rohita* (Rohu).The fishes are good indicators responding towards aquatic pollution. In the present study physiological responses like carbohydrate metabolism shows that alterations in blood glucose, red muscle, white muscle,glycogen and liver glycogen indicated that the fish was when subjected to pesticide stress under sub-lethal conditions,tissue carbohydrate levels affected at regular intervals (7th day,15th day) at the end of 30th day normalcy reached confirming the adaptation of fish to sub-lethal levels.Which shows that due to detoxifying mechanism fish may be adjusted to the sub-lethal pesticide exposure. Shembekar *et al.* (2009) recorded Hyper glycemic conditions during the exposure of Dimethoate pesticide in freshwater fish *Macrones vittatus*.

Materials and Methods

Labeo rohita fish was collected from local aqua culture Andhra Pradesh Fisheries Department Anantapur, because Labeo is Known for adaptability of laboratory conditions besides the availability and commercial value four inches length fishes were selected to be above investigation. Besides it is suitable for toxicity studies (Srinivasan and Swaminathan,Basha Mohideen 1984,Shahanawaz 1986).In order to evaluate pesticide toxicity the static bioassay is followed where the biological responses was recorded in static water (Dougroff *et al.,* 1951) LC$_{50}$ determined (Finney, 1964) and it was found the LC$_{50}$ was 197 ppm.The blood glucose was estimated by colorimetric method (Nelson and Somogy 1957). The Glycogen content in the liver and muscle at different sub-lethal exposure period in phophamidon besides control was estimated by anthrone method (Carrol *et al.,* 1956).In the above experiments fish of the same size taken and dived into six batch's six animals in each batch.

Results and Discussion

During sub-lethal exposure periods of phosphamidon of fish *Labeo rohita*.The blood glucose levels at 24 hrs there is sudden decrease, with corresponding increase in liver glycogen but blood glucose level elevated after 7th day and 15th day reaching maximal elevation but at 30th day period the above parameters reached nearer to the control. The liver glycogen followed same trend but in reverse way to blood glucose, Hence the variations in blood glucose precisely coincided with variations in liver glycogen, where as both red and white muscle glycogen also registered a fall as that of liver glycogen at different sub-lithal exposure periods of phosphamidon fig/table. These variations are found to be highly significant ($P < 0.001$) further these variations are found to be maximal and minimal at 15th day and 30th day periods in relation to control. Thus the carp exhibited fairly a good amount of percentage recovery in these parameters at 30th day sub-lethal exposure period of Phosphamidon.

It has been evident from the earlier study that Vessiolas *et al.* (1976) that carbohydrate metabolism disturbed in rats poisoned with sevin. From the studies of Renu and Drixler (1973) it has been established that organo phosphorus insecticides like Melathion could increase the blood glucose and decrease liver and muscle glycogen.similarly increased serum glucose levels were recorded in the carps after administration of Melathion (Sakaguchi and Hiromi 1972).

In the present study there is decrement in blood glucose levels of *Labeo rohita* at 24hrs sub-lethal exposure periods of phosphoamidon but at 7th and 15th day sub-lethal exposure period there is progressive elevation in blood glucose and at 30th day period,the elevated blood glucose levels lowered reaches almost to the normal levels (Figure 29.1 and Table 29.1). The maximum elevation of blood glucose level was noticed at 15th day exposure period (46,68 per cent) inversely correlated with maximum suppression in oxygen consumption opercular rate. Correlated with the above effects there is a maximum decrease in liver and muscle glycogen at 15th day, the above findings are in compromise with the reports of Koundinya and Ram Murthy (1979, 1980), Siva Prasad Rao and Raman Rao (1979), Verma *et al.* (1979) Dalela *et al.* (1981)

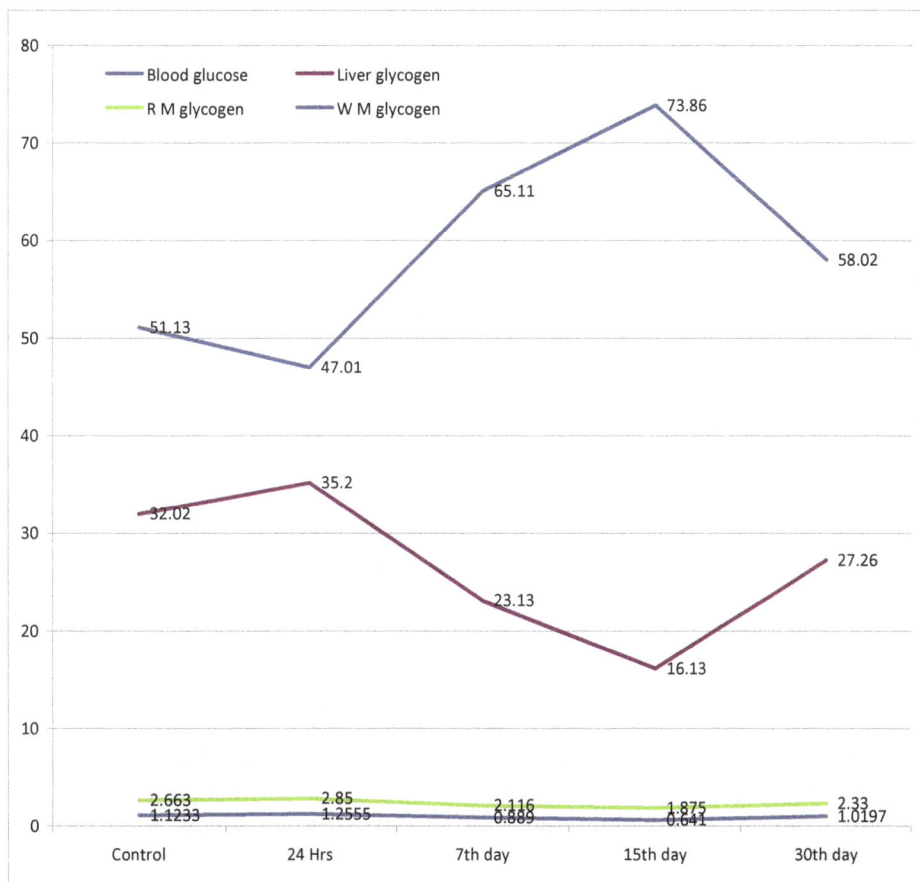

Figure 29.1: Blood Glucose (mg/100 ml blood glucose) Liver and Muscle Glycogen Measured (mg/g wet weight of the tissue).

Jayantha rao *et al.* (1982).The possible reason for hyper glycemic during phosphamidon stress is hypoplasia in Islets of langerhans of Pancreas could lead to decreased insulin resulting hyper glycemic and decreased in liver and muscle glycogen.Further increase in blood glucose level and fall in liver and muscle glycogen levels are indicative of increase in the rate of glycogenolysis, another explanation for decreased in liver and muscle glycogen content might be due to the suppression in the rate of glycogenosis or gluconeogenesis in the liver. The hyper glycemic condition noticed in this carp of the present study might be due to the stimulation of glucogon hormone secretion by pancreas (Harper, 1978).

Table 29.1: Blood Glucose mg/100 ml of Blood, Liver Glycogen, Red Muscle and white Muscle Glycogen Expressed in mg/gram Wet Weight of Tissue

Sl.No.	Parameter	Control	Sub-Lethal Exposure Period			
			24hrs	7th day	15th day	30th day
1	Blood glucose	51.13	47.01	65.11	73.83	58.2
	SD	±178.0	±179	±1.94	±1.329	±1.145
2	Liver glycogen	32.02	35.20	23.13	16.13	27.26
	SD	±1.22	±0.629	±3.82	±1.032	±4.51
3	Red muscle glycogen	2.633	2.850	2.116	1.875	2.330
	SD	±0.11	±0.083	±0.116	±0.068	±0.077
4	White muscleglycogen	1.1233	1.2555	0.889	0.641	1.0197
	SD	±0.013	±0.042	±0.18	±0.036	±0.048

References

Basha Mohideen and Shahanawaz. 1984-1986Effect of temperature on the toxicity of Melathion in Indian major carp *Labeo Rohita* M. Phil thesis S. K. U. Anantapur.

Carrol, N. V. 1956. Glycogen in the liver and muscle by use of anthrone reagent, *J. Biol. Chem.*, 22: 583-593.

Doudroff P, Anderson B. G., Burdic G. E, 1951. Environmental pollution by pesticides, p. 281(ed. C. A. Edwards, London and Newyork Plenum Press).

Finney, D. J. 1964. Probit analysis 2nd edition Cambridge University Press.

Longley, R. W. and Row, J. H. Doudroff *et al.,* 1951. *Sewage and Industrial wastes*, 23: 1930.

Rathod, M. V. Lokhande, M. V. and Shembekar, V. S. 2009. Effect of Dimethoate on blood sugar level of freshwater *Macrones Vittatus. Ecology and Fisheries.* 2(1): 101-106.

Sakaguchi and Hiromi. 1972. Nippon Suisan Gokkaishi 6, 38, 555

Siva Prasad Rao. K, 1980. Studies on some aspects of metabolic changes with emphasis on carbohydrate utilization of all free systems of freshwater teleost tilapia subjected to methyl parathion exposure Ph. D. thesis S. V. University, Tirupathi.

Sriniwasan. A., and Swaminathan. G. K, 1957. Toxicity of six Organo Phosphorous insecticides to fish. *Curr. Sci.,* 36, No. 15 397-398.

Chapter 30

Histological Studies of the Ovaries of the Teleost *Rasbora daniconius* Induced by Heavy Metals

☆ *F.I. Shaikh, Mohd. Ilyas and Shaikh Imran*

ABSRACT

Lethal and sublethal exposures to mercuric chloride and altered the histological picture of the female gonial units. This exposures coused several degenerative lisions in the ovary the impairment of vitellginesis was evident in treated fish. Mercuric chloride produced more changes in the ovary.

Keywords: Rasbora daniconius, Heavy metal toxicity.

Introduction

Action of pollutants on gonadal tissue provides clear example effects which may often unrecognized at the individual level (Resenthal and Atderdic,1976). Heavy metal pollutants are a major problem in aquatic environment because of their toxicity, their peristance, their tendency to accumulate in organism and undergo foodchain amplification (Weis and Weis,1977a) Inorganic mercury was reported as a general treatogen to the killifish, *Fundulus hetroclitus* (Weis and Weis,1977b) and cadmium was considered both as embrotoxin and teratogen to the blugill, *Leporis macrothisus* (Eaten,1974) a careful perusal of the literature reveals that not much information is available on the toxicity of these metals to fishes. The present work describes the toxicological action of mercury and cadmium on the oocyte differentitiation of the tetrost *Rasbora daniconius*.

Materials and Methods

Mature adults of *Rasbora daniconius* were collected from Bindusara River from Beed District.The fishes were feed with the live pieces of Earthworm every alternate day and allow acclimatizing in the laboratory conditions in large aquaria. For 15 days prior to the experimentation water was renewed every day to provided freshwater rich in oxygen.

The fishes were (Average weight 9±0.2 gm and length 9.5cm selected were placed in separate aquaria keeping one group as control a sub lethal dose of 0.225ppm of mercury chloride was administered for 10, 20 and 30 day. The ovary of each control and test fish was dissected and fixed in aqueous Bions. Then the Naries were dehydrated through graded series of ethanol embedded in purafin wax (56-60°C) and sectioned at a thickness of 5-7μ the sections were stained with harris haemotoxylin and counterstained with eosin.

Results and Discussion

In ovaries of control fish the oognies and mature oocytes appeared normal with well defined oocyte differentiation (Figure 30.1).Ovaries of fish exposed to mercuric chloride had different morphological appearances. The first observed toxic action of metals after 10 days in the young oocytes were extensive vacuolization in the oocortex, necrosis of oolemma and hypertrophy of follicular cells (Figure 30.2). Thereafter mercury induced more lytic changes in the oolemma. Which resulted in atresia of oocytes after 20 days (Figure 30.3). It was evident from the loosely arranged shrunken less stainable and less yolky mature oocytes (Figure 30.4). Reduction in their counts were also noticed from the histological suspense in their counts in the present work.

It seems certain that mercury chloride inhibits the transfer of nutrients across the oolemma. There by causing the inhibition of vitellogensis metals accumulation in the tissues of the adult fish would be transferred to the eggs during oogensis possibly

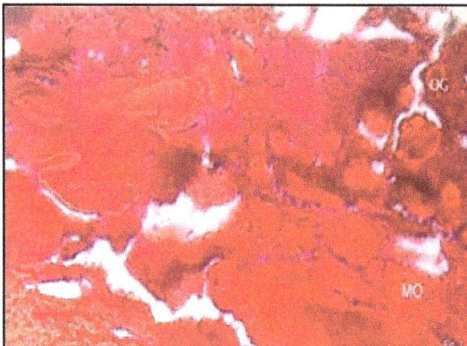

Figure 30.1: T.S. of Control Ovary X 100.　　　**Figure 30.2: T.S. of HgCl$_2$ Treated Ovary After 10 Days X 100.**

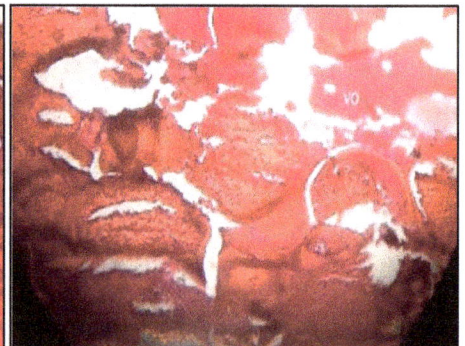

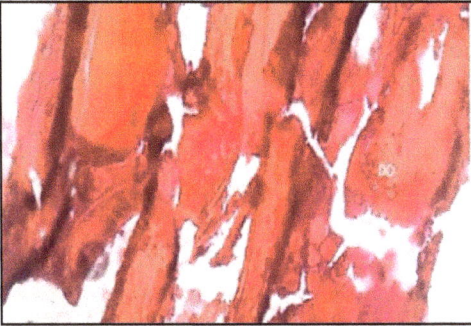

Figure 30.3: T.S. of HgCl₂ Treated Ovary After 20 Days X 100.

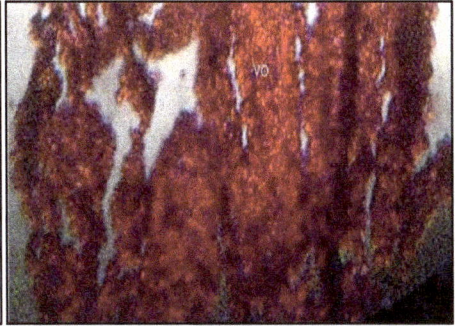

Figure 30.4: T.S. of HgCl₂ Treated Ovary After 30 Days X 100.

making deleterious effect on the embryonic development (Weis and Weis 1976b).Mackime *et al.* (1976) reiterated the same concept of the transfer of mercury from the parents to the through yolk, which caused deformities low hatchability and death of the brook trout embryos reduced productivity of sperm and ova was observed in the gold fish *Carassius auratus* injected with cadmium (Tafanelia and Summer Felt, 1975). Songalang and O'Halloran (1972) and Ahsan and Ahsan (1974) reported the interference of cadmium in the activity of *Solvelinus fontinalis* and *Clarias batrachus* respectively histological and biochemical effects of heavy metals on the ovary received little attention (James *et al.,* 2003). Deshmukh and Kulkarni (2005), Kumar and Pant (1984) reported a significant atresia to zinc on gonads. Baruah and Das (2002) also noted partial lysis swelling atresia and change in nucleus after exposure for 20 days.

Acknowledgements

The Authors are thankful to UGC, New Delhi for providing the financial assistance to carried out present study.

References

Ahsan, S. N. and Ahsan, J. 1974. Degenerative changes in the testis of *Clarias batrachus* (Linn) caused by cadmium chloride, *Indian J. Zool.* 15: 39-43.

Baruah, B. K. and Das M. 2002. Histopathological changes in ovary of fish *Hetropneustes fossilis* exposed to paper will effluent. *Aquaculture,* 3: 29-32.

Eaten J. G. 1980. Chronic cadmium toxicity to the bluegill *Lepomis macrochirus* Rafinesque trans. Am fish. SOC Washington D. C. 15th Ed.

James, R., Sampath K, and Edward, D. S. 2003. Copper toxicity on growth and reproductive potential in an ornamental fish *Xiphophorus helleri. Asian Fisheries Science,* 16: 317-326.

Kumar, S. and Pant, S. C. 1964. Comparative effects of the sublethal poisioning of zinc, copper and lead on the gonads of the teleost *Puntius conchonius* (Ham). *Toxicology Letters* 23: 189-194.

Mckim, J. M. G. F. olson. G. W. Holcombe and E. P. Hunt. 1976. Long term effects of methyl mercuric chloride on the three generation of booktrout (*Solvelinus fontinalis*) toxicity, accumulation, distribution and elimination. *J. Fish. Res. Board Can.* 33, 2726-2739.

Rosenthal H. and D. F. Alderdice. 1976. Sublethal effect of environmental stressors, natural and pollutional on marine fish and larvae. *J. Fish. Res. Board Can.* 33, 2047-2065.

Sanglang G. B. and M. K. O'halloran. 1974. Adverse effect of cadmium on brooktrout testis and on in vitro testicular androgen synthesis Bio. Reprod 9, 374-403.

Tafanelli R. and R. C. Summerfelt. 1975. Cadmium induced histopathological changes in goldfish in Pathology of fishes (Ed) W. E. Ribelin and G. Migaki, University of Wisconsin press Madison wis 613-645.

Weis J. S and Weis, J. S. 1977a. Effect of Heavy Metals on development of the killifish, *Fundulus hetroclitus. J. Fish Biol* 11, 49-54.

Weis, P. and Weis, J. S. 1977b. Methylmercury teratogensis in the killifish *Fundulus hetroclitus. Teratology.* 16, 317-326.

Chapter 31

A Study of the Toxic Effect of Detergent Surf Excel on the Muscle Protein of a Freshwater Fish *Cyprinus carpio* (Linn.)

☆ *S. Jeyakumar and S. Mala*

ABSTRACT

Detergents are surface active ingredients, once put into water they tend to remain there resisting conversion into fewer complexes into more soluble substances thereby creating foams in cesspools. Wastewater having detergent in them will bring about changes in bio-chemical constituents in the tissues of aquatic fauna especially fishes.Protein level in the muscle tissues of *Cyprinus carpio* exposed to sub – lethal concentrations of synthetic detergent surf excel for a short period was investigated. A decreasing trend was observed in the protein values on exposure to sub- lethal concentrations of Surf excel. The observed changes confirmed the impact of the synthetic detergent Surf excel on the physiology of fish.

Keywords*: Synthetic detergent, Cyprinus carpio, Bio-chemical constituents.*

Introduction

Pollution of water is an important dimension of environmental degradation. The disposal of the domestic water containing detergents directly enters the aquatic medium burdens the ecosystem these detergents manifest their toxic effect not only on aquatic organism but also on human beings through aquatic food source. In the aquatic organism the detergents cause physiological, Pathological and bio-chemical alteration (Yeragi *et al.,* 2003). The detergents are able to affect the activity of biologically active molecules such as lipids, aminoacids, co-enzymes and other protein containing compounds.(Devaraj, 1987). Liver,Kidney,Brain, Gills and muscles are the most vulnerable part of a fish exposed to medium containing any type of toxicants (Jana and Bandyopodhya, 1987). In the present study, it is aimed to investigate the toxic effects of commercially available synthetic detergent surf excel) on the protein level in the muscle tissue of *Cyprinus carpio.*

Materials and Methods

The test animal *Cyprinus carpio* (10gm ± 0.5 g live weight) were procured from manimuthar dam (longitudes: 77.6° E, Latitude 8.5° N, Altitude 11.62 m above sea level, Tirunelveli Dt,Tamil Nadu, South India) and acclimated to laboratory conditions for a period of one month in dechlorinated tap water. During the acclimation period, the fish were fed on pelleted feed with 38 per cent protein level. After the static bio – assay of the detergent surf excel, the test fishes were exposed to chosen sub-lethal levels (0.004 per cent - 0.008 per cent),which was taken in the plastic troughs of 30 litres capacity. Ten fishes were introduced into each trough and triplicates were maintained in each concentration. The test fish reared in dechlorinated tap water served as control.

Bio chemical analyses were carried out at the 30th day of exposure. Methods described by Lowery *et al.,* was followed to estimate the muscle protein content.

Results and Discussion

The toxic effect of the detergent surf excel on the bio-chemical constituent (protein) in the muscle tissue of a freshwater fish *Cyprinus carpio* were found out. The results were presented in the Table 31.1. In the experimental the level of protein is always lower than the control in the full course of the experiment, and with the increasing concentration of the detergent a gradual decrease was noted indicating the relationship of the protein level of the muscle tissue with that of the detergent concentration this indicates detergents are creating stress to the aquatic organisms

Table 31.1: Relationship of Protein Levels in the Muscle Tissue of
***Cyprinus carpio* with the Concentration of Detergent**

Sl.No.	Detergent Conc. in Per cent	Average	SD	SE	't' Value	P Value	Y=a+bx	'r' Value
1	0	15.2	0.874	0.618			16.438	
2	0.004	13.1	0.2	0.141	2.41	2.92	11.368	
3	0.005	11.6	0.361	0.255	3.85	2.92	10.101	
4	0.006	8.3	0.153	0.108	8.09	2.92	8.833	0.886
5	0.007	7.1	0.1	0.071	9.76	2.92	7.566	
6	0.008	5.3	0.153	0.108	11.62	2.92	6.295	

(Jones *et al.*, 1987). Fishes muscle tissues have high protein value they form an important trophic level in the aquatic food chain. The reduction in the protein level noted in the present study may be attributed to the fact that the detergent might have created tissue damage (kaber, *etal.*, 1981) *i.e.* necrosis or disturbances of celluar function and consequent impairment in protein synthesis machinery, this may declined protein synthesizing capacity of the muscle (Malla Reddy and Basha Mohideen, 1988) and may lead to TCA cycle through aminotransferase (Palanichamy and Potidore, 1980; Natarajan, 1983) which is a condition to cope up with the stress condition during which organs have high catabolic potency intense proteolysis which has reflected in the form of decrease in the muscle protein.

References

Devaraj, H., and Devaraj. N. 1987. Rat intestinal lipid changes in patulin toxicity. Ind. *J. Exp. Biol.* 25: 637-638.

Jana. S. and Bandyo padhya, S. 1981. Effect of metals on some bio chemical parameters in the fish, *Chanana punctatus, J. Environ. Ecol.* (5) 488-493.

Jones. K. A; Brown. S. B. and Horay. T. J. 1987. Behavioural and Biochemical studies of onset and recovery from acid stress in Aretic Char. (*Salvelinus alpinus*) *Can. J. Fish. Aquat. Sci* 441B1. 3: 381.

Kaber, A. S. I; Janardhan Rao. K. S. and Ramana Rao. K. V. 1981. Effect of Malathion Exposure on some physical parameters of whole body and on tissue of teleost *Tilapia mossambica. BioScience,* 3(1): 17-21.

Malla Reddy. P. and Basha Mohideen. M. 1988. Toxic impact of fenvalerate on the protein metabolism in the branchial tissue of a fish *Cyprinus carpio. Curr-Sci.* 57: 211-212.

Natarajan. G. M. 1983. Metasytox Effects of lethal (LC. 50h) concentration on free amino acid and glutamate dehydro genases in some tissues of air breathing fish, *Channa striatus* (Bleeker). *Comp. Physiol. Ecol.* 8: 254-256.

Palanichamy. S. R. and Potidore, T. 1980. Mechanism of heavy metal inhibition of aminoacid transport in the intestine of marine fishes. *Biol. Bull Mar. Labwoods* (II). 159: 458.

Yeragi, S. G., Rana. A. M. and Koli. V. A. 2003. Effect of Pesticide on Protein metabolism of mudskipper *Boleophthalamus dussumieri J. Ecotoxicol. Environ. Moint.* 13(3).

Chapter 32

Effect of Herb Satavari (*Asparagus racemosus*) on Growth Performance of *Cirrhinus mrigala* Fingerlings

☆ *T.K. Thakur, Dushyant Kumar Damle and Sandhya R. Gaur*

ABSTRACT

An experiment was conducted to assess the effect of Satavari (*Asparagus racemosus*) root powder as a herbal growth promoter on the growth performance of *Cirrihinus mrigala* fingerlings. Fish (3.16±0.02 g and 42±0.02 cm) were fed with five dietary treatments, each having four replications. Treatment groups had a varying levels of Satavari root powder (0.5 per cent, 1 per cent, 1.5 per cent and 2 per cent) included in their diets. The control diet was free from Satavari root powder. Diets formulated contained 30 per cent crude protein and feed was administrated @ 4 per cent live body weight twice daily for 8 weeks. Weekly growth of fishes was recorded and feeding level was adjusted accordingly. At the end of the experiment, growth performance of *Cirrhinus mrigala* fingerlings was analyzed. The results showed that all levels of Satavari administration in the diets of *Cirrhinus mrigala* produced significantly (P<0.05) higher growth as compared to the control. Fingerlings fed with the diet containing 1.5 per cent Satavari had significantly (P<0.05) higher weight gain (90.68 per cent) and higher specific growth rate (1.15 per cent) than other experimental diets. Survival per cent was also greatest in the 1.5 per cent Satavari group. Economic analysis of experimental diets also revealed that the addition of 1 – 1.5 per cent of Satavari to the diet of *Cirrhinus mrigala* fingerlings is quite beneficial. In conclusion, it can be suggested that addition of 1.0- 1.5 per cent of medicinal herb Satavari as a growth promoter to the fish diet can not only promote growth, but also reduce the cost of production, and thus can increase the profitability in Carp culture.

Keywords: *Growth performance, Carps, Herb inclusion, Satavari.*

Introduction

Aquaculture is mainly feed based industry where supplementary feed is the single most important factor affecting input cost. Feed cost is considered to be the highest recurrent cost often ranging from 60-70 per cent depending on the intensity of the culture (Pandian *et al.,* 2001). To reduce this cost various techniques have been employed *viz.*, fertilization, use of mixed feeding schedule, stocking of genetically superior fish, stocking with stunted carp fingerlings, using metabolic stimulants in diets referred to as feed additives or growth promoters etc. These include substances like hormones, antibiotics, minerals, vitamins, feed attractants and stimulants (Gangadhara and Ramesha, 2001). Although hormones and their analogues, antibiotics and a host of other substances have been used for growth promotion, the fear of residual effects and doubt regarding consumer acceptance have prompted a search for substances which are user-friendly and origin more from nature. This has resulted in the use of herbal preparations as growth stimulants/promoters. Inclusion of these growth promoters in fish feed are reported to bring measurable increase in fish growth by way of weight gain and better feed conversion efficiency. Growth promoters not only reduce the period of rearing but also cut down the cost of production which in turn increases the profitability in aquaculture farming.

Medicinal plants are the store house and rich source of safe and cheap chemicals. In Ayurveda several herbs have been prescribed as metabolic as metabolic enhancer, anti-stress, growth promoter, appetizer, tonic, immuno-stimulants and antimicrobials since ages. However, use of such medicinal plants in fish feed has attracted only limited attention of aquaculturists. The present study aims at effect of herb, Satavari (*Asparagus racemosus*) on growth performance of *Cirrhinus mrigala* fingerlings with a view to achieve maximum production in low cost as well as without any toxic and residual effects.

Materials and Methods

Stocking of Fingerlings in Aquaria

This experiment was carried out for nearly three months during July- septmber 2007. The acclimatized fingerlings of *Cirrhinus mrigala* (average length 6.42 cm. and weight 3.15 g) were stocked randomly in aquaria of equal size (75 x 37.5 x 37.5 cm). Care was taken to select healthy fishes of almost uniform size. Before stocking the fingerlings, they were given a dip treatment with 5 ppm Potassium permanganate ($KMnO_4$) in plastic container for 4-5 minutes to disinfect the fishes. The fingerlings were stocked @10 fingerlings aquarium^{-1}.

Feed Preparation

The ingredients used for the preparation of experimental feeds were mustard oil cake, rice bran, root powder of herb (Satavari - *Asparagus racemosus* Willd.), vitamins and minerals premix (Minamil, manufactured by Brihans Laboratories, Bosari, Pune, India; Table 32.1) and soybean oil procured from local market of Raipur. Carboxy methyl cellulose grade was used as a feed binder.

Feed Formulation

Basal diet containing mustard oil cake and rice bran were prepared on the basis of 30 per cent protein level with the help of Pearson's square method of feed formulation. Experimental feed – T_2, T_3, T_4, T_5 were prepared by mixing measured proportion of dry root powder of Satavari in basal diet while control feed – T_1 devoid of the herbal powder of Satavari (Table 32.1).

Table 32.1: Ingredients Composition of Various Experimental Diets

Ingredients (per cent)	Diets				
	T_1	T_2	T_3	T_4	T_5
Rice bran	8.23	8.23	8.23	8.23	8.23
Mustard oil cake	85.77	85.77	85.77	85.77	85.77
Satavari	0.0	0.5	1.0	1.5	2.0
Soybean oil	2	2	2	2	2
Vitamin-mineral mixture	1	1	1	1	1
Carboxy methyl cellulose	3	3	3	3	3

Mixing of Ingredients and Pelletization

All the ingredients were mixed and dough was prepared by adding water. Dough was pressure cooked for 25 minutes for sterilization, vitamins and minerals mix added @ 1 per cent after cooling of dough, and pellets (2 mm. diameter) prepared using a hand pelletizer.

Experimental Design

The experimental design has five distinct experimental groups (T_1, T2, T_3, T_4 and T_5). with four replications in Completely Randomized Design (CRD) were kept.

Feeding Treatments

The fingerlings of *Cirrhinus mrigala* were fed with experimental diets – T_1, T_2, T_3, T_4, T_5 in 20 aquaria @ 4 per cent per Kg body weight of the fishes for 8 weeks under suitable condition of aeration, temperature and water quality. Each daily diet was divided into two equal parts and offered to the fish twice daily at 8.00 – 8.30 a.m. and 4.00-4.30 p.m. Feed was applied in petridishes at bottom of aquaria. Feeding rate was adjusted on weekly basis based on the increment in fish weight.

Physico-chemical Parameters of Water

Certain physico-chemical parameters *viz.* temperature, pH, dissolved oxygen, free carbon dioxide, alkalinity and hardness of water in aquaria were analyzed by following the Standard Methods (APHA, 1995). Water temperature was recorded daily while other parameters were recorded at weekly intervals.

Growth Study in Fish

The growth parameter of the fingerlings were recorded on weekly bases and evaluation of growth and dietary performance done by estimating net wet gain (NWG), percentage weight gain (PWG), average daily gain (ADG), specific growth rate (SGR) and survival percentage.

Feed Consumption

The unconsumed feed was collected after three hours of feeding, by siphoning the bottom of glass aquaria and filtering the water through silk cloth No. 30. This was then dried in an oven and weighed. The quantity of feed consumed was determined by deducting the dried unconsumed feed from the total feed given. Daily consumption (twice daily) was recorded for each group and thereafter was calculated as percentage of body weight.

Proximate Analysis of Feed

Samples of feed ingredients, diets and whole body of fishes were analyzed for the proximate composition as per the methods described by the AOAC(1990).

Economic Evaluation of Diets

Economic analysis of each experimental diet was calculated on the basis of local retail rates prevalent in the market (Table 32.2). It was assumed that all other costs were constant in each of the diets, and hence the difference in costs of the diets is the reflection of how effective the diet was at achieving the fish growth.

Table 32.2: Market Price (2008) of Feed Ingredients

Sl.No.	Particulars	Cost (Rs./kg)
1.	Rice bran	3.00
2.	Mustard oil cake	9.00
3.	Satavari	450.00

Result and Discussion

Growth Performance of Fingerlings of *Cirrhinus mrigala*

Data pertaining to growth performance of *Cirrihinus mrigala* fingerlings fed with varying levels of Satavari mixed diets are depicted in Table 32.3. The average initial and final body weight of Mrigal fingerlings were ranged between 3.13 – 3.18 g and 4.49 – 6.05 g respectively among all the treatments. Statistical analysis of variance indicated that the average gain in weight in fingerlings was significantly influenced

Table 32.3: Growth Performance of *Cirrhinus mrigala* (Ham.) Fingerlings Fed on Varying Levels of Satavari Mixed Diet for Eight Weeks

Levels of Satavari Per cent of Diet	g kg^{-1} b.w. Day^{-1}	NWG (g)	PWG (per cent)	ADG (mg day^{-1})	SGR (per cent day^{-1})	FI	FCR	FCE	PER	Survival (per cent)
0.0 per cent (T$_1$)	—	1.31^{e}	41.29^{e}	23.40^{e}	0.617^{e}	3.38^{d}	2.58^{a}	0.388^{e}	1.29^{e}	97.50^{a}
0.5 per cent (T$_2$)	0.085	1.71^{d}	53.90^{d}	30.63^{d}	0.770^{d}	3.50^{c}	2.04^{b}	0.490^{d}	1.64^{d}	97.50^{a}
1.0 per cent (T$_3$)	0.170	2.48^{b}	78.65^{b}	44.25^{b}	1.036^{b}	3.65^{ab}	1.47^{d}	0.679^{b}	2.26^{b}	100.00^{a}
1.5 per cent (T$_4$)	0.248	2.88^{a}	90.68^{a}	51.38^{a}	1.152^{a}	3.67^{a}	1.28^{e}	0.784^{a}	2.61^{a}	100.00^{a}
2.0 per cent (T$_5$)	0.337	2.13^{c}	68.05^{c}	38.00^{c}	0.927^{c}	3.58^{b}	1.68^{c}	0.595^{c}	1.98^{c}	100.00^{a}
SEm ±		0.04	1.08	0.70	0.011	0.02	0.03	0.008	0.03	1.58
CD (P=0.05)		0.12	3.26	2.10	0.034	0.07	0.08	0.024	0.08	NS

Values bearing same superscripts are not significantly (P=0.05) different.

NS: Non-significant at 5 per cent.

by different levels of Satavari. The net weight gain was found significantly (P=0.05) higher in fishes fed on herbal diet T_4 (2.88 g) followed by fishes fed on herbal diet T_3 (2.48 g), T_5 (2.13 g), T_2 (1.71 g) and was observed to be lowest in fishes fed with control diet T_1 (1.31 g) respectively (Table 32.3). Further CD test indicated significant difference among all the treatments.

In this study, net weight gain markedly accelerated with increasing level of herb, Satavari to the diets of Mrigal fingerlings. The relatively low weight gain (normal) of control group can be related to various factors affecting growth of fishes (Donaldson *et al.,* 1979), however the inhibitory effects of these factors on growth are overcome with treatments receiving root powder of Satavari. These results are in agreement with those obtained by Kavitha (1996), who found that the feeding of herb, Satavari enhanced growth performance of *Labeo rohita.* Similar results were also obtained by Santiago (1991) in Nil Tilapia, Shambhu and Jayaprakash, 2001) in *Penaeus indicus,* Citarasu *et al.* (2003) and Venketramalingam *et al.* (2007) in *Penaeus monodon,* Naiyr (2004) in *Labeo rohita,* Ji *et al.* (2007) in Japanese Flaunder, Kumar (2002a), Kumar (2002b), Kour 2003), and Dhangar (2004) in *Cirrhinus mrigala.*

Highest percentage weight gain was found for the fishes fed on herbal diet T_4 (90.68 per cent) followed by T3 (78.65 per cent), T_5 (68.05 per cent), T_2 (53.90 per cent) and was found lowest for fishes fed on control diet T_1 (41.29 per cent) respectively (Table 32.3). Difference in Percentage Weight Gain (PWG) was found significant (P=0.05) among all the treatments.

Impact of herb, Satavari as supplemented diets on the growth of Indian Major Carp, *Labeo rohita,* studied by Kavitha (1996) revealed that the fish fed with Satavari mixed diets showed significantly higher PWG (119.6 per cent). In the present study the PWG increased to 90.68 per cent in the herb treated groups of fishes as observed by Sharma *et al.* (1986) also. Studies of Francis *et al.* (2002), Kour 2003), Rathore (2005) and Sharma *et al.* (2007) on various species of fishes also reported higher percentage weight gain in fishes fed on herbal diets.

The fingerlings of *Cirrhinus mrigala* fed on control diet (T_1) have an Average Daily Gain (ADG) of 23.40 mg day^{-1} as shown in Table 32.3. This was significantly increased (P=0.05) to 30.63, 44.25, 51.38 and 38.00 (mg day $^{-1}$) in diets T_2, T_3, T_4 and T_5 fed groups of fishes respectively. The difference in ADG was found to be significant among all the treatments. The highest ADG was recorded for fishes fed with herbal diet T_4 and was found lowest for fishes fed with control diet T_1.

It has been observed that application of herb in the supplementary feed has beneficial effect on Specific Growth Rate (SGR) of fishes. In the present study, SGR of fishes during the experimental period were 0.617 per cent, 0.770 per cent, 1.036 per cent, 1.152 per cent and 0.927 per cent (day^{-1}) for group of fishes on diets T_1 to T_5 respectively (Table 32.3). Significant (P=0.05) difference in SGR was found among all the treatments. The highest SGR was observed in the fishes fed with herbal diet T_4 and was found lowest in fished fed with control diet T_1.

Similar results were found by Kavitha (1996). They revealed that feeding of Satavari to *Labeo rohita* fry exhibited highest specific growth rate (1.02 per cent). In the present study the highest specific growth rate was observed 1.152 per cent in the groups of fishes fed with Satavari mixed diets. Shabmbhu and Jayaprakash (2001), Shalaby *et al.* (2006), Wang *et al.* (2006), Venketramalingam *et al.* (2007) and Yu *et al.* (2008) were also found significantly higher SGR after feeding of herbs to different species of fishes. This increase in SGR may be due to the enzymatic breakdown of food, availability of nutrients for absorption (Das *et al.*, 1987) and increased secretion of enzymes in the stomach and intestine (Maheshappa 1993 and Sadakshari 1993). Moreover it may be due to the increase in the level of RNA/DNA ratio in the muscle and hepatopancreas which lead to protein synthesis resulting to enhanced growth (Mathers *et al.*, 1992).

In the present study, weekly average individually feed intake (FI) of *Cirrhinus mrigala* fingerlings during the experimental period was presented in Table 32.3. Total Feed Intake (FI) for fishes in relation to different diets was 3.38, 3.50, 3.65, 3.67 and 3.58 (g fish^{-1}) in T_1, T_2, T_3, T_4 and T_5 respectively. The results showed that the fingerlings of *Cirrhinus mrigala* fed with herbal diets increased the feed intake significantly (P=0.05) as compared to control diet. The difference in values of feed intake for the group of fishes fed on diets T_4 and T_3 was found non-significant (P=0.05) while it showed significant difference between fishes fed on diets T_4 and T_5, T_4 and T_2, T_4 and T_1, T_3 and T_5, T_3 and T_2, T_3 and T_1, T_5 and T_2, T_5 and T_1 and T_2 and T_1 respectively (Table 32.3).

The growth rate promotion by Livol (IHF-100), a new growth promoter from Indian herbs, was studied by Jayaprakash and Euphrasia, 1996) and reported that the herbal diet increased the stimulation of appetite of *Penaeus indicus* leading to enhanced feed intake. In the present study feed intake increased with increasing levels of Satavari in the diets of *Cirrhinus mrigala* fingerlings. It may be due to the inclusion of herbal appetizer and metabolic enhancer, *Asparagus racemosus* in the diets of *Cirrhinus mrigala* (Sharma *et al.*, 2006).

Data pertaining to Feed Conversion Ratio (FCR) for Mrigal fingerlings throughout the experimental period is depicted in Table 32.3. The values of FCR for fishes were 2.58, 2.04, 1.47, 1.28 and 1.68 for the diets T_1 to T_5 respectively. The FCR values for fishes fed on different levels of Satavari mixed diets exhibited significant (P=0.05) difference among all the treatment including control. The best FCR value was observed for the fishes fed with herbal diet T_4 as compared to other experimental diets.

In the present study the food conversion ratio (FCR) improved (1.28) significantly with increasing levels of herb, Satavari. A similar improvement in FCR (2.25) was obtained by Kavitha (1996) in *Labeo rohita*. It may be due to the incorporation of Satavari as a growth promoter to the diets of *Cirrhinus mrigala* fingerlings which improve digestibility of feed and utilization of nutrients (Kumar *et al.*, 2006). Hormonal and non-hormonal growth promoters improve the nutrient digestibility is reported for several species (Gangadhar *et al.*, 1998; Keshavanath and Renuka,. 1998). Improvement in FCR were also reported by Sivaram *et al.* (2004) in Greasy Grouper, Rekhate *et al.* (2005) in Broiler, Rathore (2005 in *Cirrhinus mrigala*, Wang *et al.* (2006 in *Paralichthys olivaceus* and Ji *et al.* (2007)) in Japanese Flounder, while fed each with different species of herb.

Feed Conversion Efficiency (FCE) was also significantly influenced by different levels of Satavari mixed in the diets. The FCE values observed during the experimental period were 0.388, 0.490, 0.679, 0.784 and 0.595 for the group of fishes fed on diets T_1, T_2, T_3, T_4 and T_5 respectively (Table 32.3). Significant (P=0.05) difference was found between all the treatments and the fishes fed on diet T_4 showed the highest FCE value than other experimental diets.

The Protein Efficiency Ratio (PER) observed during the experimental period is presented in Table 32.3. It was recorded as 1.29, 1.64, 2.26, 2.61 and 1.98 for fishes fed with diets T_1, T_2, T_3, T_4 and T_5 respectively. The PER values were significantly (P=0.05) different among all the fishes fed on different levels of Satavari mixed diets including control. The highest PER value was found for fishes fed with herbal diet T_4 and was found lowest for fishes fed with control diet T_1.

Feed Conversion Efficiency (FCE) and Protein Efficiency Ratio (PER) are used as quality indicators for fish diet and amino acid balance. So, these parameters are used to assess protein utilization and turn over. These results are also in agreement with those obtained by Kavitha (1996), Kumar (2002a), Rathore (2005), Wang *et al.* (2006), Goda 2008) and Yu *et al.* (2008) who found that herbal diet increased feed efficiency ratio and protein efficiency ratio in fish as compared to their respective control diet.

Data pertaining to survival percentage of experimental fishes throughout the experimental period was depicted in the Table 32.3. The difference in survival was found Non-significant (P=0.05) among all the treatments. Fingerlings of *Cirrhinus mrigala* showed 97.5 per cent survival when fed with control diet (T_1). The same survival was also observed in the treatment group T_2. Survival was increased to 100 per cent in all the group of fishes fed with diets T_3, T_4 and T_5.

In the earlier studies of Citarasu *et al.* (2003), Shalaby *et al.* (2006), Sharma *et al.* (2007) and Sahu *et al.* (2007) on various species of herbs and fishes exhibited that fishes fed on herbal diet showed the tendency to higher survival which supports this finding. Satavari play a role in the immune system stimulation in fishes. It may be due to rise in haemoglobin, plasma total protein, serum total protein and globulin in herbal treated groups which improve blood indices and enhanced general health condition of animals (Rekhate *et al.* (2005).

Cost of Production of Fingerlings

The data pertaining to economic analysis of the experimental diets is described in Table 32.4. The cost of per kg experimental diets in the present study were found to be Rs. 7.97, Rs. 10.22, Rs. 12.47, Rs. 14.72 and Rs. 16.07 for T_1, T_2, T_3, T_4 and T_5 respectively. However, production cost for raising 1 Kg of fish in terms of feed intake (based on FCR) was found Rs. 20.54, Rs. 20.85, Rs. 18.36, Rs. 18.77 and Rs. 28.54 for the treatment T_1, T_2, T_3, T_4 and T_5 respectively. The production cost was found to be the lowest in the treatment T_3 followed by T_4 with the cost-benefit ratio of 1.72 and 1.66 respectively. The net income obtained in the case of T_3 was the highest among all the diets. Therefore from economic point of view the experimental diet T_3 was found most beneficial in the present investigation. However, as far as growth and nutrients value of *Cirrhinus mrigala* is considered, the diet T_4 was found to give the best results.

Table 32.4: Economics of Raising Fingerlings under different Experimental Diet

Treatment	Cost of Feed (Rs/kg)	Total Feed Intake (g)	Net Weight Gain (g)	Total Cost of Feed (Paisa)	Cost of Production (Rs/kg weight increase)	Gross Income (Paisa)	Net Income (Paisa)	Cost-Benefit Ratio
T1	7.97	3.38	1.31	2.69	20.54	6.56	3.86	1.43
T2	10.22	3.50	1.72	3.58	20.85	8.58	5.00	1.40
T3	12.47	3.65	2.48	4.55	18.36	12.39	7.84	1.72
T4	14.72	3.67	2.88	5.40	18.77	14.39	8.99	1.66
T5	16.97	3.58	2.13	6.07	28.54	10.64	4.57	0.75

In the present study, the highest percentage weight gain (90.68 per cent), Specific growth rate (1.152 per cent) and food conversion efficiency (0.784) obtained in group fed with diet including 1.5 per cent Satavari, clearly indicates that it plays a very imperative role as growth promoter in freshwater fishes specially carps, as also reported by Maheshappa *et al.* (1999), during their study with Livol inclusion in diets of Carp. They reported that Livol is relatively costly as it required additional cost of rupees 0.15, 0.30, and 0.45 respectively for 0.5 per cent, 1 per cent and 1.5 per cent inclusion per Kg of diet. However, from the point of view of improved nutrient digestibility and growth, the addition of Livol to the diet of carps was also found beneficial. Thus this study further proves that use of Satavari is beneficial to fishes as diet supplement.

Conclusion

The experiment conducted on *Cirrhinus mrigala* fingerlings to observe the effect of inclusion of herb Satavari, may be concluded as follows:

1. Satavari acts as a growth promoter, because its use enhanced the growth performance of fish.

2. Inclusion of Satavari @ 1.5 per cent of the diet was found optimum for the growth of *Cirrhinus mrigala* fingerlings.

3. Due to higher cost-benefit ratio the diet with 1 per cent level of Satavari was found most economical followed by 1.5 per cent level of Satavari.

4. Use of Satavari as a growth promoter is beneficial for fish culture.

References

AOAC, 1990. Official Method of Analysis. 16th Edn., Association of Analytical Chemists, Arlington, VA USA.

APHA. 1995. Standard methods for the examination of water and wastewater. *American Pub. Health Asso.,* Washington, DC, USA.

Goda, A. M. A. S. 2008. Effect of dietary Ginseng herb (Ginsana G 115) supplementation on growth performance, feed utilization, and hematological indices of Nil Tilapia, *Oreochromis niloticus* (L.), fingerlings. *J. World Aqua. Soc.,* 39(2): 205-214.

Citarasu, T., K. Venkatramalingam, M. M. Babu, R. R. J. Sekar and M. Petermarian, 2003 Influence of the antibacterial herbs, *Solanum trilobatum*, *Andrographis paniculata* and *Psoralea corylifolia* on the survival, growth and bacterial load of *Penaeus monodon* post larvae. *Aqua. International*, 11(6): 581-595.

Das, K. M., A. Ghos and A. Ghos, 1987. Studies on the comparative activity of some digestive enzymes in fry and adult of Mullet, *Liza parasia* (Ham). *J. Aqua. Trop.*, 2: 9-15.

Dhangar, D., 2004. Use of seed of an aquatic herb, Lotus (*Nelumbium speciosum* Willd.) as a growth promoter in the supplementary feed of Indian major carp, *Cirrhinus mrigala* (Ham.). M. Sc. Thesis, Maharana Pratap University of Agriculture and Technology, Udaipur, India.

Donaldson, E. M., U. H. M. Fagerlund, D. A. Higgs and J. R. McBride, 1979. Hormonal Enhancement of Growth in Fish. In: Fish Physiology, Vol. 3, Hoar, W. S. and D. J. Randall (Eds.). Academic Press, New York, USA., pp: 455-597.

Gangadhara, B. and T. J. Ramesha, 2001. Non-hormonal growth promoters: A boom to aquacultures. *Fish. Chimes*, 21: 49-50.

Francis, G., H. P. S. Makkar and K. Becker, 2002. Dietary supplementation with a *Quillaja saponin* mixture improves growth performance and metabolic efficiency in Common Carp (*Cyprinus carpio*). *Auaculture*, 203: 311-320.

Gangadhar, B., M. C. Nandeesha, T. J. Varghees and P. K. Keshvanath, 1998. Effect of feeding 19-Norethisterone on grwoth and body composition of Rohu, *Labeo rohita*. *Assian Fish. Sci.*, 11: 51-58.

Jayaprakash, V. and J. Euphrasia, 1996. Growth performance of *Labeo rohita* (Ham.) Livol (IHF-1000), a herbal product. *Proc. Indian Nat. Sci. Acad. B*, 63: 1-10.

Ji, S. C., G. S. Jeong, G. S. Im, S. W. Lee, J. H. Yoo and K. Takii, . 2007. Dietary medicinal herbs improve growh performance, fatty acid utilization, and stress recovery of Japanese Flounder. *J. Fish. Sci.*, 73 (1): 70-76.

Kavitha, K., 1996. Impact of two herbs supplemented diets on the growth of Indian Major Carp, *Labeo rohita* (Ham.). M. Sc. Thesis, RAU, Bikaner, India.

Keshavanath, and Renuka, . 1998. Effect of dietary L-carnitine supplements on growth and body composition of *Labeo rohita* (Ham.) fingerlings. *Aqua. Nutrition*, 4(2): 83-87.

Kour, D., 2003. Use of herb "Bala" (*Sida cordifolia*, Linn.) as growth promoter in the supplementary feed of an Indian Major Carp, *Cirrhinus mrigala* (Ham.). M. Sc. Thesis, Maharana Pratap University of Agriculture and Technology, Udaipur, India.

Kumar, A., 2000a. Use of Ashwagandha, *Withania somnifera* (L.) Dunal as growth promoter in the supplementary feed of and Indian Major Carp *Cirrhinus mrigala* (Ham.). M. Sc. Thesis, Maharana Pratap University of Agriculture and Technology, Udaipur, India.

Kumar, R., 2002b. The effect of two herbs, mulethi (*Glucyrrhiza glabra,* Linn.) and kali Musli (*Curculigo orchioides,* Gaertn.) as growth promoter in the supplementary feed of an Indian Major Carp, *Cirrhinus mrigala* (Ham.). M. Sc. Thesis, Maharana Pratap University of Agriculture and Technology, Udaipur, India.

Kumar, A., N. Gupta and D. P. Tiwari, 2006. Effect of herbs as feed additive on In vitro and in sacco dry matter digestibility of paddy straw. *Indian J. Anim. Sci.,* 76: 847-850.

Maheshappa, K., 1993. Effect of different doses of Livol on growth and body composition of Rohu, *Labeo rohita* (Ham.). M.F.Sc. Thesis, University of Agricultural Science, Banglore, India

Maheshappa, K., Ramesha, T. J., Gangadhar, B. and Varghese, T. J. 1999. Growth performance and biochemical composition of Rohu, *Labeo rohita* to Livol incorporated diets. *Indian Journal of Animal Sciences* (communicated).

Mathers, E. M., D. F. Haulihan and M. J. Chunnigham, 1992. Nucleic acid concentrations and enzymes activities as correlates of growth of Saithe, *Pollachius virens*: Growth rate estimates of open sea fish. *Mar. Biol.,* 112: 363-369.

Naiyr, P., 2004. Use of herb, Gokhru (*Pedalium murex,* Linn.) as growth promoter in the supplementary feed of an Indian Major Carp, *Labeo rohita* (Ham.). M. Sc. Thesis, Maharana Pratap University of Agriculture and Technology, Udaipur, India.

Pandian, M. R., S. N. Mohanty and S. Ayyappan, 2001. Food Requirements of Fish and Food Production in India. In: Sustainable Indian Fisheries, Pandian, T. J. (Ed.). *National Academy of Agricultural Sciences,* New Delhi, India, pp: 153-165.

Rathore, L. K., 2005. Use of Vidari Kand (*Pueraria tuberose*) as a herbal growth promoter in the supplementary feed of an exotic carp, *Cyprinus carpio* var. communis L. M. Sc. Thesis, Maharana Pratap University of Agriculture and Technology, Udaipur, India.

Rekhate, D. H., S. Ukey and A. P. Dhok, 2005. Performance of broilers supplemented with Shatavari (*Asparagus racemosus* willd.) root powder. Department of Animal Nutrition, Post Graduate Institute of Veterinary and Animal Sciences, Akola-444 104 (MS), Maharashtra, Animal and Fishery Sciences University, Nagpur, Source: IPSACON-2005.

Sadakshari, G. S., 1993. Effect of Bioboost forte, Livol and Amchemin AQ on growth and body composition of Common Carp, *Cyprinus carpio* (Linn.). M. F. Sc. Thesis, University of Agricultural Sciences, Bangalore, Karnataka, India.

Sahu, S., B. K. Das, B. K. Mishra, J. Pradhan and N. Sarangi, 2007. Effect of *Allium sativum* on the immunity and survival of *Labeo rohita* infected with *Aeromonas hydrophila. J. Applied Ichthyol.,* 23: 80-86.

Santiago, C. B., 1991. Feed, survival and food conversion of Nil Tilapia fingerlings fed diet containing Bayo-n-ox, a commercial growth promoter. Isr. *J. Aqua.,* Bamidgeh, 42 (2): 77-81.

Shambhu, C. and Jayaprakash, V. 2001. Livol (IHF-1000), a new herbal growth promoter in White Prawn, *Penaeus indicus* (Crustacea). *Ind. J. Marine Sci.*, 30 (1): 38-43.

Shalaby, A. M., Y. A. Khattab and A. M. Abdel Rahman, 2006. Effect of Garlic (*Allium sativum*) and Chloramphenicol on growth performance, physiological parameters and survival of Nile Tilapia (*Oreochromis niloticus*). *J. Venom. Anim. Toxins incl. Trop.*, 12(2), Botucatu Apr. /June.

Sharma, S. K., L. L. Sharma and D. Kour, 2006. Use of herbs as metabolic enhancer in freshwater fish. *Fish. Chimes*, 26: 25-28.

Sharma, C., L. L. Sharma and B. K. Sharma, 2007. The herb Kaunch (*Mucuna pruriens*) as growth promoter in supplementary diet of common carp fingerlings. *Fish. Chimes*, 26: 69-71.

Sharma, S., S. Dahanukar and S. M. Karandikar, 1986. Effect of long-trem administration of the root of Ashwagandha (*Withania somnifera*) and Shatavari (*Asparagus racemosus*) in Rats. *Indian Drugs*, 23: 133-139.

Sivaram, V., M. M. Babu, G. Immanuel, S. Murugadass, T. Citarasu and M. P. Marian, 2004. Growth and immune response of juvenile Greasy Groupers *(Epinephelus tauvina)* fed with herbal antibacterial active principle supplemented diets against *Vibrio harveyi* infections. *Aquaculture*, 237 (1/4): 9-20.

Venketramalingam, K., J. G. Christopher and T. Citarasu, 2007. *Zingiber officinalis* an herbal appetizer in the Tiger Shrimp *Penaeus monodon* (Fabricius) larviculture. *Aqua. Nutrition*, 13 (6): 439-443.

Wang, J. Q., Y. X. Sun and J. C. Zhang, 2006. Effect of Chinese herb additives on growth, digestive activity and non-specific immunity in Flounder *Paralichthys olivaceus*. *J. Fish. China*, 30: 90-96.

Yu, M. C., Z. J. Li, H. Z. Lin, G. L. Wen and S. Ma, 2008. Effects of dietary Bacillus and medicinal herbs on the growth, digestive enzyme activity and serum biochemical parameters of the shrimp *Litopenaeus vannamei*. *Aquacul. Int.*, 16: 471-480.

Chapter 33

Studies on Nutritional Composition of some Edible Fishes (Cyprinids) Found in Markets of Raipur District of Chhattisgarh

☆ *Neha Chandrawanshi, Dushyant Kumar Damle*
and Sandhya R. Gaur

ABSTRACT

Biochemical composition of five cyprinid fishes namely Catla (*Catla catla*), Rohu (*Labeo rohita*), Mrigala (*Cirrhinus mrigala*), Silver carp (*Hypophthalmichthys molitrix*) and Common carp (*Cyprinus carpio*) found in markets of Raipur district of Chhattisgarh was studied in order to evaluate their nutritional values. The mean value of moisture, protein, lipid, carbohydrate and ash content was found as 78.99 ± 6.56, 18.68 ± 0.87, 2.6 ± 0.92, 0.86 ± 0.60, 1.11 ± 0.06 in Catla; 75.10 ± 9.88, 16.48 ± 0.31, 2.17 ± 0.81, 1.12 ± 0.89, 1.53 ± 0.27 in Rohu; 72.4 ± 7.69, 16.8 ± 0.2, 1.65 ± 0.75, 0.78 ± 0.70, 1.77 ± 0.22 in Mrigala; 69.66 ± 7.95, 16.26 ± 0.79, 2.6 ± 0.63, 0.77 ± 0.98, 1.52 ± 0.30 in Silver carp and 65.02 ± 11.39, 15.36 ± 0.75, 2.77 ± 0.68, 0.81 ± 0.50, 0.95 ± 0.18 in Common carp. In Chhattisgarh, rohu has got the most preference due to its taste and palatability, but this study shows that catla is having equally ample nutrients along with higher content of protein than rohu and lesser content of fat than common carp. Moreover its cost is also less than rohu, hence can be referred as a good source of animal protein in people's daily diet.

Introduction

In Chhattisgarh, fish is highly consumed by the people. A number of varieties of fish species are found in Chhattisgarh and all types of them are preferred by the people. Out of different varieties of fishes, carps are most popular and liked by the people. Fish has been recognized as an excellent food source for human beings for centuries and is preferred as a perfect diet not only due to its excellent taste and high digestibility but also because of having higher proportions of unsaturated fatty acids, essential amino acids and minerals for the formation of functional and structural proteins (Kumar, 1992).

Body composition is a good indicator of the physiological condition of a fish but it is relatively time consuming to measure (Hernandez *et al.,* 2001). Proximate body composition is mainly the analysis of water, fat, protein and ash contents of fish. Carbohydrates and non-protein compounds are present in negligible amount and are usually ignored for routine analysis (Cui and Wootton, 1988). Knowledge of biochemical composition of fish muscle is of great help in evaluating not only its nutritive value but also the quality assessment and ultimately optimum utilization of the inhabitant natural resources. The nutritive value of fish is determined either by the ratio between the edible and non-edible parts of the body, or by their chemical composition, on the basis of which the caloric value of their meat can be evaluated.

Studies of carp fishes under the agro climatic condition of Chhattisgarh may be quite significant as it will relate to the nutritional value of these fishes. The information obtained on fat, protein and minerals and their percentages in relation to size are very important for fish used as food to the consumers. It makes it easy to select the most suitable species having optimum protein content and size for human consumption. Therefore, present study was proposed to evaluate the nutritional composition of some species of cyprinids namely *Catla catla, Labeo rohita, Cirrhinus mrigala, Hypophthalmichthys molitrix, Cyprinus carpio* which are found in Raipur district.

Materials and Methods

From a lot of freshly caught carps under study, of around 1 kg were collected from different market sites. The identification of fishes was done as per the key given by Day (1889) and Jayaram (2002). Before proceeding for analysis of flesh head, viscera, bones, fins, scales and tails of these fishes were removed carefully. The muscle of carp fishes was removed with the help of knife. Collected muscle parts were then taken into the mortal and are crushed with the help of pestle. The grinded muscle part was then preceded for analysis. Estimation of moisture, protein, lipid, carbohydrate and ash were done by APHA, 1998, Lowry's method, 1951, Soxhlet method, Anthrone reagent (Travelyan and Harrison, 1952) and (AOAC, 1995). Statistical analysis was done by ANOVA single factor, data were analyzed and the significance differences between the treatments were determined by CRD.

Results and Discussion

Moisture Content

In catla, Hussain *et al.* (2011) reported moisture content to be 54.91 per cent - 63.06 per cent, Hassan *et al.* (2010) showed 75.43 - 76.45 per cent and Ali *et al.* (2005) found it to be 68.84 per cent. In present study the moisture content was found to be 78.99 per cent. In case of rohu, Memon *et al.* (2010) analyzed moisture content to be 75.65 per cent in summer season and 72.91 per cent in spring season. Ali *et al.* (2005) reported it to be 72.81 per cent. In the present study the moisture content was found to be 75.10 per cent. Similarly in mrigala, the present study showed the moisture content to be 72.4 per cent whereas, (Ali *et al.*, 2005), have reported moisture to be 69.5 per cent, present study shows silver carp moisture content to be 69.66 per cent. Common carp moisture in present study was 65.02 per cent whereas Farhoudi *et al.* (2011) reported it to be 82.0 – 88.4 per cent and Ali *et al.* (2005) to be 65.60 per cent. This result is quite similar to the results obtained by previous workers, this may be due to the fact that these fishes were also collected mainly from rivers of Raipur district.

Protein Content

Protein was found to be the most dominant biochemical constituent in the muscle of catla. The present study showed that protein content in 5 carps is significantly different from each other. Protein values were observed to be 18.68 per cent in catla, 16.48 per cent in rohu, 16.85 per cent in mrigala, 16.26 per cent in silver carp and 15.36 per cent in common carp. Ashraf *et al.* (2011) reported protein content in silver carp to be 16.61 and 16.01 per cent whereas, Abdullah *et al.* (2011) and Ahmed *et al.* (2012) reported protein content of catla, rohu and mrigala and found it to be 109.6, 127.5, 110.2 mg/gm of tissue and 16.90, 17.49, 17.16 per cent respectively. Hadjinikolova (2008), Hadjinikolova *et al.* (2008), Stolle *et al.* (1994) and Jeyaraj *et al.* (2011) reported the protein content of common carp and recorded it to be 77.0 – 84.5 per cent, 66.0 per cent, 13.0 – 21.9 per cent and 60 per cent respectively. The range of protein content in carp during the present investigation *i.e.* 15.36 – 18.68, clearly shows that growth of fishes is directly proportional to the optimum temperature, as the temperature was between 25°C– 32°C supporting maximum growth during the study period.

Lipid Content

Present investigation was focused on the nutritional composition of some edible cyprinids fishes and the results obtained show that lipid content of catla, rohu, mrigala, silver carp and common carp ranged between 1.65 – 2.77 per cent. Hassan *et al.* (2010) reported lipid of catla to be 3.32 – 4.72 per cent. Swapna *et al.* (2009) observed lipid class in freshwater fishes and reported catla lipid content as 1.2 per cent. Shekhar *et al.* (2008) studied seasonal changes in flesh quality of catla and reported lipid content of catla to be 2.1 to 2.4 per cent Ahmed *et al.* (2012) also reported lipid content of catla to be 2.01 per cent. In case of rohu, Abdullah *et al.* (2011), Ali *et al.* (2005) found lipid content to be 18.0 per cent and 17.33 per cent. Memon *et al.* (2010), Bakhtiyar *et al.* (2011), Mahboob *et al.* (2004) and Ahmed *et al.* (2012) also estimated lipid content of rohu and found it to be 0.95 – 1.95 per cent, 1.34 per cent, 1.94 – 2.32 per cent and

3.16 per cent respectively. Present study shows lipid content of rohu to be 2.17 per cent. Similarly in mrigala lipid content was found to be 1.65 per cent in present study whereas, other workers have reported lipid content of mrigala to be 12.4 per cent, 20.00 per cent and 1.8 per cent by Abdullah *et al.* (2011), Ali *et al.* (2005) and Swapna *et al.* (2009) respectively. In case of common carp, Swapna *et al.* (2009) and Farhoudi *et al.* (2011) reported lipid content to be 3.8 per cent and 4.2 – 10.8 per cent respectively. Ashraf *et al.* (2011) observed lipid content in silver carp and reported that lipid content ranges from 1.93 – 2.28 and in present study it was reported to be 2.6 per cent.

Several workers have related lipid content with moisture content and moisture was found to be inversely proportional to lipid content. The present study also elucidated the same, as in all the carps the lipid and moisture content were estimated to be 2.6 per cent lipid and 78.99 per cent moisture in catla, 2.17 per cent lipid and 75.10 per cent moisture in rohu, 1.65 per cent lipid and 72.4 per cent moisture in mrigala, 2.6 per cent lipid and 69.66 per cent moisture in silver carp and 2.77 per cent lipid and 65.02 per cent moisture in common carp. Maximum lipid was found in common carp, where the moisture content was lowest.

Carbohydrate Content

Shekhar *et al.* (2008) reported carbohydrate content to be 2.3 – 2.9 per cent. Present investigation shows catla carbohydrate content to be 0.86 per cent. In rohu, Memon *et al.* (2010) and Jabeen and Chaudhry (2011) observed carbohydrate content and found to be 0.99 – 1.58 per cent and 23.71 – 33.50 per cent respectively, whereas in present study it is 1.12 per cent. Similarly in mrigala, Abdullah *et al.* (2011) also reported carbohydrate content to be 13.6 mg/gm of tissue whereas present study shows its content to be 0.70 per cent. Jabeen and Chaudhry (2011) found carbohydrate content to be 24.51 – 37.25 per cent in common carp. Thus range is quite lower than the observations done by previous authors. The low values of carbohydrates recorded in the present study could be because glycogen in many fishes does not contribute much to the reserves in the body (Jayasree *et al.*, 1994). Ramaiyan *et al.* (1976) reported similar findings in 11 species.

Ash Content

In catla, Shekhar *et al.* (2008) reported ash content to be 1.20 – 1.40 per cent. In the present study the ash content was found to be 1.11 per cent. Shekhar *et al.* (2008), who has done his work in seasonal changes in flesh quality of *Catla catla*. In case of rohu, Memon *et al.* (2010) estimated the ash content and found to be 1.02 – 1.15 per cent. According to present study the ash content was estimated to be 1.53 per cent, which shows the similar result in correspondence to Memon *et al.* (2010),who estimated nutritional aspects and seasonal influence on fatty acid composition of rohu from Indus river, Pakistan. Similarly in mrigala, in present study the ash content was found to be 1.77 per cent, whereas khan *et al.* (2012) reported ash content to be 3.5 – 4.1 per cent, whereas Ahmed *et al.* (2012) observed ash content of mrigala and found to be 1.66 per cent. Farhoudi *et al.* (2011) and Jabeen and Chaudhary (2011) observed ash content of common carp and found its content to be 7.6 – 16.4 per cent and 9.66 per cent. However, in present study ash content of common carp was found only 0.95 per cent. In silver carp, ash content was found to be 1.52 per cent.

Table 33.1: Biochemical Composition of Cyprinid Fishes found in Markets of Raipur District

Treatment	Moisture (Per cent)	Protein (Per cent)	Lipid (Per cent)	Carbohydrate mg/gm of tissue	Ash (Per cent)
Catla (T_1)	78.99 ± 6.56	18.68 ± 0.87	2.6 ± 0.92	8.6 ± 0.60	1.11 ± 0.06
Rohu (T_2)	75.10 ± 9.88	16.48 ± 0.31	2.17 ± 0.81	11.2 ± 0.89	1.53 ± 0.27
Mrigala (T_3)	72.4 ± 7.69	16.8 ± 0.2	1.65 ± 0.75	7.08 ± 0.70	1.77 ± 0.22
Silver carp (T_4)	69.66 ± 7.95	16.26 ± 0.79	2.6 ± 0.63	7.72 ± 0.98	1.52 ± 0.30
Common carp(T_5)	65.02 ± 11.39	15.36 ± 0.75	2.77 ± 0.68	8.12 ± 0.50	0.95 ± 0.18
SEm (±)	3.96	0.28	0.34	0.34	0.10
CD (0.05)	11.70	0.85	1.01	1.00	0.35
CV (per cent)	12.27	3.87	32.01	8.89	16.15
Level of significance(.05 per cent)	NS	S	NS	S	S

Through the current study it is clear that cyprinid fishes are good source of good quality lipid, protein and essential minerals in habitats of Chhattisgarh and each species of fish has its own nutritional value parameters due to their different food preferences and ecological conditions.

Conclusion

Based on the above results of the present research, it can be concluded that the nutritional body composition of selected fishes, including nutrients, is within nutritional ranges required by human beings in their diet. The percentages of protein content ranged between 15.36 – 18.68 per cent, which is quite high. The moisture and lipid content were found to be inversely proportional to each other. From the present study it is clear that cyprinid fishes are good source of good quality lipid, protein and essential minerals in habitats of Chhattisgarh and each species of fish has its own nutritional value parameters due to their different food preferences and ecological conditions. The range for ash content (0.95 – 1.77 per cent) gave an indication that the fish samples may be good sources of minerals such as calcium, potassium, zinc, iron and magnesium. Among all carps, rohu has got most preference due to its taste and palatability in Chhattisgarh, but this study shows that catla is having equally ample nutrients along with high content of protein (more than rohu), less content of fat (lesser than common carp) and its cost is also less than rohu, hence can be referred as a good source of animal protein in people's daily diet.

References

Ahmed, S., Rahman A., Ghulam, M., Hossain, B. and Nahar, N., 2012. Nutrient composition of indigenous and exotic fishes of rainfed waterlogged paddy fields in Lakshmipur, Bangladesh. *World J. of Zoology* 7 (2): 135-140.

Ali, M., Iqbal, F., Salam, A., Iram, S. and Athar, M. 2005. Comparative study of body composition of different fish species from brackish water pond. *International J. of Enviornmental Science Technology,* 2(3): 229-232.

AOAC, 1955. Official method of analysis of AOAC International, vol. 1, 16th edn. (ed. Cunniff,

P. A.). AOAC International, Arlington, USA.

APHA, 1998. Standard methods for examination of water and wastewater. 20th Ed. American Public Health Association, AWWA, WPCA, Washington, D. C., U. S. A. 1193.

Ashraf, M., Zafar, A. and Naeem M. 2011. Nutritional values of carnivorous fish species. *International. J. of Agriulture and Biology,* 13(5): 701-706.

Ashraf, M., Zafar, A., Rauf, A., Mehboob, S. and Qureshi Naureen, A. 2011. Nutritional values of wild and cultivated silver carp *(Hypophthalmichthys molitrix)* and grass carp (*Ctenopharyngodon idella*). *International J. of Agriculture and Biology*, 13(2): 210-214.

Bakhtiyar, Y., Langer S., Karlopia S. K. and Ahmed I., 2011. Growth, survival and proximate body composition of *Labeo rohita* larvae fed artificial food and natural food organisms under laboratory condition. *International J. of Fisheries and Aquaculture* 3 (6): 114-117.

Cui, Y. and Wootton, R. J., 1988. Bioenergetics of growth of Cyprinids, Phoxinus, the effect of the ration and temperature on growth rate and efficiency. *J. of Fish Biology,* 33: 763-773.

Day, F., 1889. The fauna of British India, including Ceylon and Burma fishes, 1 and 2. Taylor and Francis, London.

Farhoudi, A., Abedian kenari, A. M., Nazari R. M., and Makhdoomi, C. H. 2011. Study of body composition, lipid and fatty acid profile during larval development in Caspian sea carp (*Cyprinus carpio*). *J. of Fisheries and Aquatic Science,* 6(4): 417-428.

Hadjinikolova, L. 2008. Investigations on the chemical composition of carp (*Cyprinus carpio* L.), bighead carp (*Aristichthys nobilis* rich.) and pike (*Esox lusius* L.) during different stages of individual growth. *Bulgarian J. of Agricultural Science,* 14 (2): 121-126.

Hadjinikolova, L., Nikolova, L., and Stoeva, A. 2008. Comparative investigations on the nutritive value of carp fish meat (*Cyprinidae*), grown at organic aquaculture conditions. *Bulgarian J. of Agricultural Science,* 14(2): 127-132.

Hassan, M., Shahid Chatha, S. A., Tahira I. and Hussain B. 2010. Total lipids and fatty acid profile in the liver of wild and farmed *Catla catla* fish, GRASAS Y ACEITES, ENERO-MARZO, 61 (1): 52-57.

Hernandez, M. D., Martinez, F. J. and Garcia B., 2001. Sensory evaluation of farmed sharpsnout seabream (*Diplodus puntazzo*). *Aquaculture International,* 9: 519-529.

Hussain, B., Mahboob, S., Hassan, M., Liaqat, F., Sultana, T. and Tariq H. 2011. Comparitive analysis of proximate composition of head from wild and farmed catla catla. *The J. of Animal and Plant Science,* 21(2): 207-21.

Jabeen, F. and Chaudhry, A. S., 2011. Chemical compositions and fatty acid profiles of three freshwater fish species. *Food Chemistry,* 125: 991–996.

Jayaram, K. C., 2002. Fundamentals of fish taxonomy. Narendra Publishing House, New Delhi. 174.

Jeyaraj, N., Rajan, M. R. and Santhanam, P., 2012. Dietary requirement of vitamin a and biochemical composition of common carp *Cyprinus carpio* var. communis. *J. of Fisheries and Aquatic Science,* 7: 65-71.

Jayasree, V., Parulekar, A. H., Wahidulla, S. and Kamat, S. Y., 1994. Seasonal changes in biochemical composition of *Holothuria vleucospilota* (Echinodermata). *Indian J. of Marine Science,* 23: 17-119.

Khan, N., Ashraf, M., Qureshi, N. A., Sarker, P. K., Vandenberg, G. W. and Rasool, F., 2012. Effect of similar feeding regime on growth and body composition of Indian major carps (*Catla catla, Cirrhinus mrigala* and *Labeo rohita*) under mono and polyculture. *African Journal of Biotechnology,* 11(44): 10280-10290.

Kumar, D. 1992. Fish culture in un-drainable ponds. A manual for extension F. A. O. Fisheries Technical paper. No. 235. Rome: 239.

Lowry, O. H., Rosebrough, N. J., Farr, A. L. and Randall, R. J. 1951. Protein measurement with the folin phenol reagent. *J. Biol. Chem,* 193: 265-275.

Mahboob, S., Liaquat, F., Liaquat, S., Hassan, M. And Rafique, M., 2004. Proximate composition of meat and dressing losses of wild and farmed *Labeo rohita* (rohu). *Pakistan J. Zoology,* 36(1): 39-43.

Memon, Nusrat N., Talpur, Farah N. and Bhanger, Muhammad I. 2010. Nutritional aspects and seasonal influence on fatty acid composition of carp (*Labeo rohita*) from the Indus River, Pakistan. *Polish J. of Food and Nutritional Sciences*, 60(3): 217-223.

Ramaiyan, V., Paul, A. L. and Pandian, T. J., 1976. Biochemical studies on the fishes of the order Clupeiformes. *J. of Marine Biology Association India*, 18(3): 516-524.

Shekhar, C., Rao, A. P. and Abidi A. B. 2008. Season changes in flesh quality of freshwater fish Catla catla (Ham.). *The Allahabad farmer J.,* 63(2): 1-6.

Stolle, A., Sedlmeier, H., Nassar, A., Eisgruber, H., Youssef, H. and Lotfi, A. 1944. The nutritive value of carp (*Cyprinus carpio*). Tierarztl Prax, 22(6): 512-4.

Swapna, H. C., Kumar Rai, A., Bhaskar, N. and Sachindra, N. M., 2009. Lipid classes and fatty acid profile of selected Indian freshwater fishes. *J. of Food Science Technology*, 47(4): 394–400.

Travelyan, W. E. and Harrison, J. S. 1952. Studies on yeast metabolism, Fractionation and microdetermination of cell carbohydrates. *Biochem. J.,* 50: 298-303.

Chapter 34

Antibacterial Activity of Extract of *Chaetomorpha aerea* against Fish Pathogens

☆ *G. Sattanathan and R. Rajesh*

ABSTRACT

The antibacterial activity of different extracts of marine algae of *Chaetomorpha aerea* was determined against a wide variety of pathogenic bacteria. The extracts were against fish pathogenic bacteria like *Pseudomonas fluorescens, Pseudomonas punctata, Bacillus subtilis, Staphylococcus aureus* and *Escherchia coli* by cup plate method. Minimum Inhibitory Concentration (MIC) values of each active extract were determined. It is concluded that petroleum ether extract and ethanolic extract of *Chaetomorpha aerea* algae exhibited antibacterial activity.

Keywords*: Chaetomorpha aerea, Labeo rohita, Macro algae, Antibacterial activity, Cup plate method.*

Introduction

Over the last few decades the great advances in our understanding of the causes of transmission, treatment of prevention of infectious diseases have fostered complacency about infections in a society which is well nourished and has access to vaccines, antibiotics and other drugs (Vadhini, *et al.,* 2002). Alternative systems of medicine *viz.* Ayurveda, Siddha and Traditional Chinese medicine have become more popular in recent years. According to one estimate, more than 700 mono and poly-herbal preparations in the form of decoction, tincture, tablets and capsules from more than 100 plants are in clinical use (Chakraborthy G.S., 2008). In fact; plants produce a diverse range of bioactive molecules, making them a rich source of different types of medicines. Plants with possible antimicrobial activity should be tested against an appropriate microbial model to confirm the activity and to ascertain the parameters associated with it (Ali, *et al.,* 2007).

India has huge wealth of seaweeds approximately 700 species of marine algae are found in the region of the Indian cost among these nearly 60 species are commercially important (Reichelt J.L.,and Borowitrika M.A., 1984). The *Chaetomorpha aerea* is a marine algae belonging to family Cladophorales. Algae have been used in traditional medicine for a long time. The biological activity of many of the liquid constituent from marine algae has been studied and reported on several aspects of *Chaetomorpha aerea* (Valchos., 1997 and Salvador, *et al.,* 2009).The *Chaetomorpha aerea* have antimicrobial, anti-inflammatory, astringent, diuretic and hypoglycemic properties (Rajasulochana, *et al.,* 2009).

Due to the fast that the marine algae *Chaetomorpha aerea* is very useful found by above mentionedreports and there is a need to find out more about the potentially of this algae as an antimicrobial agent. The present study is, therefore, designed to assess the potency of ethanolic and petroleum ether extracts of *Chaetomorpha aerea* on some fish pathogens.

Materials and Methods

Collection of Algae

The marine algae of *Chaetomorpha aerea* were obtained from Parankipettai, Tamilnadu in India Figure 34.1. The algae was collected in a sterile container and maintained at room temperature.

Extraction of Algae

Preparation of the extract algae were collected from Parankipettai, India (Figure 34.1) and dried. Dried algae were crushed in pestle and porter. The algae powdered (500gm) were exhaustively extracted with 70 per cent ethanol and then successively with petroleum ether. The extracts were then re dissolved in 10 per cent DMSO (v/v) to yield solutions containing 100.0mg of extract per ml.

Collection of Fish

Labeo rohita were collected in lower Anaicut, Tamil Nadu. The fish was collected in sterile container maintained at lab conditions and excess of fish preserved in 50 per cent of formalin solutions.

Isolation of Bacteria

The fish gut region was removed carefully and then cut in to small pieces. The 1gm of gut pieces was weighed aseptically, crushed and added to the dissolved in 99ml of distilled water. From this each dilution 1ml each added in to petriplates in duplicates. The samples was poured in to petriplates and mixed the medium and allowed to solidify. After the pour plate technique the plates were incubated at 37°C for 48 hours. After incubation period the isolated colonies were streaked in to the

Figure 34.1 : *Cheatomorpha aerea.*

nutrient agar plates for purification and identification. The isolated colonies were transformed to nutrient agar slants and stored for further studies.

Identification of Bacteria

Identification of the bacteria was done based on the bacterial identification chart (Surendran and Gopakumar, 1981). The isolated cultures were streaked onto the Nutrient agar and incubated at 37°C for 48 hours.

Antibacterial Activity

Invitro antibacterial activity was performed with ethanol and petroleum ether extracts of algae of *Chaematomorpha aerea* against the fish pathogenic bacteria by the disc diffusion method (Barry, 1980 and Rios, *et al.,* 1988). Each Petri dish was inoculated with one of the bacterial cultures suitably diluted to contain above 10^6 cells/ml by spreading 0.1ml suspension of the organism with a sterile cotton swab. In each plate cups of 6mm diameter were made at equal distances using sterile cork borer. One cup was filled with 0.1 ml of standard drug, another with 0.1 ml of DMF, and others were filled with 0.1ml of samples in sterile DMF. The Petridis were incubated at 37°C for 48 hours. The diameter zone of inhibition in mm was recorded after incubation.

The extracts that showed antibacterial activity were subjected to minimum inhibitory concentration (MIC) assay by serial two fold dilution method (Florey, *et al.,* 1989). MIC was interpreted as the lowest concentration of the sample, which showed clear fluid without development of turbidity.

Results and Discussion

The antibacterial activities of the ethanolic and petroleum ether extracts of *Chaetomorpha aerea* showed significant variations as shown in table 1. Among the two extracts tested, ethanolic extract had greater antibacterial potential, followed by petroleum ether extracts. *Fluorescens* (17mm) and *Pesudomonas punctata* (15mm). Petroleum ether extract was very effective against E. coli (16mm). Antimicrobial potency of the algae extract of *Chaetomorpha aerea* against the tested bacteria strains were expressed in MIC as presented in Table 34.1 and Figure 34.2 respectively.

Table 34.1: Antibacterial Activity of different Extracts of Algae of *Chaetomorpha aerea*

Micro-organism	Zone of Inhibition (mm)			MIC (µg/ml)	
	Ethanolic Extract	Petroleum Ether	Gentamycin	Ethanolic Extract	Petroleum Ether
P. fluorescens	17	10	21	15.21	29.65
P. punctate	15	9	23	15.30	29.36
B. subtilis	14	12	25	30.41	31.50
S. aureus	11	13	28	30.43	25.92
E.coli	09	15	26	29.38	26.75

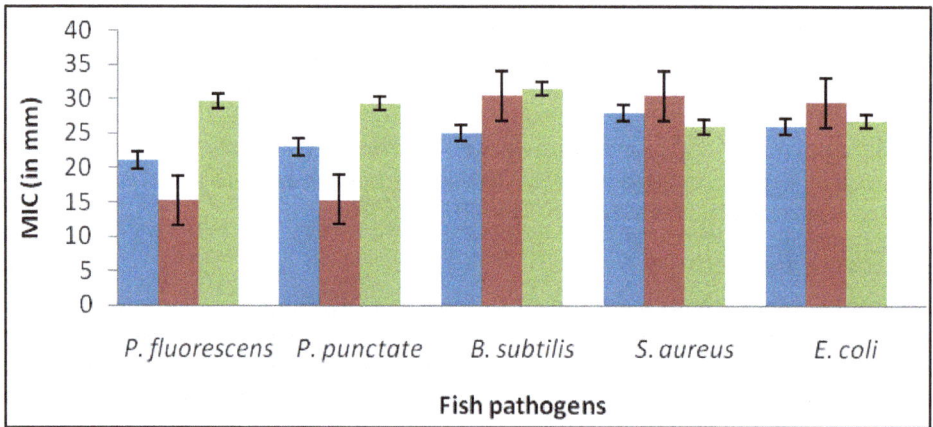

Figure 35.2: Antibacterial Activity of *C. aerea* against Fish Pathogens.

Many studies have highlighted that molecules from algae showed original biochemical compositions, behind nutritional (Fleurence, 1999), medical and antibacterial properties (Chakraborty, *et al.,* 2010). Seaweeds contain various compounds as polysaccharides, proteins, lipids, amino-acids, sterols or phenolic molecules which show bioactivity against microorganisms (Wong *et al.,* 1994 and Phang *et al.,* 1986) or virus (Nirmal Kumar *et al.,* 2010). The other hand, sulfated polysaccharides extracted from marine algae can be used for their anticoagulant and antithrombotic properties (Abrantes *et al.,* 2010). The main goal of this study was in a first time to find a valorization path of the macroalga *Chaetomorpha aerea*, which is an ecological problem for French west coast and especially oyster ponds. The extracellular polysaccharides of this green alga were firstly extracted and their compositions were characterized. Indeed, certain authors have highlighted that green algae as *Chaetomorpha* were composed of interesting polysaccharides (arabinogalactans), sometimes sulfated (Mao *et al.*, 2005; Venkata Rao and Sri Ramana, 1991; Anand Ganesh *et al.*, 2009).

In our present study, a wide range of fish pathogenic microorganisms were examined, the fish pathogenic bacteria. This may indicate that the *Chaetomorpha aerea* extracts have broad inhibitory activities to pathogenic microorganisms and promising to act as potential antibacterial agents from natural plant sources.

References

Abrantes, J. L., J. Barbosa, D. Cavalcanti, R. C. Pereira, C. L. Frederico Fontes, V. L. Teixeira, T. L. Moreno Souza, and I. C. P. Paixao., 2010. The effects of the diterpenes isolated from the Brazilian brown algae *Dictyota pfafii* and *Dictyota menstrualis* against the herpes simplex type-1 replicative cycle. *Planta Med.* 76: 339-344.

Ali M. A., Alam N. M., Yeasmin S., Khan A. M and Sayeed M. A., 2007. Antimicrobial Screening of different extracts of *Piper longum* L., *Research Journal of Agriculture and Biological Sciences.* 3(6), 852-857.

Anand Ganesh, E., S. Das, G. Arun, S. Balamurugan, R. and Ruban Raj., 2009. Heparin like compound from green alga *Chaetomorpha antennina* - as potential anticoagulant agent. *Asian J. Med. Sci.* 1: 114-116.

Barry A. L., 1980. Procedures for testing antimicrobial agents in agar media. In antibiotic in laboratory medicine, Williams and Wilkins Co., Baltimore, USA, 1-23.

Chakraborthy G. S., 2008. Antimicrobial activity of the leaf extracts of *Calendula officinalis*L., *Journal of Herbal Medicine and Toxicology.* 2(2), 65-66.

Chakraborty, K., A. P. Lipton, R. Paul Raj, K. K. Vijayan, 2010. Antibacterial labdane diterpénoïdes of *Ulva fasciata Delile* from southwestern coast of the India Peninsula. *Food Chem.* 119: 1399-1408.

Fleurence, J. 1999. Seaweed proteins: biochemical, nutritional aspects and potential uses. *Trends Food Sci. Technol.* 10: 25-28.

Florey H. W., Chain E., and Florey M. E., 1989. The Antibiotic, Vol I, Oxford University Press, New York.

Mao, W., X. Zang, Y. Li, and H. Zhang, 2005. Sulfated polysaccharides from marine green algae *Ulva conglobata* and their anticoagulant activity. *J. Appl. Phycol.* 18: 9-14.

Nirmal Kumar, J. I., R. N. Kumar, M. K. Amb, A. Bora, and S. Kraborty, 2010. Variation of biochemical composition of eighteen marine macroalgae collected from Okha Coast, Gulf of Kutch, India. *Electron. J. Environ. Agric. Food Chem.* 9: 404-410.

Phang, S, M., Y. K. Lee, M. A. Borowitzka, and B. A. Whitton (eds.). Algal biotechnology in the Asia-Pacific region. University of Malaya, Kuala Lumpur, Malaysia.

Rajasulochana, R., Ahamotharan and Krishnamoorthy P, 2009. Antimicrobial activity of *Chaetomorpha aerea* Marine algae chloroform: Methanol (2: 1 v/v extract), *Journal of American Science*. 5(3): 20-25.

Reichelt, L., and Borowitrika M. A., 1984. Antimicrobial activity of from Marine Algae result of a large-scale screening programme, *Hydrobiology*, 116/117: 158-167.

Rios J. J., Reico M. C., and Villar A., 1988. Antimicrobial screening of natural products, *Journal of Ethnophrmacology*. 23: 127-149.

Salvador, N., Gomez Garreta A., Lavelli L., and M. A Ribera, 2007. Antimicrobial activity of *Chaetomorpha aerea* macro algae, *Iberian Journal of Science*. 71: 101-103.

Surendran, P. K and K. Gopakumar, 1981. Scheme for identification of bacterial cultures. *Fishtech*, 18: 133-141.

Vadhini H., Kamalinejad M and Sedaghati N., 2002. Antimicrobial properties of *Croccus sativa* L., *Iranian Journal of Pharmaceutical Research*. 1, 33-35.

Valchos, V. A. T and C. Von Holy., 1997. Antimicrobial activity of extract from selected Southern African Marine Macro Algae. *S. Af, Res J. Sci* 93: 328-332.

Venkata Rao, E., and K. Sri Ramana., 1991. Structural studies of a polysaccharide isolated from the green seaweed *Chaetomorpha antennina*. *Carbohydr. Res*. 217: 163-170.

Wong, W. H., S. H. Goh and S. M. Phang., 1994. Antibacterial properties of Malaysian seaweeds. pp. 75-81.

Chapter 35

Avifauna of Ekrukh wetland (Hipparga Lake) of Solapur District, Maharashtra

☆ *P.V. Darekar, S. H. Chougule*
and A.C. Kumbhar

ABSTRACT

Ekrukh wetland is one of the man made water bodIES located close to Solapur city in North Solapur Tahasil. It is constructed in British reigning. Its total storage capacity is 3330 M.Cu.Ft. This reservoir commands a gross area of 17.152 acres. The water from Ekrukh water tank is currently used for irrigation and drinking. This water body contains a diverge assemblage of resident and migratory birds.

An attempt is made in this investigation to prepare a checklist of avifauna from the Ekrukh wetland over the period of 12 months. *i.e.* from May 2013 to April 2014. They are categorized as resident, local and migratory. Further more, the birds which are noticed in the study period are classified as aquatic and terrestrial. More than hundreds of species belonging to 41 families have been recorded in this reservoir.

Keywords: *Avinfauna, Ekruh wetland, Solapur, Maharashtra.*

Introduction

The present research paper reports the check list of Avifauna of Ekrukh wetland (Hipparga Lake) of North Solapur Tahasil Solapur of District. (M.S.).The Ekrukh wetland is located at Tale Hipparaga village close to Solapur city. Ekrukh wetland is one of the man made water reservoirs constructed very close to Solapur city in Maharashtra and it has a total capacity of 3,330 M.Cu.Ft. This reservoir commands a gross area of 17,152 acres. Ekrukh wetland is a historical man made water body. Earlier the water from Ekrukh wetland is utilized for the purpose of irrigation and drinking, but the water of Ekrukh water reservoir is now supplied to Solapur city for drinking as well as for industrial purposes. The water body contains a diverse assemblage of resident and migratory birds. Till now, different researchers have studied avifauna of Solapur in both aquatic and terrestrial habitat in the last two decades. Avifauna of Solapur was studied by Rahmani (1989) and Mahabal (1989) and Kumbhar *et al.* (2013). Anecdotes of avifauna were discussed by Prasad (2003). Checklist of birds of Solapur was published by Gaikwad *et al.* (1997). Survey of impact of changes in Wetland Ecosystem on the population and breeding status of avifauna of Ekrukh lake was conducted by Hippargi *et al.* (2012).

Materials and Methods

The regular survey camps were arranged at morning and evening time of every fortnight, scheduled from May 2013 to April 2014 to find out relative abundance of the birds at study site. For this survey, Olympus binocular was used for bird watching. Photography was done with the help of SLR camera (Canon 1100D) and zoom lenses (55-210mm and 70-300). The observed birds were identified with the help of field guides and pictorial literature. The record of observed birds is maintained by basic bird count and point count method.

Results and Discussion

In the present investigation, totally more than hundreds of bird species belonging to 41 families have been recorded at Ekrukh wetland for the period of 12 months *i.e.* from May 2013 to April 2014. As the study period includes monsoon, winter and summer season, the variability in the climate is observed which resulted in the different bird count at the study site. The recorded birds are categorized into resident, local and migratory birds, and furthermore, aquatic and terrestrial. The availability of variety of food and the nesting behavior of many birds was also studied.

Key to Abbreviations

First Suffix: (Status)
R: Resident
M: Migrant
RM: Local Migrant

Second Suffix: (Occurrence)
A: Aquatic
T: Terrestrial

Table 35.1: Checklist of Avifauna of Ekrukh Wetland

Sl.No.	Common name	Scientific Name	Family	Residential status	Occur-rence
1.	Common quail	*Coturnix coturnix*	Phasianidae	R	T
2.	Indian peafowl	*Pavo cristatus*		R	T
3.	Lesser whistling duck	*Dendrocygna javanica*	Anatidae	LM	A
4.	Ruddy shel duck	*Tadorna ferruginea*		M	A
5.	Comb duck	*Sarkidiornis melanotos*		M	A
6.	Bar headed goose	*Anser indicus*		M	A
7.	Spot billed duck	*Ana poecilorhyncha*		R	A
8.	Northern shoveller	*Anas poecilorhyncha*		M	A
9.	Common teal	*Anas clypeata*		M	A
10.	Garganey	*Anas querquedula*		M	A
11.	Northern pintail	*Anas acuta*		M	A
12.	Red chested pochard	*Rhodonessa rufina*		M	A
13.	Common pochard	*Aythya ferina*		M	A
14.	Indian Grey Hornbill	*Ocyceros birostris*	Bucerotidae	R	T
15.	Small Kingfisher	*Alcedo atthis*	Alcedinidae	R	T
16.	White breasted Kingfisher	*Halcyon smyrnensis*		R	T
17.	Pied kingfisher	*Ceryle rudis*		R	A
18.	Green bee eater	*Merops orientalis*	Meropidae	R	T
19.	Hoopoe	*Upupa epops*	Upupidae	R	T
20.	Indian Roller	*Coracias bengalensis*	Coraciidae	R	T
21.	Asian koel	*Eudynamys scolopacea*	Cuculidae	R	T
22.	Brain fever bird	*Hierococccyx varius*		LM	T
23.	Crow pheasant	*Centropus sinensis*		R	T
24.	Alexandrine parakeet	*Psittacula eupatria*	Psittacidae	R	T
25.	Rose ringed parakeet	*Psittacula krameri*		R	T
26.	House swift	*Apus affinis*	Apodidae	R	T
27.	Spotted owlet	*Athene brama*	Strigidae	R	T
28.	Indian nightjar	Caprimulgus asiaticus	Caprimulgidae	R	T
29.	Rock pigeon	*Columba livia*	Columbidae	R	T
30.	Laughing dove	*Streptopelia senegalensis*		R	T
31.	Eurasian collared dove	*Streptopelia decaocto*		R	T
32.	White breasted water hen	*Amauromis phoenicurus*	Rallidae	R	A
33.	Common moorhen	*Gallinula chloropus*		R	A
34.	Common coot	*Fulica atra*		R	A
35.	Chestnut bellied sandgrouse	*Pterocles atra*	Pteroclidae	R	T

Contd...

Table 35.1–*Contd...*

Sl.No.	Common name	Scientific Name	Family	Residential status	Occurrence
36.	Common snipe	*Gallinago gallinago*	Rostratulidae	M	A
37.	Greater painted snipe	*Rostratula bengalensis*		R	A
38.	Spotted redshank	*Tringa erythropus*	Scolopacidae	M	A
39.	Common greenshank	*Tringa nebularis*		M	A
40.	Green sandpiper	*Tringa ochropus*		M	A
41.	Wood sandpiper	*Tringa glareola*		M	A
42.	Common sandpiper	*Actitis hypoleucos*		R	A
43.	Pheasant tailed jacana	*Hydrophasianus chirurgus*	Jacanidae	R	A
44.	Little ringed plover	*Charadrius dubius*	Charadriidae	R	A
45.	Kentish plover	*Charadrius alexandrinus*		R	A
46.	Red wattled lapwing	*Vanellus indicus*		R	A
47.	Yellow wattled lapwing	*Vanellus malabaricus*		R	T
48.	Black winged stilt	*Himantopus himantopus*	Recurvirostridae	R	A
49.	Indian courser	*Cursorius coromandelicus*	Glareolidae	R	T
50.	Pratincole	*Glareola lactea*		R	T
51.	Brown headed gull	*Larus brunnicephalus*	Laridae	M	A
52.	River tern	*Sterna aurantia*		M	A
53.	Black bellied tern	*Sterna acuticauda*		M	A
54.	Brahminy kite	*Haliastur indus*	Accipitridae	R	T
55.	Black kite	*Milvus migrans*		R	T
56.	Shikra	*Accipiter badius*		R	T
57.	Eurasian marsh harrier	*Circus aeruginosus*		LM	A
58.	Pallied harrier	*Circus macrourus*		M	A
59.	Little grebe	*Tachybaptus ruficollis*	Podicipedidae	R	A
60.	Little cormorants	*Phalacrocorax niger*	Phalacrocoracidae	R	A
61.	Indian cormorants	*Phalacrocorax fuscicollis*		R	A
62.	Great cormorants	*Phalacrocorax carbo*		R	A
63.	Little egret	*Egreta intermedia*	Ardeidae	R	A
64.	Large egret	*Ardea alba*		R	A
65.	Cattle egret	*Bubulcus ibis*		R	T
66.	Grey heron	*Ardea cenerea*		R	A
67.	Purple heron	*Ardea purpurea*		LM	A
68.	Indian pond heron	*Ardeola grayii*		R	A
69.	Glossy ibis	*Plegadis falcinellus*	Threskiornithidae	LM	A
70.	White ibis	*Threskiornis melanocephala*		R	A
71.	Black ibis	*Pseudibis papillosa*		R	T

Contd...

Table 35.1–*Contd...*

Sl.No.	Common name	Scientific Name	Family	Residential status	Occurrence
72.	Eurasian spoonbill	*Platalea leucorodia*		R	A
73.	Painted stork	*Mycteria leucocephala*	Ciconidae	R	A
74.	Asian open bill	*Anastomus oscitans*		R	A
75.	Wooly necked stork	*Ciconia episcopus*		R	T
76.	Rufous tailed shrike	*Lanius isabellinus*	Laniidae	M	T
77.	Bay backed shrike	*Lanius vittatus*		R	T
78.	House crow	*Corvus splendens*	Corvidae	R	T
79.	Jungle crow	*Corvus macrorhynchos*		R	T
80.	Indian robin	*Saxicoloides fulicata*	Muscicapidae	R	T
81.	Oriental magpie robin	*Capsychus saularis*		LM	T
82.	Common stone chat	*Saxicola torquata*		LM	T
83.	Pied bush chat	*Saxicola caprata*		LM	T
84.	Brahminy myna	*Sturnus pagodarum*	Sturnidae	R	T
85.	Rosy starling	*Sturnus roseus*		M	T
86.	Common myna	*Acridotheres tristis*		R	T
87.	Red rumped swallow	*Hirundo daurica*	Hirundinidae	R	T
88.	Greater flamingo	*Phoenicopterus ruber*	Phenicoptreridae	M	A
89.	Red vented bulbul	*Pycnonotus cafer*	Pycnonotidae	R	T
90.	Ashy prinia	*Prinia socialis*	Cisticolidae	R	T
91.	Purple rumped sunbird	*Nectarinia zeylonica*	Nectariniidae	R	T
92.	Common tailorbird	*Orthotomus sutorius*	Sylviidae	R	T
93.	Yellow eyed babbler	*Chrysomma sinense*		R	T
94.	Large grey babbler	*Turdoides malcolmi*		R	T
95.	Ashy crowned sparrow lark	*Eremopterix grisea*	Alaudidae	R	T
96.	Rufous tailor lark	*Ammomanes phoenicurus*		R	T
97.	House sparrow	*Passor domesticus indicus*	Passeridae	R	T
98.	White wagtail	*Motacilla alba*		M	T
99.	Yellow wagtail	*Motacilla flava*		M	T
100.	Grey wagtail	*Motacill cinerea*		M	T
101.	Paddy field pipit	*Anthus rufulus*		R	T
102.	Baya weaver	*Ploceus philippinus*		R	T
103.	Indian silverbill	*Lonchura malabarica*		R	T
104.	Scaly breaseted munia	*Lonchura puntualata*		R	T
105.	Black headed bunting	*Emberiza melanocephala*	Fringillidae	M	T
106.	Red headed bunting	*Emberiza bruniceps*		M	T

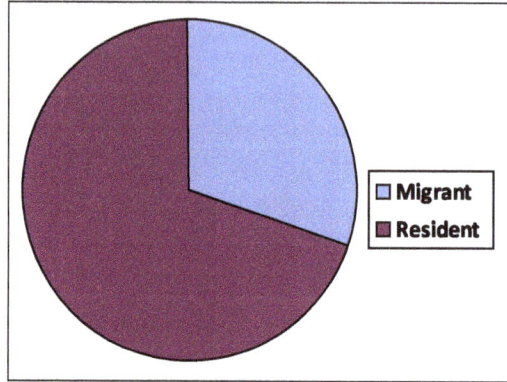

Figure 35.1:The Proportion of Resident, Migrant Birds that Visits the Ekrukh Wetland.

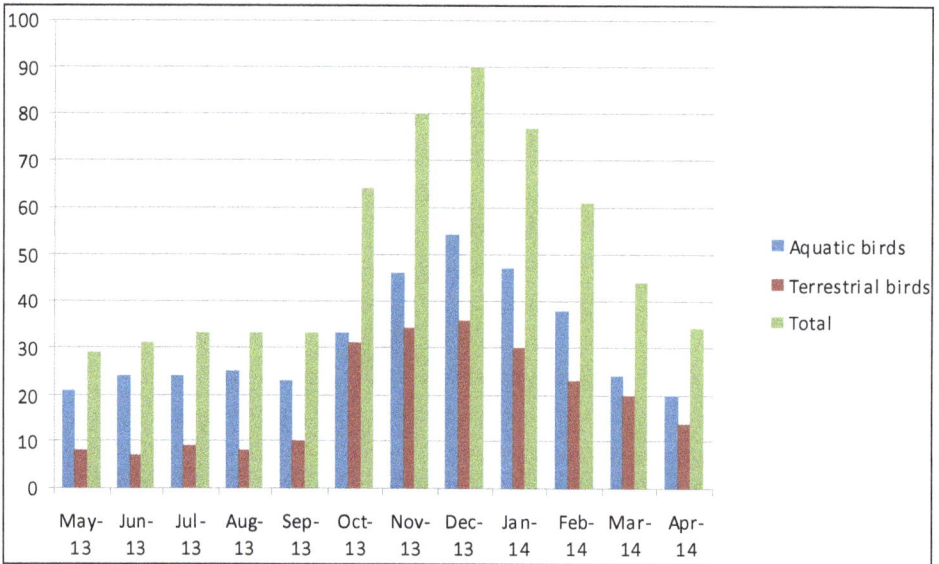

Figure 35.2: Record of Aquatic and Terrestrial Birds at Ekrukh Wetland.

Conclusion

Total 106 bird species belonging to 41 families have been recorded at Ekrukh wetland during the period of 12 months from May 2013 to April 2014. The number of bird species was observed maximum during winter season as different migratory birds visit this wetland. It can be concluded that this wetland is rich in variety of food that attracts different types of birds. The number of migratory birds is also remarkable and they are recorded in maximum population in winter season. The water level of wetland is moderate with vast area for feeding habitat. The poaching is not noticed in the study period. The Ekrukh wetland is an excellent water body for the winter visitors with local birds. It must be protected from human interference and to be kept free from effects of pollution.

Table 35.2: Record of Aquatic and Terrestrial Birds at Ekrukh Wetland

Month	Species of Aquatic Birds	Species of Terrestrial Birds	Total
May-13	21	8	29
	20	9	29
Jun-13	24	7	31
	22	12	34
Jul-13	24	9	33
	28	7	35
Aug-13	25	8	33
	27	10	37
Sep-13	23	10	33
	25	8	33
Oct-13	33	31	64
	30	23	53
Nov-13	46	34	80
	53	32	85
Dec-13	54	36	90
	51	31	82
Jan-14	47	30	77
	49	29	78
Feb-14	38	23	61
	35	25	60
Mar-14	24	20	44
	24	22	46
Apr-14	20	14	34
	22	10	32

References

Banerjee, Anand. 2008. ommon birds of Indian Subcontinent, Rupa Publications Pvt Ltd, New Delhi

Kumbhar, A. C. 2013. Avifauna of Girazani Irrigation Tank of Malshiras Tahasil of Solapur District, Maharashtra. *Ecology and Fisheries*, 6(1): 95-100.

Krays Kazmierezas. 2000. Om Book, International Press.

Salim Ali. 2012. Indian Birds, Bombay Natural History Society, Oxford University press

Satish Pande. Pramod Deshpande and Sant, Niranjan. 2013. Birds of Maharashtra, Ela Foundation, Pune, Maharashtra.

Chapter 36

Ecology of Purple Moorhen (*Porphyrio porphyrio*) in Dharamaveer Sambhaji Tank in Solapur City, Maharashtra

☆ *A.C. Kumbhar, B.N. Ghorpade and A.L. Deshmukh*

ABSTRACT

Purple Moorhens are among the most beautiful birds that inhabit the wetlands of Western Maharashtra and form an important component of Dharmaveer Sambhaji Water Tank of Solapur city. The availability and abundance of food, habit conditions and precipitation rates are the important limiting factors of Sambhaji tank. Construction of new societies of human population and associated disturbances are found to be a detrimental factor in the distribution pattern of this wetland. Monthly variations of these birds are studied from June -2011 to July- 2012. The population varies significantly from month to month in this water body. The highest numbers are noted in January 2012.

Keywords: *Beautiful birds, Human society, Detrimental factor, Wetland.*

Introduction

Purple Moorhens *(Porphyrio porphyrio)* are the beautiful birds with purplish colour and long red legs. They have a short heavy red beak. While walking they jerk their stumpy tail up and down. These birds very noisy during the breeding season (July and August). The present study deals with the monthly variations in pattern of habitat. This study was conducted between June -2011 to May – 2012.

Materials and Methods

The Dharmaveer Water tank (Kambar Talav) is located inside the city. Earlier it was out side of the city. It was constructed for the source of water for grazing animals and because of speedy spread of city area now the tank is occupied in the heart of city. Geographically it is situated on 18°–04'-0" latitude and 75°–07'-0" longitudes. Basically the tank is divided into three regions. *i.e.* the region towards the postal colony, the middle region and the region towards ex- serviceman's colony. The site for present study is restricted at the third region of tank. This region of tank is rich in the vegetation of typha, eichornia and other water weeds.

The study of various activity patterns was arranged. The brids were observed individually in every month for the duration of fourteen months. The study activity patterns were classified as feeding, walking, bathing, swimming, chasing and preening.

Results and Discussion

The purple moorhens were observed throughout the study period. They were found maximum number during Jan – 2012. During the post monsoon period they were minimum in number. In all month they spent more time in walking around the field after feeding. They were also found occasionally on tar road where the waste was dumped at the bank of tank. The maximum time of the day was spent in feeding in these birds. It was noted that the variations were observed in the activities pattern from month to month. The wing expansion or wing flapping were also interestingly noted during feeding for every few minutes.

Figure 36.1: Feeding of Purple Moorehen at Study Area.

Figure 36.2: Study Site: Dharamaveer Sambhaji Tank (Kambar Taav).

Figure 36.3: Feeding of Purple Moorehen on Dumped Waste at the Bank of Tank.

Table 36.1: Record of Noticed Purple Moorhens in Study Period at Sambhaji Lake

Sl.No.	Month	No. of Birds Noticed	Sl.No.	Month	No of. Birds Noticed
1.	Jun-2012	45	7.	Dec-2012	70
2.	Jul-2012	18	8.	Jan-2013	80
3.	Aug-2012	15	9.	Feb- 2013	75
4.	Sept-2012	10	10.	Mar-2013	50
5.	Oct-2012	20	11.	Apr-2013	51
6.	Nov-2012	23	12.	May-2013	50

Figure 36.4: Feeding Behaviour of Purple Moorehen with Street Dog Indicated Disturbed Ecosystem.

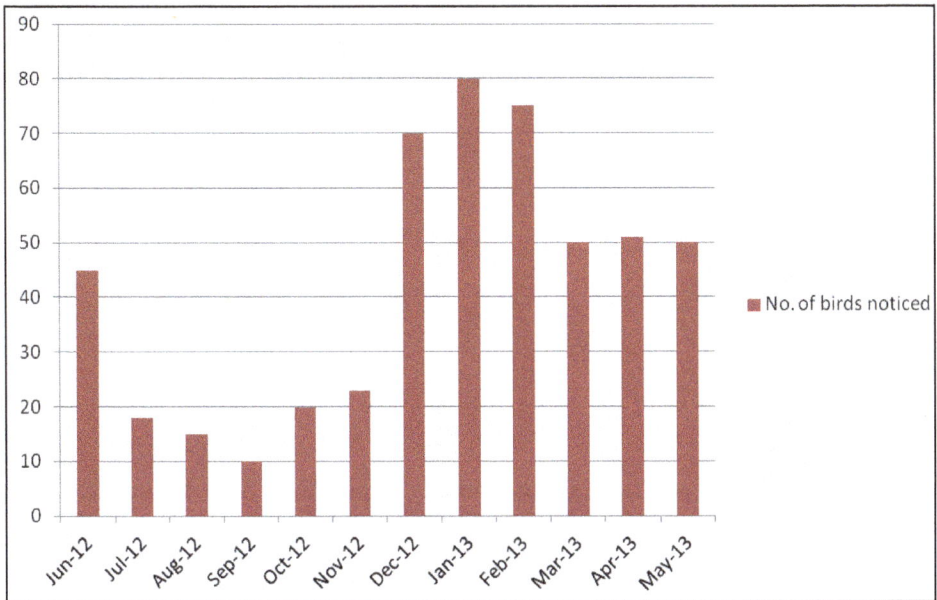

Figure 36.5: Record of Noticed Moorhens in Study Period at Sambhaji Lake.

Conclusion

The wetlands are important components to maintain the food chain and food web of biotype. The wetlands must be preserved to avoid decline in bird population. The newly constructed roads, colonies of human population, vehicle disturbances and water pollution are major causes to district such water bodies and this automatically threaten to the ecosystem on which in turn into the gradual decrease in bird's population. The present study site is under threatened. Earlier this water body was very famous for lotus vegetation. But it is now totally polluted from human interference. This water body should be preserved and protected for the natural habitat of varieties of birds. Appropriate conservative strategies should be applied to protect such a natural habitat.

References

Ali. S and Replay, S. D. 1983. Hand Book of Birds of India and Pakistan. Bombay Natural History Society, Oxford University Press. New Delhi

Ali, S. and Riplay, S. D. 1995. A Pictorial Guide to the Birds of Indian Subcontinent. Bombay Natural History Society, Oxford University Press. New Delhi

Ali, S. 1996. The Book of Indian Birds. Oxford University Press. New Delhi

Birds of Lonawala and Khandala 2009. ELA Foundation, Pune

Manjula Menan, 2004, Ecology of Purple Moorhen in Azjinhillam Wetland, Kerala.

Chapter 37

Traditional and Synthetic Materials Used in Fishing Gears of Inland Fisheries in Marathwada, India

☆ *S.P. Chavan, P.M. Kannewad, S.V. Poul and M.S. Kadam*

ABSTRACT

Inland capture fishery of Marathwada Region of Maharashtra State in India mainly consists of river Godavari and its tributaries and small, medium, large reservoirs constructed on various tributaries of river Godavari. Up to 1970's decade there was not much modernization in the fishing crafts and gears used in the inland capture fishery of this region. The use of locally available materials and plant based fibers to construct the nets and boats for fishing was common. The fish catch efforts were more but fish catch was very less. Therefore fisher communities from SC, ST, NT and VJNT classified cast and tribe in the constitution of India were economically and subsequently in other sectors of their life were poor. Since1980's decade polyester, polyethylene and polyvinyl synthetic fibers are being used to produce the water proof, durable and high strength threads and wires. The availability of these materials for the fisher communities to construct the nets and boats has created a boom in inland capture fisheries in Marathwada Region. Less efforts of fishing, more fish catch in less time, cost effective and durable designs of nets and gears are the major improvements helped for the developments in the living standards and economy of fishermen from this region. Nonconventional fishing, offseason fishing, non-expertise fishing are the emerged problems along with the alcoholism as their age old problem, all these are threat to sustainable use of fisheries resource of this region is the most sticking negative reflection of this improvement in fishing equipments.

Keywords: *Synthetic, Nets, Boats, Inland fishing, Marathwada, Maharashtra.*

Introduction

In the organized and unorganized inland fisheries sector of Marathwada region of Maharashtra State about 0.65 million people are dependent on inland capture fisheries. 80-85 per cent people have their part-time involvement in fisheries sector (Niture and Chavan, 2010), along with this business the fisher communities are also involved in the agriculture in their own farm, labor work in building construction, to run a small shop, as a labor in the clay-brick preparation business and other miscellaneous business to earn the money for their livelihood. After the long term review of data from year 1970-2011 on the working pattern of fisheries cooperative societies, life and living standard of fisher communities in this region, fish catch by these communities and the marketing; it was concluded that this sector is much unorganized. There are many reports on the analysis of physical and chemical characteristics of reservoirs as a part of fisheries management research in recent time from this region but it was found that, not much attention is being given for the study of fishing gears and nets, crafts and boats (Niture and Chavan, 2009). The fishing nets, gears and crafts used by the fisher communities are one of the important components in deciding the catch composition, catch quantity and ultimately the income they get from this entire business. It is essential to review the type of materials which were traditionally being used from the year 1970's to year 2000 and in recent decade up to 2013-2014. Naturally available materials of plant origin like timber and bamboo, Jute, cotton thread, coconut coir, teak wood, locally available light weight wood, iron hooks and rings were in use to construct the different kinds of nets and boats up to year 1970-1975. Later on with the use of plastics in early 1980's decade the use of various synthetic materials started in the inland fisheries of this region. There are more benefits rather than losses to the fisher communities due to use of synthetic materials like thermocol for one man boat; nylon, rayon, polyester threads to construct different kinds of nets for effective and effortless fishing in the rivers and reservoirs with many other indirect benefits is a positive aspect. While the involvement of non-expert, unskilled fishermen with poor knowledge about the fish life especially distribution, biology, breeding and migration and having opportunistic approach has resulted in non-sustainable utilization of natural aquatic resource. Spread of synthetic non-degradable waste of nets and rafts in the rivers and reservoirs is the main hazardous impact on inland fisheries sector of Marathwada region was also observed as an emerging pollution problem.

Materials and Methods

The study is based on the data collected about the fishing gears and craft used by the fisher communities distributed in the study area; their traditional knowledge and the comparison, confirmation with the data available in the standard literature. The data were analyzed using the Union model by Marques, JGW (1991). According to this model, all available information on the surveyed subject is to be considered. Local information provided by the participants was compared with those from the specialized academic literature. Based on synchronic and diachronic interviews, the controls were performed through verification tests of consistency and validity of responses (Marques, JGW, 1991). All ethnographic material (Recordings, transcriptions, field notes and photographs) is stored at the laboratory of Zoology and Regional Centre for Network of Indian Universities on Cultural and Biological Diversity (NIUCBD), Department of Zoology, School of life Sciences, Swami Ramanand Teerth Marathwada University, Nanded, Maharashtra State, India. The fisher tribes and casts distributed in around 2200 villages (DIC, 2011; 2012; 2013) and all over Marathwada region living in groups of 10-25 families on the medium, small or large reservoirs in Marathwada Region during the month of September to late May. All these groups either take the contract of fishing from the lease owners of the reservoirs or under the title of fish cooperative society, they are the lease owners of the reservoirs. For the study of type of synthetic materials used, all the sites of fishing were observed for a period of five years from 2009-2014. All these issues concerning the current status of their life and the benefits they are getting from the use of fishing gears as compared to earlier situation when the natural material was being used is discussed. Nearly 2000 fishermen were personally interviewed and 126 various synthetic materials, accessories, equipments, gears, crafts and boats were examined to reach the conclusion. Fishing equipments and their efficiency to increase the income of fishermen is a criteria selected to evaluate the status of fishery of this region.

Study Area

The study was conducted for the status of inland fisheries sector of Marathwada Region having 64717.91 Km^2 area in 08 different districts Aurangabad, Parbhani, Jalna, Beed, Latur, Osmanabad, Hingoli and Nanded in 76 talukas of all districts in the region. The study was conducted for 0.65 million population of fishermen working part-time or full time on 918 different reservoirs from which 10 were major reservoirs (1000 ha. and above) *e.g.* Nathsagar (Jaikwadi) near Paithan district Aurangabad, Yelderi and Siddheshwar on river Purna near Jintur and Aundha Nagnath cities in Parbhani and Hingoli district respectively; Manjra in district Latur and Terna in district Osmanabad; Majalgaon in district Beed; Karadkhed, Barul, Loni, Manyad, Vishnupuri, Penganga in Nanded district. It was an observation on the fishing equipments and discussion with the fisher communities distributed in the coastal villages along the 2361 Km long total stretch of river Godavari in the region. Visited to the fisher families living in villages in the coastal region of rivers like Godavari, Purna, Dudhna, Kayadhu, Manyad, Terna, Manjra, Masoli, Lendi, Sindaphana, Bindusara, Karpara of Godavari River basin (Figure 37.1).

Fishing Nets and Crafts Used

The fishing nets used to catch the wild and cultivable species of fishes from the reservoirs, lakes and rivers are explained as below with the details of synthetic and natural or traditional materials used in their construction are given in Tables 37.1 and 37.2.

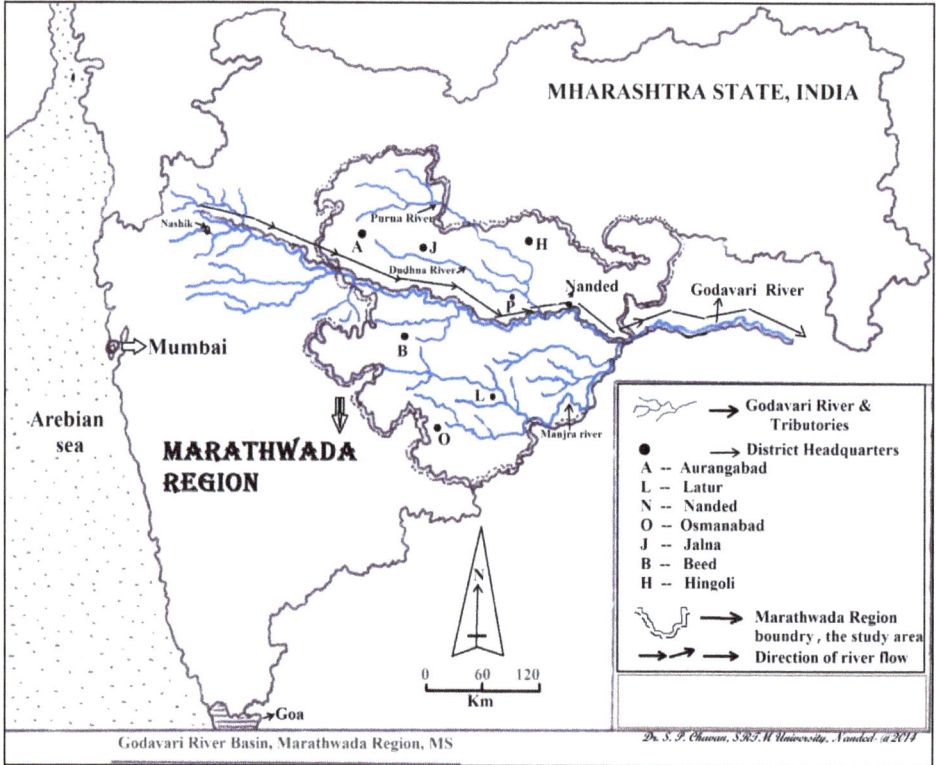

Figure 37.1: Map of Maharashtra State in India Showing Study Area of Marathwada Region Godavari River Basin.

1. Gill Net

It is one of the most common fishing net used to catch the fishes from all kinds of freshwater habitats in Marathwada region and also found in major parts of the world (Ahmed *et al.,* 2008). The mesh size of this net varies from 2.0 cm. to 15.0 cm. Indian Major Carps including *Labeo rohita, Cirrihina mrigala, Catla catla* and *Labeo calbasu* were found in the catch by using this net by the fishermen in this region. The net can be operated in various habitats of this region.

2. Cast Net

It is one of the most ancient and traditional type of one man operated net used in freshwater and marine water capture fishery (Emmanul *et al.,* 2008) (Figure 37.2). In the inland fishery of Marathwada in the 1970's decade the cast net was prepared by using the cotton thread hence it was of small size and having less catch efficiency, less durability and needed more efforts to operate the net, also it was not possible to operate in the weedy and thorny bottom (Tables 37.1 and 37.2). Recently used cast net is totally made from all synthetic materials including nylon rope, nylon 66 thread, Iron weights etc. Hence the net is more durable, litre in weight, easy to carry and operate, large in size, with high efficiency of catch and also it can be used in the weedy and thorny surface of the habitat due to good strength of threads used in the

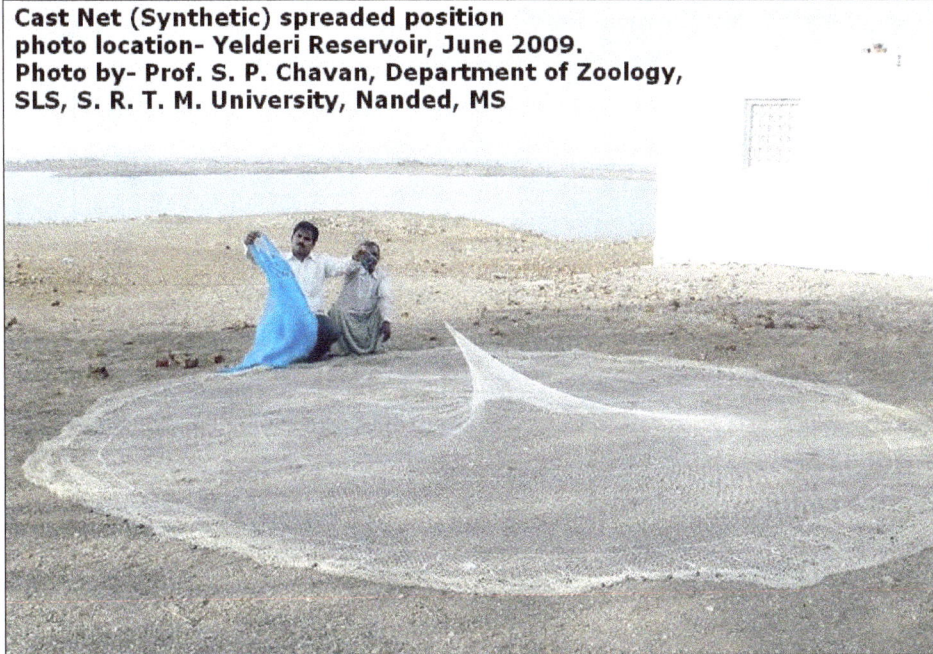

Figure 37.2: Cast Net (Synthetic material) Used in Reservoir Fishery of Marathwada Region (Photo at Yelderi reservoir, Parbhani District).

construction of this net. All details of the net for improvement in the fish catch in comparison is given in Tables 37.1 and 37.2.

3. Bag Net

The net is called as Bag Net because it has centrally fixed bag between two lateral net pieces as guiders for fishes to trap in the bag. The net is named locally as 'Pandi'. It can be used as a fixed bag net (passive net) or it may be dragged by two groups of fishermen (3-4 in each group) during the net operation (Active net). The central bag has terminal opening and closing arrangement to remove the trapped fishes (Figure 37.5) from the bag net. There is a preventive net-fold at the mouth of net to prevent the back escape of fishes from the net. It is of large size, completely made from synthetic fabrics, net pieces, ropes and strings.

4. Drag Net

It is a drag net made from synthetic nylon threads. It has strong upper and lower line ropes useful for dragging. Synthetic floats fixed to upper line at regular intervals and the iron sinkers fixed to lower line. The length of net is about 1-2 km, it require group of 8 fishermen. This net is useful to catch specific fish species like *Cyprinus carpio, Catla catla,* river prawns of *Macrobrachium* species etc.

5. Drag-Lift Net

It is made from knight fabric and three bamboo sticks. The frame of bamboo is triangular in shape. This net is operated in shallow and weedy water. The fishermen

Table 37.1: Natural and Synthetic Material Used in Fishing Gears of Marathwada Region

Sl.No.	Type of Net/Fishing Gear/Boat/Craft	Particulars	Fishermen Required to Operate	Synthetic Material Used in Construction	Catch Capacity/ Day
1.	Cast Net	Circular Net 10-15 Mdiameter(after spreading)	1-2	Nylon, Crypton	5-110 kg
2.	Gill Net	50-1000 M	1-10	Nylon, Polyester	10-100 kg
3.	Drag net	20-25 M	8-10	Nylon 66,Reyon	
4.	Bag Net	8-10 M side walls, 15 M Central Bag	2-6	Nylon and Reyon	200-300 kg
5.	Trap Net	5-6 M Side walls, 5-6Bags fixed in series one after other	2-6	Polyester Knight cloth, Nylon	30-40 kg
6.	Lift Net	3 m triangular shaped frame of bamboo, central bag 3m	1-2	Polyester	5 kg
7.	Hook and Line	100-150 hooks with 1/2 – 1 M wire fixed to Nylon long line with floats and baits to each hook	1-2	Steel Metal and Nylon.	20 kg
8.	Hook and angle	1-5 hooks fixed to single line with float, wt. and bait	1-2	Steel and Nylon	5 kg
9.	Thermocol Rafts (one Man carrier)	2 x 0.8 x 0.2 mL x W x HThermocol sheet	1	Thermocol	150 kg
10.	Thermocol Raft (2-4 Men and net Carrier)	1.8 x0.8 x0.3 m Thermocol sheets, 2-4 sheets fixed with bamboo to form large platform.	2-4	Thermocol	4 kg
11.	Rounded Iron Vessel	2-3 M diameter 1 -1.5 m height, Iron, boat shaped vessel	1-2	Iron plates Nut- Bolt.	1000 kg
12.	Masula type of Centrally Keeled boat	2.8 x 1 x1 M iron sheet vessel, with central narrow keel.	2	Mango tree wood blogs	1500 kg
13.	Catch carry Nets	Knight cloth of Nylon, 1 x 1M bag	1-2	Nylon thread, Knight fabric	50 kg
14.	Catch storage baskets	from Bamboo, 30-40 Kg. Capacity	1-2	Bamboo strips Coconut leaves.	20 kg
15.	Catch carry Baskets	from bamboo, 20-30 Kg. Capacity	1-2	Bamboo strips	20 kg
16.	Catch storage Thermocol Box	3 x 2x 2 ft. Thermocol Boxes with crushed ice	2-3	Bamboo baskets Thermocol	50 kg

M: Meter; ft: Feet; kg: Kilo Gram.

drag the net with hand while the fisher women drag it with the reverse movement with the drag line fixed around her waist (Figure 37.4). It is specially used to catch the freshwater prawns from the shallow and coastal region of river or reservoir. The catch capacity, the design etc. are given in Tables 37.1 and 37.2.

Table 37.2: The Cost of Natural and Synthetic Materials Used in the Fishing Gears and Crafts of Marathwada Region

Sl.No.	Fishing Net/Gear, Boat, Craft Other Equipment	Natural = N, Synthetic = S	Durability	Unit Cost (INR)
1	Fiber Plastic Floats	S	8-10 Years	10/Piece
2	Casting Iron or Alloy Weights or sinkers	S	Life long	15/Piece
3	Upper Line of Nylon	S	3-4 Years	300/kg
4	Lower Line of Nylon	S	3-4 Years	400/kg
	Upper line of Cotton	N	1-2 Years	100/kg
	Lower line of Cotton	N	1-2 Years	200/kg
5	Thread for net weaving			
	a. Nylon 66	S	8-10 Years	
	b. Polyester	S	7-8 Years	
	c. Rayon	S	7-8 Years	
	d. Crypton	S	9-10 Years	500/kg
	e. Cotton	S	2-3 Years	
	f. Nylon 1	S	9-10 Years	
	g. Nylon 4	S	8-9 Years	
	h. Nylon 6	S	8-9 Years	
	i. Indian Jute	N	1-2 Years	
6.	Wooden logs	N	10-15 Years	6000/raft
7.	Bamboo	N	10-15 Years	100/piece
8.	Thermocol Sheets 5x2x1	S	4-5 Years	600/piece
9.	Wooden sticks	N	1-2 Years	150/pece
10.	Dried weed bundles	N	½- 1 Year	Free naturally
11.	Metal casting hooks			
	a. ½ inch	S	All Hooks	
	b. 1 inch	S	10-15 Years	250/12 pieces
	c. 1 ½ inch	S		
12.	Knight cloth of Nylon	S	5-6 Years	30/M
13.	Cotton Cloth	N	1-2 Years	50/M
14.	Synthetic transparent nylon net	S	3-4 Years	350/kg
15.	Metal sheets for boats	S	10-14 Years	40/kg
16.	Waterproof ropes of Nylon	S	7-8 Years	600/kg
17.	Ceramic weight (Sinkers)	N	10-12 Years	10/Piece
18.	Concrete weights (Sinkers)	S	10-14 Years	20/Piece
19.	Stone as weight (Sinkers)	N	Many Years	Freely available in river
20.	Hooks of stone	N	5-6 Years	20/Piece
21.	Dried fruits of Pumpkin as floats	N	1 Year	2/Piece

INR: Indian Rupees; S: Synthetic; N: Natural.

Figure 37.3: Large Drag-Bag Net (Pandi) at Yelderi Reservoir.

Figure 37.4: Drag Lift net (Synthetic Net) Operation in Coastal Part of a Reservoir for Prawn Collection by Fisher Women in Marathwada Region, Maharashtra State.

Figure 37.5: Drag-Bag Net of Synthetic Material, Operated by Fishermen.

Figure 37.6: Thermocol Raft Used in Freshwater Capture Fishery of Marathwada Region.

6. Hooks and Lines

This is passively operated fishing gear. Series of hooks of metal are hanged from main line with 1-3 Ft. distance between two hooks and each hook is fixed at the tip of nylon thread of 1-3 ft. distance from the main supporting rope. Terminals of supporting main rope are fixed to the supporting bamboo sticks which are fixed 2-3 ft. in the bottom and kept 1-2 ft. above the water surface. When the hooks of the line are adjusted to 10-15 cm. below the water surface then the live insects and small size frog species (*Rana tigrina*) are fixed as a bait to create the movement on the water surface which attract the predatory fish species like *Channa striatus, Channa gachua, Wallago attu, Notopterus kapirat etc.* (FAO, 1995). Predatory fishes which may get hook-trapped. When the hook-line is fixed to reaching the bottom of a water body in the river or reservoir, in this situation the bait applied is earthworm, small dead fishes, small prawns, wheat flour roasted balls then the bottom dwelling fishes like *Clarias batrachus* may get hook trapped.

7. Thermacol Raft

Thermocol made rafts has created a new revolution in the freshwater capture fishery of Marathwada Region (Niture and Chavan 2009, 2010). Thermocol is a synthetic material. l02x0.8x0.2 M size thermocol piece is the preferred size used to prepare one man carrier raft. Two sheets of thermocol of this dimension are fixed in the knight cloth bag to prepare a raft. It carries 1-2 fishermen and the catch (Figures 37.6 and 37.7). 4-6 sheets of thermocol are fixed together with wooden or bamboo

Figure 37.7: Thermocol Raft Used in Freshwater Capture Fishery of Marathwada Region.

Figure 37.8: Thermocol Raft (Large, Modified) Carries Synthetic Gill Nets, Bag Net and 2-3 Fishermen in Reservoir Fishery of Marathwada.

support to prepare a large raft which may carry 200-300 kg of fish catch and 2-3 fishermen with 1-2 large gill nets and other accessories (Figure 37.8). It is most convenient type of material due to litre weight, waterproof qualities, high strength and more durability, easy to operate and carry from one to next place, also it is cost effective.

Wooden Log Raft

Straight wooden logs from stem part of locally available tree *Zizipus* sp. locally named as 'Bori' are cut down in to pieces of 2-3 M length and 20-25 cm diameter after well drying. All the wooden logs are fixed together by applying specific knot of the nylon rope so as to form a floating large platform which act as a wooden log raft. It may carry load of about 800-1000 kg.

Results and Discussion

Traditionally used fishing nets are not changed with their basic design, therefore there is no need to get the training for the operation of newly equipped fishing nets. The accessories required to construct the nets are changed from naturally available local material to synthetic materials.The durability of all types of nets used in inland fishery of Marathwada has increased to 7-8 years from earlier durability of 1-2 years.

It is only due to the use of synthetic materials in the nets and gears of fishing. The details are shown in Figures 37.12 and 37.13. Locally made cotton threads and ropes were used before the introduction of synthetic materials was having water soaking property therefore after fishing the moisture remained in the cotton is favorable for decomposition process. The use of cotton threads and ropes was found nearly eliminated from the construction of nets and gears of the inland fishery sector of this region.

Table 37.3: Durability of Synthetic and Natural Materials Used in Crafts and Nets of Inland Fishery in Marathwada Region of Maharashtra

Sl.No.	Name of Material for Net/Gear/ Craft construction	Natural: N/ Synthetic: S	Durability in Years
1.	Fibro-plastic float	A-S	15
2.	Dried fruits of Cuccurbitaceae	A-N	01
3.	Synthetic weight (Metal)	B-S	20
4.	Stone pieces/Pebbles	B-N	30
5.	Thermocol Sheets	C-S	10
6.	Metal Sheets	C-N	20
7.	Synthetic fibers	D-S	12
8.	Cotton Thread	D-N	01
9.	Knight Cloth	E-S	01
10.	Cotton Fabric	E-N	01

As: Fibro-plastic float; AN: Dried fruits of Cuccurbitaceae; BS: Synthetic weight (Metal); BN: Stone pieces/Pebbles; CS: Thermocol Sheets; CN: Metal Sheets; DS: Synthetic fibers; DN: Cotton Thread; ES: Knight Cloth; EN: Cotton Fabric.

Table 37.4: Types of Net and their Efficiency to Catch the Fish from Inland Water of Marathwada Region

Sl.No.	Type of Net	Abbreviation	Catch Efficiency in kg/day
1.	Natural Gill net	NG	10
2.	Synthetic Gill Net	SG	100
3.	Natural Drag net	ND	20
4.	Synthetic Drag net	SD	400
5.	Natural Bag Net	NB	50
6.	Synthetic Bag net	SB	200
7.	Natural Cast net	NC	20
8.	Synthetic Cast net	SC	60

NG: Natural Gill net; SG: Synthetic Gill Net; ND: Natural Drag net; SD: Synthetic Drag net; NB: Natural Bag Net; SB: Synthetic Bag net; NC: Natural Cast netl SC: Synthetic Cast net.

The upper line and lower line of the gill net, drag net lift net and cast net are the supporting ropes of these nets. These ropes are also used to drag/lift the nets during their operations. The replacement of synthetic materials like Nylon 66, Nylon 6, Rayon in place of cotton threads and ropes has improved the strength of these ropes so that during dragging and lifting operation of these nets there is very rare or no chance of damage or breaking of these ropes.

The floats used in upper line of various nets till 1970's decade were the dried fruits of family - cucurbitacea of tender plants especially bitter pumpkin. The naturally available floats were of un-even size, less durable than synthetic floats. Naturally available floats may get crack, spoil, sink down easily due to less buoyancy water soaking and colorless hence difficult to identify while recently used synthetic floats are colorful red, white; yellow etc. therefore easy to identify the location on water surface. Synthetic floats are fibro-plastic, waterproof, available in even size, high buoyant, 50-60 time durable than specially designed for easily fitting in to the net line (Figure 37.9). Due to use of synthetic materials in the floats the overall efficiency of net is increased, the efforts of fisherman to operate the net are decreased, the synthetic floats are cost effective (Marques *et al.,* 1991, Meecan *et al.,* 2001).

The weights or sinkers used in recent time to fix the lower line are specially designed to fix in to the net easily (Figure 37.11). The weights are made from alloy metal or cast iron, noncorrosive, available in any fixed weight capacity from 10-15 gm. The synthetic weights are superior than natural weights which were used earlier in the period of 1970's decade. The naturally available weights like stone pieces,

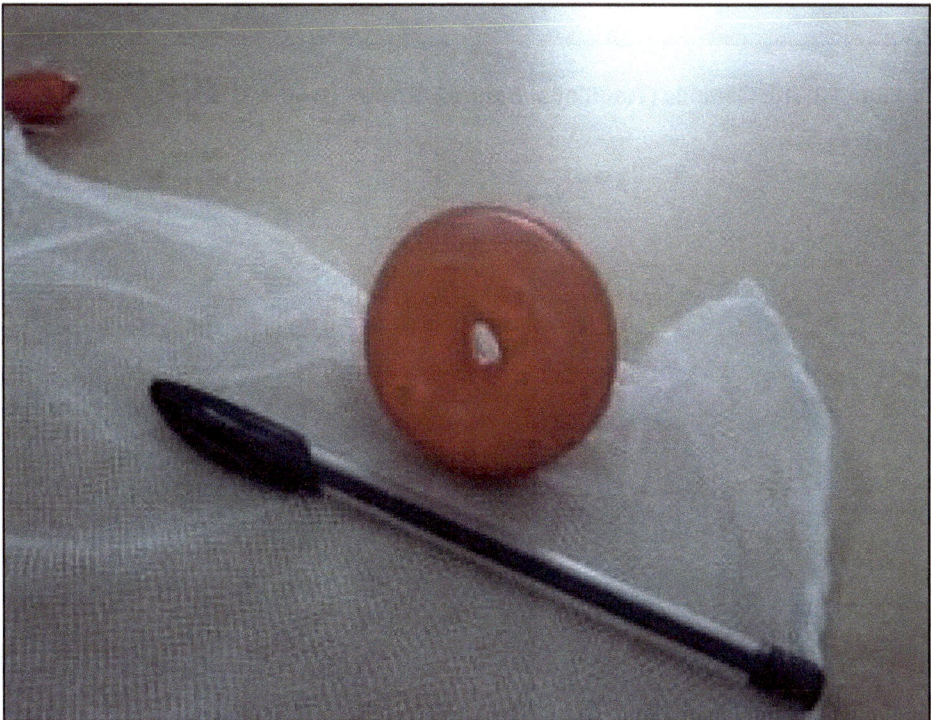

Figure 37.9: Fibro-Plastic Synthetic Float Used to Fix through its Central Opening in to the Upper Line of Net.

Figure 37.10: Ceramic (Traditional/Natural) Sinkers Used in Lower Line of Nets.

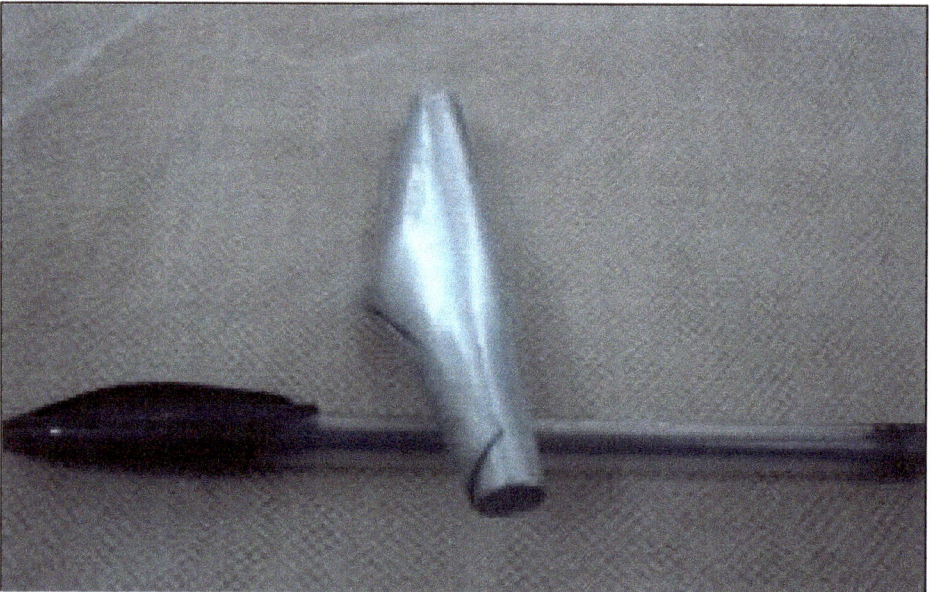

Figure 37.11: Galvanized Metal Sheet Weight to Fix in the Bottom/Lower Line of Nets.

pebbles etc. were of uneven size and weight, they were difficult to fix in to the lower line of the net, there is chance of detachment from the net hence create net operation problem. Concrete material, Metals, Metal alloy (Figure 37.11), lead, ceramics (Figure 37.10) is commonly used as weight material now days.

The size of gill nets made entirely from cotton was 50-60 ft. in length, hand made; having 1-2 years durability was used in Inland fishery of this region. The weight of cotton net was high as compare to its length and size. It was with high maintenance required for drying, preventing it from decomposition. Recently used gill nets made of all synthetic material are 10-15 time longer than traditional nets, with less weight, with high catch efficiency, more durable, with very less maintenance and easy to operate. This improvement in gillnet has increased the fish catch 10-15 time.

The cotton thread used for net weaving and rope preparation is found easily visible in water to fishes due to attachment of the silt and plankton to the net in water hence the fishes avoid the net. The synthetic material like nylon is invisible in water for fishes due to mixing or matching of the thread coloration with water color, hence there are high chances of fish trapping. In this regard to the synthetic material is superior to natural material for increasing the fish catch.

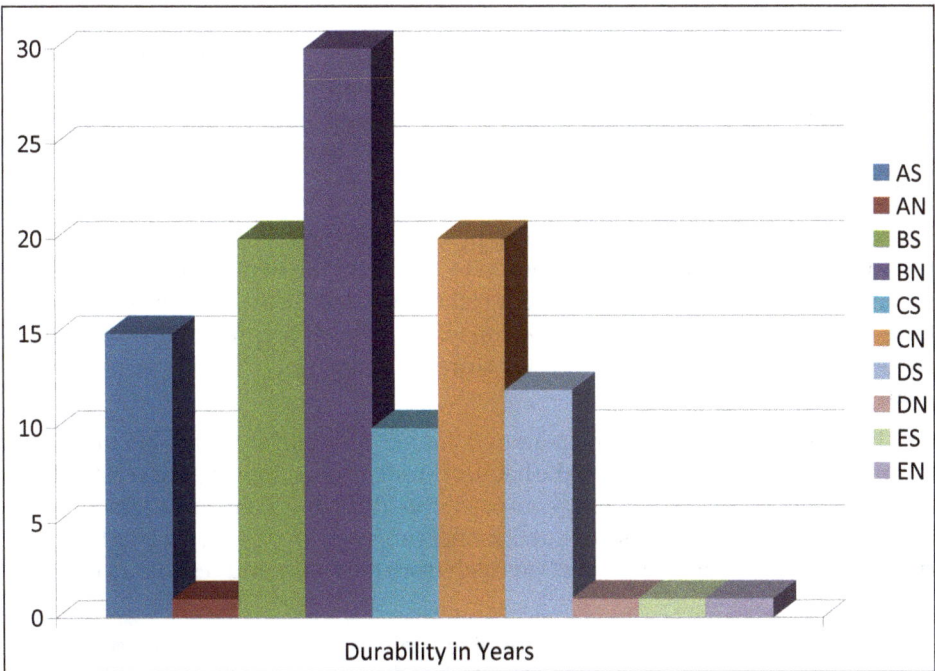

Figure 37.12: Durability of Traditional/Natural and Synthetic Materials Used in Nets and Crafts of Capture Fishery of Marathwada Region, Maharashtra.

As: Fibro-plastic float; AN: Dried fruits of Cuccurbitaceae; BS: Synthetic weight (Metal); BN: Stone pieces/Pebbles; CS: Thermocol Sheets; CN: Metal Sheets; DS: Synthetic fibers; DN: Cotton Thread; ES: Knight Cloth; EN: Cotton Fabric.

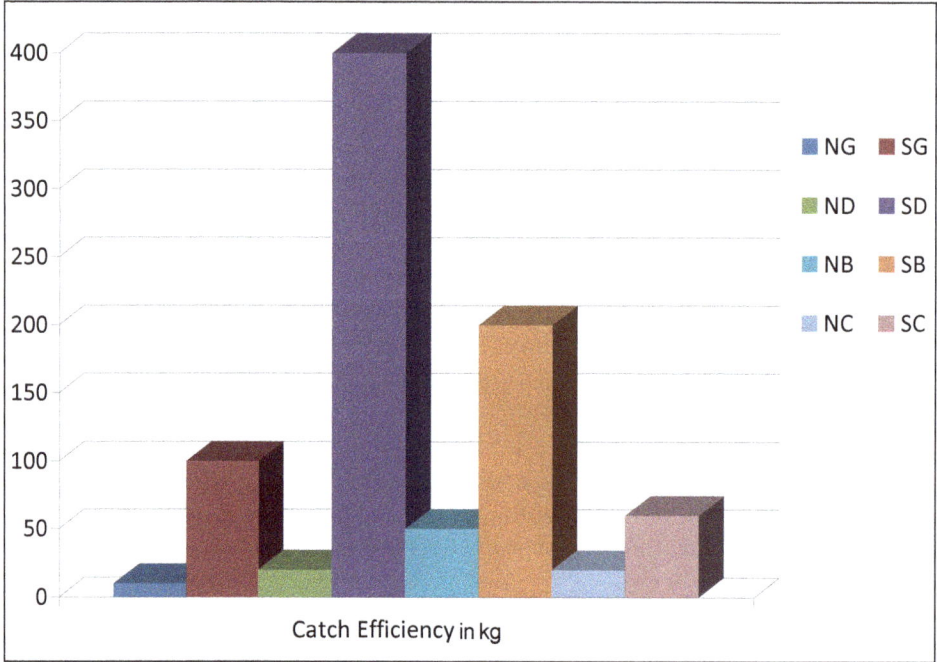

Figure 37.13: Catch Efficiency of Traditional/Natural and Synthetic Nets Used in Inland Capture Fishery of Marathwada Region.

NG: Natural Gill net; SG: Synthetic Gill Net; ND: Natural Drag net; SD: Synthetic Drag net; NB: Natural Bag Net; SB: Synthetic Bag net; NC: Natural Cast net! SC: Synthetic Cast net.

The most durable, heat stable, fire-proof, water-proof, high strength material used in very recent time in 2000 decade to construct the nets is 'Bakelite' this has made the revolution in the fishing gear technology, because some time the rivals of fishermen burn the nets, but the Bakelite made nets are fireproof. Other beneficial properties of the Bakelite thread are, these are more superior to the all synthetic fibers in all qualities.

In recent time the synthetic materials for net weaving are available in various colors to match the water color of the habitat during fishing, for *e.g.* green colored nets are used in pond and lake fishery (Campose *et. al*, 2010) due to greenish color of water and faint yellow colored nets of synthetic fibers are used for fishing in turbid water of rivers. This has also increased the fish catch from the rivers and reservoirs in recent time.

Average income of fisherman involved in capture fishery of this region has increased from Rs. 10-20 Rs./Day in 1970's decade to Rs. 400-500/day to this decade 2000. Due to application of synthetic and high quality durable materials to construct the nets and different gear (Zar *et al.,* 1999; Zhou *et al.,* 1995) but it was also found that, 70 per cent fishermen were not involved in the business for full time and daily but they were part time fishermen. Gill nets, Cast nets, Drag nets are unable to hold the load of occasional heavy catch.

Due to easily available low cost and durable readymade fishing gears and nets any one who is unemployed but having interest in fishing are found involved in the fishing process. All these accidentally entered people in this business start fishing in variety of habitats but these people don't have knowledge about the fish life, brood fish and juvenile catch prevention resulted in unconventional fishing causing considerable damage to fishery resources of this region (Agyennin-Boateng, 1989; Ahmed *et al.,* 2008; FAO, 1995). The expert fishermen use the body scale observation method to detect the age of fishes which is not done by opportunistic people.

Conclusion

From the study it can be concluded that the annual income of fishermen has increased 10 times in last three years as compared to earlier. The synthetic materials formed mainly from synthetic fibers and polymers are cost effective, easily available and durable as compared to natural material which has all inferior qualities compared to synthetic materials, this has made easy fishing with fewer efforts and less time with low income in the 1980's decade. 100 per cent fishermen population gets benefitted to raise their income in 2000 decade (Ricker, 1973; Conover, 1980). Few inexperienced fishermen and with little knowledge about fisheries and fish catch there is considerable damage to fishery resource of this region. Catch of Brood fishes during breeding and spawning, catch of the young and juveniles, off season fishing are the major damages to the Inland fisheries sector (King, 2007) of Marathwada has to be taken in to serious consideration.

Framing and implementing the guidelines for fishing and control, monitoring the activities of fishermen for sustainable utilization of the fisheries resource is suggested.

Non-decomposing or late decomposing synthetic wastes found spreaded all over the rivers and their tributaries due the drifting through floods and damaged or left-out nets is a serious concern over the damage to biodiversity living along the river basins. These materials especially nets may entangle or wrap with the legs, hooves of animals which visit the rivers for water drinking during night or day. Trawling operation for fishing (Arkley, 1990) is possible in the large reservoirs of this region but it was not been carried out.

References

Agyennin-Boateng C. E. 1989. Report on the socio-economic conditions in the fishing communities in the Yeji area of the Volta lake. FAO/IDAF Project, GHA/88/004. Field document, Rome, p. 89.

Ahmed, Y. B, Inpinjolu, J. K. 2008. The gillnet selectivity of Citharinus citharus (Pisces: Citharinidae), Hydrocynus forskalii (Pisces: Characidae), Distichodus rostratus (Pisces: Distichodontidae) and Synodontis membranaceus (Pisces: Mochokidae) in kainji lake, Nigeria. *Nig. J. Fish.* 5(1): 46-62.

Arkley, K. 1990. Fishing trials to evaluate the use of square mesh selector panels fitted to Nephrops trawls-MFV Heather spring November/December 1990. Sea fish Industry Authority Report N° 383 pp: 21.

Bagenal, T. B. and Tesch, F. W. 1978. Age and growth in: Bagenal, T. (Ed), Methods for assessment of fish production in freshwater. Oxford, Blackwell Scientific Publication: pp. 101-136.

Carlander, K. D. 1969. Hand Book of freshwater fishery Biology. Vol. I, Iowa state University Press, Ames, Iova, pp. 752.

Conover, W. J. 1980. Practical non parametric statistics. 2nd Edn. Texas Technique University, New York, USA. pp. 493.

Campos, A. and P. Fonseca and V. Henriqnes. 2003. Size selectivity for four fish species of the deep ground fish assemblage of the Portughese South West coast: evidence of Mesh size, mesh configuration and cod-end catch effects. *Fish. Res.* 63: 213-233.

DIC 2011; 2012; 2013. Nanded District, Collector Office, District Information Office, Senses.

Emmanul BE, Chukwu LO, Azeez LO, 2008. Cast net design characteristics, catch composition and selectivity in tropical open lagoon. *Afr. J. Biotechnol.* 7(12): 2081-2089.

Froese, R. D. Pauly. 2011. Fish Base. World wide web electronic Publication. Version 06/2011. http: //www. fish base. org/search. php.

FAO. 1995. Code of conduct for responsible fisheries. Rome, FAO, P. 41. DIC Nanded senses, 2011 of Maharashtra State.

King, M. 2007. Fisheries Biology, Assessment and Management. Blackwell Scientific Publication Ltd., Oxford, ISBN-13: 978-1-4051-5831-2, pp: 400.

Marques, J. G. W. 1991. Aspectos ecologicos na etnoictiologia dos Pescadores do Complexo Estuarino-Lagunar Mundau-Manguaba. Campines, Campines: Alagoas. ThesisUniversidade Estadual de Campaneas.

Meekan, S. M., A. Wilson, A. Halford and Retzel. 2001. A comparison of catches of fishes and invertebrates by two light trap designs, in tropical NW Australia. *Mar. Biol.*, 139: 373 -381.

Niture, S. D. and Chavan, S. P. 2010. Crafts and gears used in Yelderi Reservoir, Maharashtra. In: Advances In Aquatic Ecology (Vol. 4), Edited by V. B. Sakhare and Patricio Rene De Los Rios Escalente. Daya Publishing House, New Delhi, 136-142.

Ricker, W. E. 1973. Linear regressions in fishery research. *J. Fish. Res. Board Can.*, 30: 409 - 434.

Tesch, F. W. 1968. Age and Growth. In: Methods for assessment for Fish Production in freshwater, Ricker W. E. (Ed.). Blackwell Scientific Publications, Oxford, UK., pp: 93-120.

Natural Research Council. 1993. Nutrient requirement of fish committee on Animal Nutrition Board on Agricultural, N. R. Council. National Academy Press Washington DC, USA.

Niture, S. D. and Chavan, S. P. 2009. Crafts and gears used in Yeldari reservoir, Maharashtra, *Ecology and Fisheries*, 2(1): 113-120.

Zar, J. H. 1999. Biostatistical analysis. 4ᵗʰ Edition., Prentice- Hall, New Jersy, USA., pp: 469.

Zhou, M. Q. Shen and J. Li. 1995. Study on population resources and environment in relation with the development and management of lake Taihu, Waxi city Res. (In Chinese) pp: 57-63.

Previous Volumes–Contents

— Volume 1 —

2007, xvi+194p., figs., tabls., ind., 25 cm   Rs. 950

ISBN 81-7035-483-8

— Volume 2 —

2008, xvi+143p., col. plts., figs., tabls., ind., 25 cm   Rs. 750

ISBN 81-7035-559-5

— Volume 3 —

2010, xiv+176p., col. plts., figs., tabls., ind., 25 cm   Rs. 800

ISBN 978-81-7035-633-2

— Volume 4 —

2010, xvii+182p., figs., tabls., ind., 25 cm Rs. 750

ISBN 978-81-7035-657-8

— Volume 5 —

2011, xviii+231p., col. plts., tabls., figs., ind., 25 cm Rs. 1200
ISBN 978-81-7035-697-4

— Volume 6 —

2012, xv+345p., col. plts., figs., tabls., ind., 25 cm   Rs. 1800

ISBN 978-81-7035-782-7

— Volume 7 —

2013, xv+345p., col. plts., figs., tabls., ind., 25 cm   Rs. 1800

ISBN 978-81-7035-782-7

— Volume 8 —

2014, xv+303p., plts., figs., tabls., ind., 25 cm   Rs. 1895

ISBN 978-93-5124-283-3

Index